THE
COATING
PROCESSES

Prepared by the Coating Process Committee
of the Coating and Graphic Arts Division
Committee Assignment No. 1571-.870110.02

Jan C. Walter, Chairman Emeritus

TAPPI

ATLANTA, GEORGIA

TAPPI PRESS

© 1993

2nd Printing – 1994
Copyright © 1993 by:

TAPPI PRESS
Technology Park/Atlanta
P.O. Box 105113
Atlanta, GA 30348-5113 U.S.A.

TAPPI Keywords: Coating (Processes); Coated Boards; Drives; Drying; Materials Handling; Process Control; Raw Materials; Rolls; Screening; Surface Treatment.

ISBN: 0-89852-058-4 • TAPPI PRESS Item No. 0102-B057

Printed and Bound in the United States of America

Library of Congress Cataloging-In-Publication Data

The Coating Processes / Jan C. Walter, editor.
Includes bibliographical references.
ISBN 0-89852-058-4

CIP application on file.

Preface

This book's predecessor, the 1965 TAPPI book titled *The Pigmented Coating Processes* and co-chaired by George Booth and Nathan Millman, still serves as a valuable reference for early coating technology. Many of the technological steps since 1965 along with today's coating technology are documented in this first edition of *The Coating Processes*. The publication is divided into eight chapters which emphasize the process of making coated paper, the preparation, application and drying of coating, and the treatment of the dried coated surface.

The book reflects the knowledge base of the TAPPI Coating Process Committee members from 1987 to 1992. A different group of committee members working in the same five-year period might create an entirely different publication. You will notice variations in writing styles, organization, points of view and illustrations, as well as some information overlap. The contributors have served as their own editors, giving you a broad exposure to the differing views of people working in the industry.

To supplement your reading, we encourage you to seek out the additional sources for information on coating offered through the products and meetings of the Coating and Graphic Arts Division of TAPPI. For information on new technology and general tutorial information, attend the many seminars and conferences offered by the division including the Coating Conference and Coated Board, Blade Coating, Binders, and Basestock Seminars. If you cannot attend these meetings, published proceedings are available. The division also offers other books on related topics.

Jan C. Walter
Ronningen-Petter

Acknowledgements

This book is the product of a five-year collaborative effort between the members (and a few invited experts) of the Coating Process Committee of the Graphic Arts and Coating Division of TAPPI. It all began on an airplane trip between the United States and Asia on which **Ronald Johns** of Engelhard Corp., who was the Coating Process Committee chairman at the time, completed an outline which formed the foundation for this book. The outline defined where we wanted to go; all we had to do was figure out how to get a committee from the then 87 members to get us there.

To form the organizational structure, the committee decided to create Committee Assignments (CAs) for each chapter. The CA chapter chairman managed the authors contributing to each chapter. These chapter managers include **Lars Andersson,** J. M. Huber Corp., who had 13 authors to orchestrate into the production of the coating preparation chapter. **Ernest DeSanti,** The Black Clawson Co., also had special challenges working with nine contributors in the drying chapter. **George Booth,** retired from The Black Clawson Co., and **Theodore Vanya** and **William Goetz,** Appleton Papers Inc., worked together as co-chairmen along with four other contributors on the coating application chapter. Authors that managed their own chapters include **Gordon Stout,** Simons Eastern Consultants, for his work on the basestock chapter; **David Damato,** Rust International Corp., for statistical process control; and **Sanford Shapiro,** Honeywell Inc., for process control.

Also behind the scenes were many people that deserve special mention for loaning this project their specialized skills. These people include **Donald McCormick** for his review of the coating preparation chapter in its infancy to help lead it to its finished form; **Charles Klass,** Klass Associates, for his expert TAPPI and Coating Process Committee liaison skills in addition to his contribution as an author; **Ralph Loehning,** Champion International, and **Jack Koval,** National Starch and Chemical Co., for their help with the starch section. **Lawrence Gaspar,** National Starch and Chemical Co., a contributor to the drying chapter, helped along with the Basestock Committee on the basestock chapter. **Michael Paczkowski,** BTG Inc.; **Albert Lankford,** Valmet; and **David Oliver,** Westvaco Corp., helped with the blade coating chapter and **Thomas Ebel,** Lamb Grays Harbor Co., with the Finishing and Converting Division provided guidance in the finishing chapter.

The author of the sections on Mixing and Dispersion, and Flow and Material Measurement, wishes to express his appreciation to **Dave Damato,** Rust Engineering; **Terry Kellogg,** J. M. Huber; **Al Barton,** Mead Coated Board; **Elizabeth Farrar,** Biosphere; and **Joel Barker,** J. M. Huber, for their review assistance and comments. An additional special thank you to **Dave Damato** for his coordination of the figures in both these sections.

The authors of the Screening section wish to express their appreciation to the following companies for their contributions: Brookfield, Cellier Corp., CRS Sirrine Engineers, Cuno, Durion, GAW, Jylhäraisio Oy, ProGuard Filtration Systems, Ronningen-Petter, Sprout Bauer, and SWECO.

The author of the Blade Coaters section conveys his appreciation for the contributions of Beloit Corp., Black Clawson, BTG, Jagenberg, and Valmet.

Finally, a special thank you to all the contributors and editors listed for sharing their expertise. All the people listed in this book have volunteered their time, many of whom without the support of their employer would not have been able to have had this experience.

Jan C. Walter
Ronningen-Petter

List of Contributors

Robert J. Alheid
Beloit Corp.
Chapter 6, Editor

Lars G. Andersson
J. M. Huber Corp., Clay Division
Chapter 2

George L. Booth
The Black Clawson Co. (retired)
Chapter 3

William L. Bracken
Bracken Resources Inc.
Chapter 4

Richard H. Bublitz
Minnesota Corn Processors
Chapter 2

G. Gordon Bugg
Mead, Mahrt Mill
Chapter 2

C. E. Coco
Protein Technologies
International, Inc.
Chapter 2

Robert A. Daane
Consultant
Chapter 4

David J. Damato
RUST International Corp.
Chapter 8

Ernest A. DeSanti
The Black Clawson Co.
Chapter 4

Leroy E. Deters
Grain Processing Corp.
Chapter 2

Dale R. Dill
Protein Technologies
International, Inc.
Chapter 2

Larry E. Fitt
Corn Productsa unit of CPC
International Inc.
Chapter 2

Ronald L. Fox
Converters Paperboard Co.
Chapter 7

Lawrence A. Gaspar
National Starch and Chemical Co.
Chapter 2

Robert V. Hershey
Potlatch Corp.
Chapter 2, Editor

James Y. Hung
Hung International
Chapter 4

Terry L. Kellogg
J. M. Huber Corp., Clay Division
Chapter 3

Gerald I. Kheboian
Kheboian & Associates
Chapter 6

Charles P. Klass
Klass Associates Inc.
Chapter 3

Herbert B. Kohler
Kohler Coating Machine Corp.
Chapter 3

Donald W. Lawton
Advance Systems Inc.
Chapter 4

John D. McInnes
JDM Associates, Inc.
Chapter 2

Dale Midyette
CRS Sirrine Engineers, Inc.
Chapter 2

John F. Munce
IR Application Consultant
Chapter 4

Robert E. Peiffer
Beloit Corp.
Chapter 4

Jack A. Perry
Q-Jet Systems Inc.
Chapter 7

E. C. Porter Jr.
Repap Wisconsin Inc.
Chapter 4

Morion R. Ricks Jr.
Impact Systems, Inc.
Chapter 4

Sanford I. Shapiro
Honeywell Inc.
Chapter 5

Douglas K. Stinebaugh
National Starch and Chemical Co.
Chapter 2

Gordon L. Stout
Simons Eastern Consultants Inc.
Chapter 1

Theodore C. Vanya
Consultant
Chapter 3

Raymond O. Wiener
Simons Eastern Consultants Inc.
Chapters 2 and 7

E. William Wight
Beloit Corp.
Chapter 3

Michael Wunderlich
Infrared Consultant
Chapter 4

The Coating Processes
©1993 TAPPI PRESS
Atlanta, Georgia

Table of Contents

Chapter 1
Basestock / 1

Gordon L. Stout, Simons Eastern Consultants Inc.

Chapter 2, Section I
Typical Coating Components / 9

Lars G. Andersson, J. M. Huber Corp., Clay Division

Chapter 2, Section II
Preparation of Starch for Pigmented Coatings / 21

Douglas K. Stinebaugh, National Starch and Chemical Co.
Lawrence A. Gaspar, National Starch and Chemical Co.
LeRoy E. Deters, Grain Processing Corp.

Chapter 2, Section III
The Preparation of Soy Polymer,
Isolated Soy Protein, and Casein Coatings / 31

Dale R. Dill, Protein Technologies International
C. E. Coco, Protein Technologies International

Chapter 2, Section IV
Mixing and Dispersion in Coating Operations / 35

G. Gordon Bugg, Mead, Mahrt Mill

Chapter 2, Section V
Continuous Coating Make-down System / 43

Richard H. Bublitz, Minnesota Corn Processors
Editor, Robert V. Hershey, Potlatch Corp.

Chapter 2, Section VI
Screening of Paper Coatings and Their Ingredients / 49

Dale Midyette, CRS Sirrine Engineers, Inc.
John D. McInnes, JDM Associates, Inc.

Chapter 2, Section VII
Pumps and Circulation Systems / 57

Raymond O. Wiener, Simons Eastern Consultants Inc.

Chapter 2, Section VIII
Flow and Material Measurement in the Coating Process / 63
G. Gordon Bugg, Mead, Mahrt Mill

Chapter 2, Section IX
Raw Material Quality Control / 69
G. Gordon Bugg, Mead, Mahrt Mill

Chapter 3, Section I
Blade Coaters / 71
Theodore C. Vanya, Consultant
Editor: E. William Wight, Beloit Corp.

Chapter 3, Section II
Rod Coaters / 83

George L. Booth, The Black Clawson Co. (retired)

Chapter 3, Section III
Air Knife Coaters / 107

Herbert B. Kohler, Kohler Coating Machinery Corp.

Chapter 3, Section IV
Transfer Roll Coaters, A History / 115

George L. Booth, The Black Clawson Co. (retired)

Chapter 3, Section VII
Coated Board / 145

Terry Kellogg, J.M. Huber Corp., Clay Division

Chapter 4
Drying / 153

William L. Bracken, Bracken Resources Inc.
Robert A. Daane, Consultant
Ernest A. DeSanti, The Black Clawson Co.
James Y. Hung, Hung International
Donald W. Lawton, Advanced Systems Inc.
John F. Munce, IR Application Consultant
Robert E. Peiffer, Beloit Corp.
E. C. Porter Jr., Repap Wisconsin Inc.
Marion R. Ricks Jr., Impact Systems, Inc.
Michael Wunderlich, Infrared Consultant

Chapter 5
Process Control / 175

Sanford I. Shapiro, Honeywell Inc.

Chapter 6
Web Handling and Off-machine Coater Drives / 181

Gerald I. Kheboian, Kheboian & Associates
Editor, Robert J. Alheid, Beloit Corp.

Chapter 7, Section I
Brush Polishing of Mineral Coated Surfaces / 219

Jack A. Perry, Q-Jet Systems Inc.

Chapter 7, Section II
Steel-to-steel Calendering / 223

Ronald L. Fox, Converters Paperboard Co.

Chapter 7, Section III
Soft-nip Calendering of Coated Paper and Paperboard / 225

Jack A. Perry, Q-Jet Systems Inc.

Chapter 7, Section IV
Supercalender / 231

Raymond O. Wiener, Simons Eastern Consultants Inc.

Chapter 8
Statistical Process Control for the Coated Paper Process / 235

David J. Damato, RUST International Corp.

List of Figures and Tables

Chapter 1

Chapter 2

Chapter 3

Chapter 4

Chapter 7

Chapter 8

Chapter 1

Basestock

Gordon L. Stout, Simons Eastern Consultants Inc.

Introduction

This discussion of basestock relative to the final coated product is general and broad in scope. The primary objective is to create additional awareness of the complexity of the process of basestock manufacture and the need for control of the variables which enter into that process. A secondary objective is to create awareness of the need for additional study in the reader's specific areas of interest. Two basic resources to consult are the TAPPI PRESS publications and various university paper science libraries.

Sheet Integrity and Runnability Characteristics

It is apparent that the basesheet as well as the coating are critical in the final product, and they come together most importantly under the criteria of "runnability." In order to get the web or mat of fibers through the entire conversion process, the substrate must first have uniform strength. Uniformity is essential in both cross-machine direction (CD) and machine direction (MD) in areas of caliper, basis weight, moisture, and absorbency. This uniformity is necessary to avoid wrinkles, ridges, and corrugations and to assure uniform coating pickup. In order to achieve uniform moisture, caliper, and finished basis weight profiles, the paper web is frequently dried to less than 1% moisture at the size press and 3-4% at the reel. This of course wastes energy, reduces production, reduces strength, and can lead to a shocking experience from static discharge at the paper machine reel; it is nonetheless standard operating procedure in many mills. Mechanical requirements of no holes of any size, a very minimal number of splices, and a cleanly slit roll edge free of dust and cracked edges are critical elements for an off-machine coater operation and any additional converting operations. The roll must also be of uniform hardness and free of corrugations, wrinkles, and dirt. It seems to require an open-minded papermaker to accept this fact when his supervisor's traditional reward comes from tons per hour off the reel.

Chapter 6 on coater drives, in this book, addresses concerns regarding web runnability. The interrelationship of web quality and web handling, in addition to the attributes that must be controlled for a successful operation, are well illustrated in this chapter.

Papermaking and Coating Interdependencies

A perception frequently exists that the coating can cover minor flaws in the substrate provided another increment of coating weight is applied. This perception is not reality. The basesheet makes up the majority of the coated product, in many cases 80% or more of the final product. Multiple high coating weight applications on unbleached kraft will dramatically improve surface smoothness and color, but even this process cannot overcome serious defects. Attempts to control color on lightweight basesheets by tinting coating without tinting the basesheet provide a graphic example of this fact. A simple exercise which will bear this out follows: Secure a beaker of standard pigmented and untinted coating, and use this in the lab to apply a uniform 6-8 g/m^2 coating layer to a pink or goldenrod sheet. I don't believe you will be satisfied with the results.

Pick strength requirements, although dependent on coating formulation and application, require the interfiber bonding of the basesheet. A slitter dust problem may be caused by an abrasive filler in the basesheet such as calcium carbonate or calcined clay. This interdependency of basesheet and coating illustrates the need for communication of needs and capabilities between the papermaker and the coater crew. This need for communication exists throughout the mill and finally beyond the mill to printing and converting operations. The people doing the day-to-day hands-on operations are the experts on the actual process capability, and they need the support of management and technical groups in order to produce a consistently high-quality product.

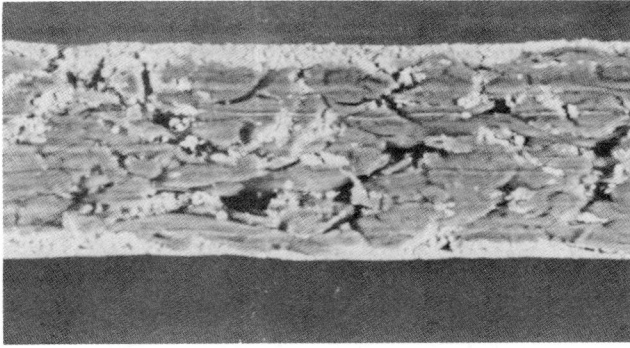

50μm

Fig. 1.1. Cross section of coated paper with moderate to high coating weight

50μm

Fig. 1.2. Cross section of lightly coated paper

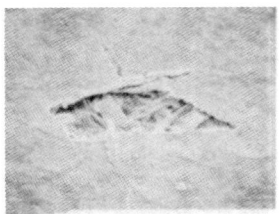 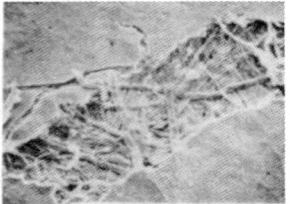

Fig. 1.3. Photomicrographs of craters in coated paper which illustrate the contrast between coated and uncoated surfaces of a paper substrate

The trend toward converting from acid to alkaline papermaking often leads to discussions related to runnability. When a major process change such as this occurs, there will be problems until new skills are learned and operational parameters are defined. Some excellent coated sheets utilize an alkaline basesheet, while equal quality coated sheets are produced using an acid basesheet.

We could list the advantages and disadvantages of an acid or alkaline basesheet, but again we need to look at its end use. It is well to realize, especially in the case of an alkaline system, that a lot of fiber can be replaced with filler. However, carried to excess, the filler will reduce sheet strength for a given basis weight. An alkaline sheet will not require as much coating to obtain the same smoothness, brightness, and opacity and will respond well to calendering because of the higher filler content. A well managed system, whether acid or alkaline, will be a very clean system, produce few holes, and present a very uniform surface to the coater head with a good chance of the result being an excellent coated product. Another consideration is that the longevity of an alkaline sheet is almost infinite while an acid sheet deteriorates over time. Even so, both acid and alkaline systems will be important in the foreseeable future because of different needs. To date, wood-containing sheets are acid or neutral by necessity. Mills have traditionally been acid, and the incentive to change based solely on economics is questionable for many mills. There is, however, a clear trend toward alkaline—and neutral-sized sheets which seems likely to continue.

Often there is a question of which side of the sheet should be coated first on coated two side (C2S) grades or which side should be coated on coated one side (C1S) grades. The answer to these questions depends more on the equipment involved than on the raw materials. With the advent of twin-wire formers the two-sided issue has generally improved, but factors such as wet press configuration also have a significant effect on two sidedness and smoothness. Which side to coat is best determined by reflecting first on the product end-use requirements and then determining the attributes of a given machine. It is common to coat the bottom side first because it is often the most closed side of the sheet. The first coater will normally apply a higher coating weight than the second coater given the same operating parameters. On C1S grades, a reasonable approach is to coat the smoother of the two sides. In the case of a machine-glazed (MG) sheet, this is the MG side; in other sheets, it is generally the felt side or top side from a traditional fourdrinier.

Occasionally, a decision to coat one side or the other is based on the curl requirements. This is usually not a good choice and is better addressed by controlling post-coating differential drying, adding a moisturizing unit, or both. A moisturizing unit utilizing a variable-speed hydrophilic roll to precisely meter a uniform film of water or starch solution to the uncoated side after coating will effectively eliminate curl problems.

I would emphasize the answers to the points raised here are found in intimate knowledge of the end user's requirements and not in what we believe the requirements are. To illustrate this story: You are in the shower, and it feels just right. Your friend passes by and sees steam and believes your welfare is at stake, then reaches in and turns up the cold water. Do we ever do this to our customers?

Smoothness—An Attribute Necessary for a Quality Product

The final coated product is generally more smooth than the basesheet and has additional gloss, opacity, and brightness. The key words are "more" and "additional" as the basesheet is enhanced by the coating layer. Application of coating to a sheet with surface defects such as dirt, shieves, lumps, etc. will negatively enhance these defects.

Smoothness is an attribute commonly sought as a primary objective in most coating operations. Smoothness is related to gloss which also is generally sought after. In order to view the attribute of smoothness from another perspective, let us compare common mineral coating to latex house paint. Except for application technology, these products are similar in function. To pursue the analogy further, have you ever painted an unsanded board or piece of rusty metal? If so, you probably observed it wasn't very attractive or very smooth. The comment "the surface wasn't properly prepared," is of course true. At 8 g/m^2 of coating weight, a thickness of dried coating of less than 5 micrometers is applied to an interwoven mat of fibers. These individual fibers range in diameter from 20 micrometers to 65 micrometers even after refining, pressing, and drying. Add to this that, due to fiber swelling and debonding, every wet and dry cycle increases sheet roughness and we can see the need for a smooth basesheet.

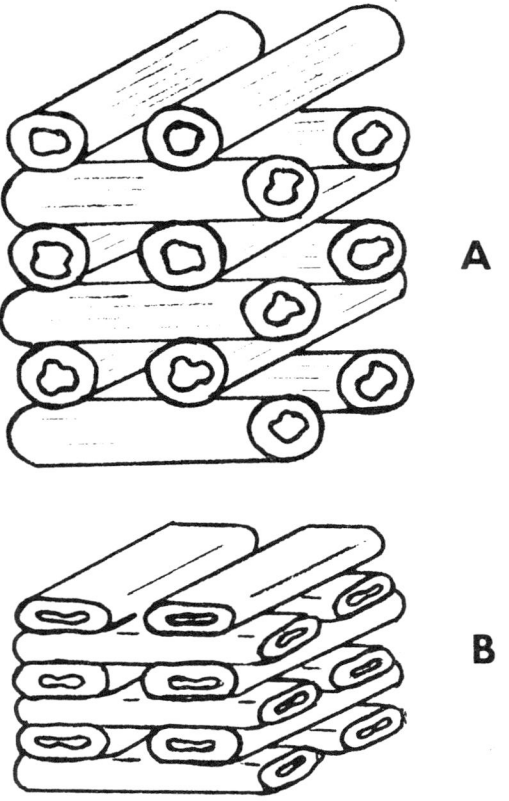

Fig. 1.4. Idealized fiber structures. The thick-walled fibers (A) are less conformable than the thin-walled fibers (B)

The micro smoothness needed for rotogravure printing requires a very smooth basesheet. Several methods of measurement of smoothness exist, but a Parker Print Surf (PPS) instrument is easily used and quite accurate. In addition, the PPS used properly can predict final coated sheet smoothness, following a standard coating application based on basesheet smoothness. This predictive test needs to be based on specific equipment, i.e., a specific paper machine and a specific off-machine coater, but is very useful within the mill. Sheffield Smoothness can predict many smoothness requirements quite well, but the

Table 1.1. Properties of North American pulpwoods

Species	Fiber Length mm	Fiber Diameter μm	Fiber Wall Thickness, μm Earlywood Latewood	Ratio L/T	Coarseness mg/100 m
Birch	1.8	20-36	3-4	500	5-8
Red Gum	1.7	20-40	5-7	300	8-10
Black Spruce	3.5	25-30	3-4 (70%) 6-7 (30%)	700	14-19
Red Cedar	3.5	30-40	2-3	1400	15-17
Southern Pine	4.6	35-45	2-5 (50%) 8-11(50%)	700	20-30
Douglas Fir	3.9	35-45	2-4 (60%) 7-9 (40%)	700	25-32
Redwood	6.1	50-65	3-4	1700	25-35

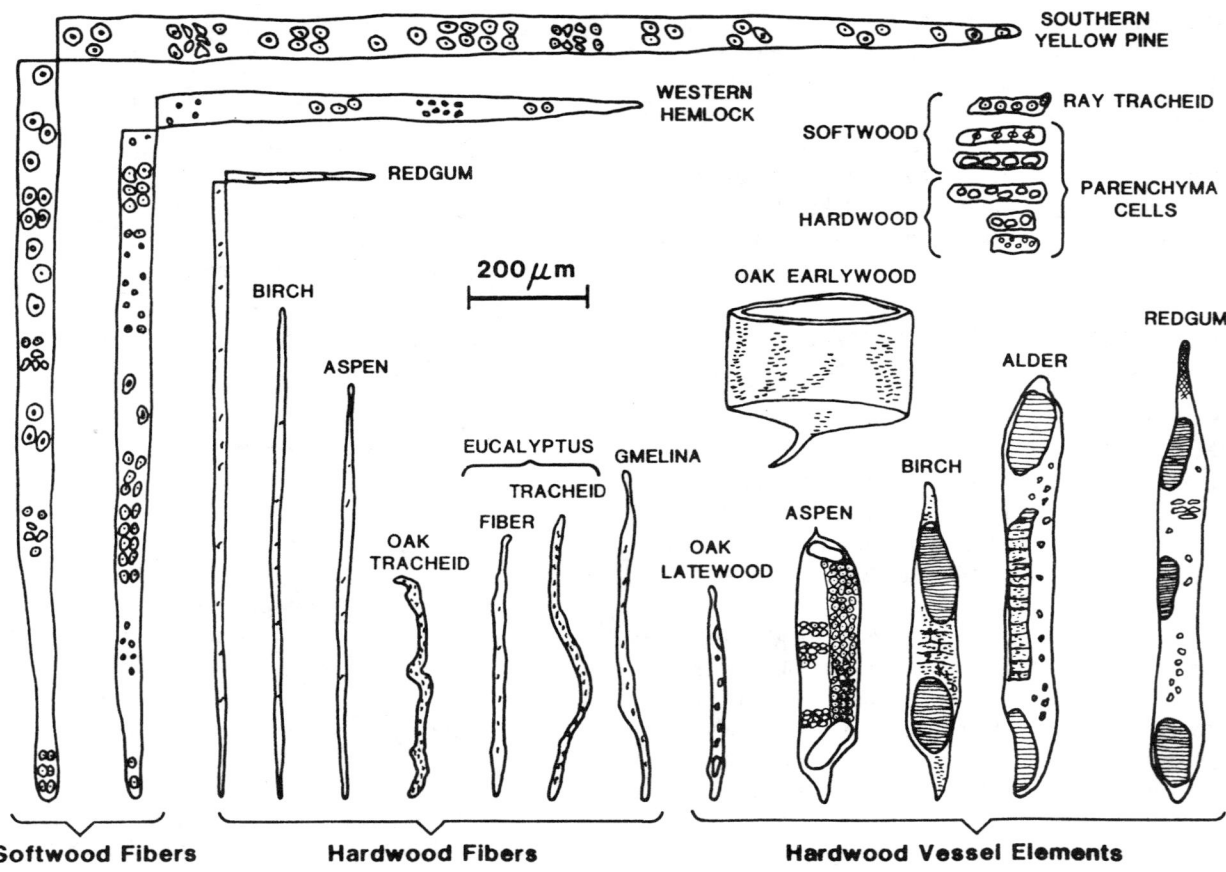

Fig. 1.5. Diagrammed major cell types in softwoods and hardwoods. All diagrams are at the same magnification to show the relative sizes of these elements

PPS is superior for prediction of rotogravure requirements.

Higher coating weight, followed by effective supercalendering, will improve smoothness but will reduce bulk and stiffness. Multiple applications of relatively low coating weights will improve smoothness, and the coating of a sheet from a machine with a properly used Yankee dryer will improve smoothness. The interdependency of the coating and the basesheet must be recognized and balanced to achieve specific end-use requirements.

Relationship of Product Requirements to the Manufacturing Process

A few basestock requirements have been discussed now. Assuming we are looking at a paper machine supplying an off-machine coater, let's walk through the process beginning with fiber selection and ending with the reel to help understand the process in more detail.

Starting with fiber selection, we have a choice first of

all of regional varieties designated NHW for northern hardwood, SHW for southern hardwood, NSW for northern softwood and SSW for southern softwood. Generally speaking, HW contains ten times the number of fibers per unit volume compared to SW. Northern fibers tend to be smaller than southern fibers. Softwoods yield longer fibers and impart more strength to the sheet. This sounds pretty straight forward, but species selection also enters into the equation. If you have observed Douglas-fir fibers microscopically, you have seen very large-diameter softwood fibers (especially compared to a eucalyptus fiber). Even the smallest fiber is larger in diameter than the coating layer thickness. As these fibers are refined, many "fines" are produced which will fill in voids of the fiber mat and improve smoothness, formation, opacity, and generally enhance the surface presented to the coater.

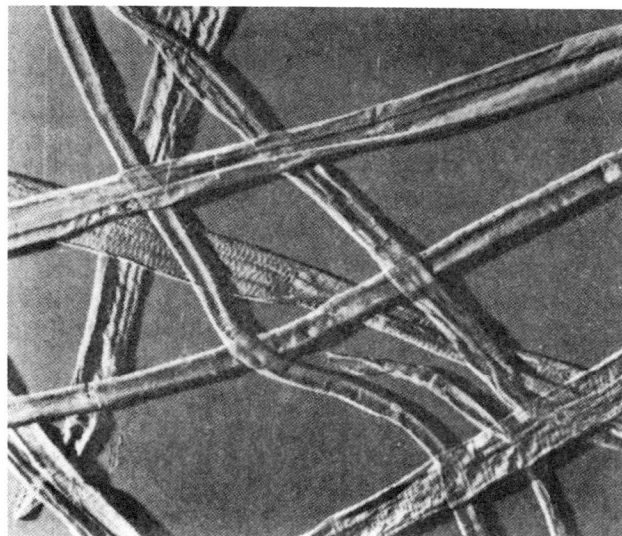

Fig. 1.6. Douglas-fir fibers have relatively thick cell walls and do not collapse as readily as other softwoods. The "spiral thickenings" enable Douglas-fir fibers to be easily identifiable (Weyerhaeuser Pulp Division)

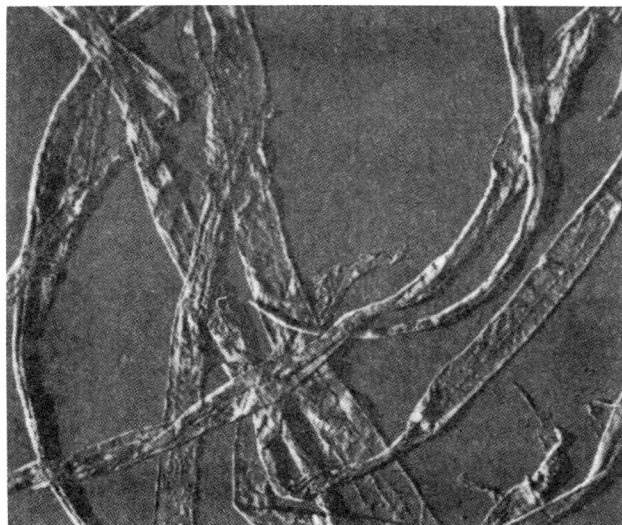

Fig. 1.7. Western red cedar has a very thin ribbon-like fiber. Average fiber length is 3.2 mm (Weyerhaeuser Pulp Division)

Along with the primary fiber source selection, another consideration has to do with the basic process of fiber separation from the roundwood or chips. In a chemical process, the lignin or glue that holds the fibers together is dissolved during a cooking process, removed during washing, and the fibers finally bleached to yield bleached pulp. If these fibers are used exclusively, we have a woodfree or "freesheet." Other options include stone grinding of certain species to produce stone groundwood and combination pulp processes such as thermomechani-

cal pulp or chemithermomechanical pulp (TMP or CTMP). These processes are based on economics and yield pulps which are quite desirable for specific end uses. Many lightweight coated (LWC) grades contain a high proportion of these pulps and yield a product which is very acceptable for the intended use. Generally speaking, for strength and quality considerations in specialty grades and coated Grades 1 through 3, the furnish is 90-100% "free," while Grade 5 will be less than 50% "free."

Another fiber source which is finding more and more favor comes in the form of broke from the mill and purchased secondary fiber. Clean recycled fiber can add a lot to the furnish and make a smoother and more opaque sheet, but contaminants can cause serious quality problems, especially on a blade coater. Precautions must be taken to keep plastic and other contaminants out of the system. A good cleaning system is necessary, but—because of the reality of economics and human nature being prone to failure—a good ongoing and concentrated effort to avoid the contaminant before it has an opportunity to get into the system is essential. Simple awareness of the consequences can help tremendously.

Unlike the furnish, a lot of attention is given to the variables in coating preparation. A coating recipe is formulated, published, and adhered to. Substitution of one manufacturer's clay for another is studied thoroughly before the change is allowed. This is true to an even greater extent in the binder system. Although raw material cost is considered, it is not the key issue in selection. When it comes to fiber selection, especially in the areas of species selection and broke source, economics is a very strong determinant. This is not hard to understand since the traditional reward for managers is based on short-term profitability rather than on long-term growth based on product quality. There should be detailed fiber specifications and rigid furnish recipes for stock prep, and these parameters should be adhered to as are the material specifications for materials applied to the paper in the coating process.

Additives to Improve Optical Properties

Fibers properly treated have a reasonable degree of strength and allow the opportunity to enhance optical properties of the finished coated product.

The addition of colorants such as dyes or pigments to enhance brightness, or to color forms bond grades yellow or pink and other variations, is a common practice. An important point to note is that any colorant added requires similar treatment and matching in the coating. Today's closed-loop systems that adjust colorant addition are quick payback projects. A very important consideration regarding colorants is the 80:20 ratio of basestock to coating. For this reason, it is not feasible to tint only the coating and expect it to carry all the color. Neither will adjusting coat weights control color variations in the basestock. Although in some specialty or board grades, it is possible to obtain desired color values with colorant addition only to the coating, it is desirable to adjust both basestock and coating colorant addition precisely in most cases.

By adding inexpensive filler clay to improve opacity and brightness, there is the additional benefit of a smoother sheet after the calender. Titanium dioxide is very effective in optical improvements even though the cost is very high. Calcined clay or large particle size clay are effective fillers in many cases. In an alkaline system, calcium carbonate adds great benefits including additional brightness at low cost. There is, however, a limit on filler use as the pigment particles do not bond as the fibers do, resulting in decreasing sheet strength as the filler amounts increase.

Because fiber and filler is suspended in water, it follows that the beneficial fines and fillers may be the first to leave in the water removal process at the forming area. Something is required to retain these fine particles so retention aids, cationic starches, and alum are added for this purpose. When this wet-end chemistry is balanced, a proper attraction exists between particles and fibers for good retention, without excessive flocking resulting in poor formation.

Refining, Forming, Pressing, and Drying

Next we need to address the process of refining. This process can enhance the coating basestock a great deal when combined with proper furnish selection. When a relatively low amount of energy has an impact on the fibers several times, the fibers tend to fibrillate, unwind, and become more flexible without being excessively shortened by cutting. If there are fewer impacts, but these impacts are more severe, then excessive cutting resulting in reduction of fiber length occurs. Sometimes there is an inclination to refine with high-impact energy when the primary fiber is course such as with Douglas-fir or southern pine. It would often be more economical and better for final product quality if more highly refined hardwood (HW) or eucalyptus were used to produce a less porous, well formed sheet. Serious communication and quality planning needs to take place between the coating people, the stock prep people, the pulp mill, and the purchasing people for a synergistic combination of the parts.

This refined slurry of pulp and filler passes through cleaners to remove remaining contaminants, then goes into the headbox and onto the wire of the paper machine. There have been headbox upgrades in the past 15 years which have resulted in greatly improved machine and cross-machine profiles and general sheet quality. The tools to correct variations are there if the feedback and response are timely. An accurate and precise combination of electronic and human interaction can improve quality a great deal in this area.

Forming fabric selection can certainly be an important aspect of how well the sheet is formed. There are fine fabrics, coarse fabrics, and multi-layered fabrics, and each suits a purpose. The fine-strand multi-layered forming fabric will tend to produce a smooth wire-side surface and inhibit water removal. Conversely, an excessively rapid drainage can be produced by coarse fabrics. Coarse fabrics and the configurations of foils and vacuum boxes on a paper machine can allow higher machine speeds but can promote pulling of fines and fillers out of the sheet to the point of producing voids or pin holes. This kind of basesheet will negatively affect a lightweight basesheet in areas of coating holdout, uniformity of product, and other similar attributes.

It is evident that a balance between speed and drainage is important, and to this end the twin-wire former was conceived. Properly balanced, the twin-wire former will produce a high-quality basesheet by removing water from both sides of the paper web. A conventional long forming table can also produce a smooth and closed sheet. Water removal is the primary function of the wet pressing section, but wet press design must address issues of two sidedness, smoothness, and caliper in addition to water removal.

Contact drying on cylinder dryers at high steam pressures, after the press, will result in fibers in the sheet raised, pulled, or sticking to the dryer cylinder surface, a phenomena referred to as dryer picking. Pre-drying with infrared dryers or using release coatings on the first few dryer cylinders helps to minimize problems associated with picking.

As the web continues through the dryer section, loose fibers may be redeposited. When the sheet gets to the size press, some of these fibers become part of the size and are reglued to the web. Others are retained in a filter or continue through the process. Even though a 200-mesh screen is capable of allowing 80-micrometer particles to pass, good size press screens of 100-150 mesh are invaluable in removal of contaminants that would otherwise be on the basesheet surface. You will recall the normal fiber diameter range is 20-65 micrometers, but greater lengths trap many fibers on the screen.

Surface debris, made up not only of fiber and scale but sometimes including precipitated amylose from the size solution and other contaminants, affects the coating stations in a blade coater. At the blade, fiber and other contaminants are pulled from the sheet and become a part of the coating; again some are removed by the coating filters, and some redeposit on the web. Some contaminants lodge under the blade, resulting in streaks and scratches. The next stop for these contaminants is often the printing press, resulting in a justifiably irate printer.

The basesheet is frequently dried to less than 1% to enter the size press. This moisture level limits the pickup and facilitates uniform pickup of size solution. It is well to remember that if the sheet is formed of uniform weight, then caliper and moisture will tend to be level and uniform also. Sizing is often a very essential part of the basesheet manufacturing process, while in other basestock machines there is no size press. It is well to consider that each rewetting and drying of the basesheet results in increased sheet roughness due to fiber swelling and debonding.

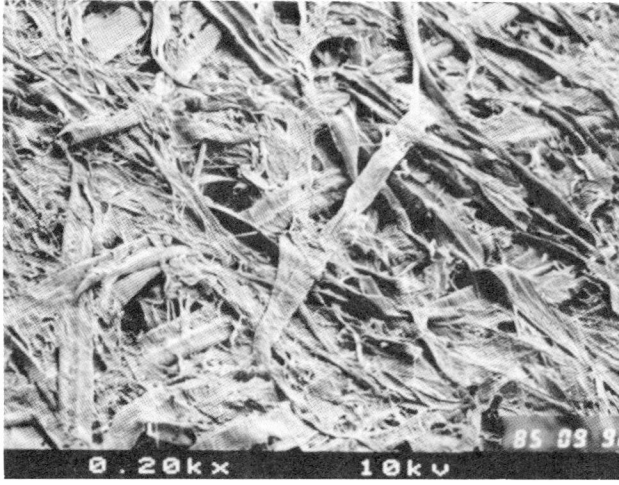

Fig. 1.8. Scanning electron micrographs of basestock before and after pretreatment with water on the blade coater. This illustrates the debonding and resultant rough surface caused when the fiber mat is rewet

A size press in itself will not produce a smoother sheet; however, the size press coupled with a properly formulated pigmented size and subsequent calendering can produce a sheet well suited for certain printing applications. This is especially true with a film metering-type size press. Following the size press with noncontact drying prior to contact drying will greatly reduce or eliminate picking in that area. Infrared drying is effectively used to level the cross-direction moisture profile as well as minimize picking in this location. The size press can apply starch size penetrating the web in a way that will greatly improve pick resistance in offset printing applications.

Another useful aspect of the size press is the use of inexpensive size solution to reduce expensive coating absorption by keeping the coating on the surface. The nature of the size press and its capability to either add or detract from sheet quality requires very close communication between the machine crew and the size preparation people.

On-machine calendering enhances sheet smoothness, but even with hard roll controlled zone calenders, reduction of caliper and bulk frequently is more than is desired. An example is producing a sheet for rotogravure printing where that attribute called "compressibility" is necessary for good performance. A well formed and uniformly dried sheet passed through a soft calender will produce an excellent sheet for many applications and generally result in very uniform caliper. This in turn results in a well constructed reel of paper free of corrugations, ridges, and wrinkles.

Fig. 1.9. Surface of a supercalendered clay starch coating, 4400X (courtesy of The Institute of Paper Science and Technology)

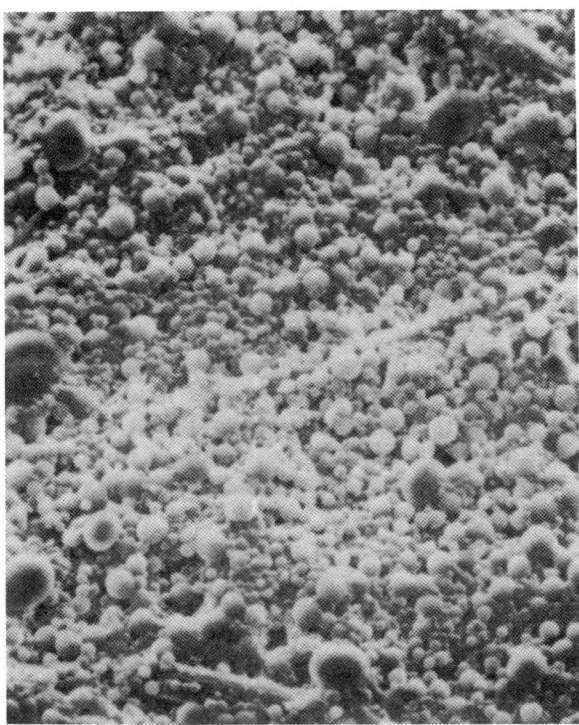

Fig. 1.10. Surface of a micro-encapsulated, CB sheet of a carbonless copy paper system, 330X (courtesy of The Institute of Paper Science and Technology)

The last steps of the process of basestock production include an effective hole detector to detect any holes which can be subsequently patched or removed. The primary value of a hole detector is to alert the crew so immediate action can be taken to eliminate the *source* of the holes. Moisture, basis weight, and caliper are monitored; and some operations include closed-loop control, both MD and CD, of these attributes. This illustrates a trend toward quality awareness and product uniformity. Daily monitoring of this equipment by concerned, properly trained personnel is essential for lasting quality improvement. Distributed control systems (DCS) can produce and process endless data leading to either confusion or excellent process control.

Proper tension control at the reel is essential to present a uniform wrinkle-free and corrugation-free sheet to the winder or rereeler operation and ultimately to the off-machine coater. One mechanical defect sometimes forgotten which results in a break at the off-machine coater (OMC) or in a subsequent converting step is called a "burst." Bursts are caused by entrained air at the nip of the reel, winder, or rereeler.

Summary

The manufacture of basestock has many components and is influenced by many people in both staff and operating positions. Raw material and equipment selection are of prime importance, and certainly we can say the manufacture of basestock consists of many variables. Certain variables are very critical to the process and therefore cannot be left to chance. If we are to produce a quality or satisfactory basesheet, the papermaker must first be familiar with the end use. The key variables in basestock manufacture must be identified, a means to measure them established, and a means to adjust or control them must be made available to the papermaker to ensure a quality basesheet is presented to the coater.

Bibliography

References

1. Hamilton, J.W., Condon, M.C., Reeves, R.H., Dalquist, R.W., Janes, R.W., Walker, D.A., 1989 TAPPI Coating Conference Basestock Quality Roundtable, audio-taped proceedings.

2. Scott, W.E. and Trosset, S., *Properties of Paper*, TAPPI PRESS, Atlanta, 1989.

3. Beatty, J.E., et al., 1988 Coating Basestock Seminar, TAPPI PRESS, Atlanta, 1989.

4. Smook, G.A., *Handbook for Pulp and Paper Technologists*, Joint Textbook Committee TAPPI and CPPA, TAPPI PRESS, Atlanta, 1986.

5. Kocurek, M.J. and Stevens, C.F.B., *Pulp and Paper Manufacture, Vol. 1: Properties of Fibrous Raw Materials and Their Preparation for Pulping*, Joint Textbook Committee, TAPPI and CPPA, TAPPI PRESS, Atlanta.

6. Kheboian, G.I., "Off-Machine Coater Drive Systems Back to Basics," TAPPI PRESS, Atlanta.

Chapter 2, Section I

Typical Coating Components

Lars G. Andersson, J. M. Huber Corp., Clay Division

Introduction

Aqueous coatings for paper or board applications contain three distinctive groups of materials besides water: pigments, binders, and additives. Any mixture not containing these three elements is really not a true paper or board coating, but can be a size press solution, a water box solution, or possibly a wash coat. Coatings that do not utilize water as the fluidizing media are usually called solvent coatings. These belong to a small, highly specialized group of coatings and will not be covered in this section. We will now take a closer look at the three material groups mentioned.

Pigments

The following pigments are commonly used in today's paper and board coatings: kaolin clay, calcium carbonate, titanium dioxide (TiO₂), and alumina trihydrate. To a lesser extent, there are also some satin white, amorphous silica and silicate, and synthetic as well as structured pigments being used. For more details, see TAPPI Monograph No. 38, *Paper Coating Pigments*.

A. Kaolin Clay

Kaolin clay is by far the largest group of coating pigments used today, and its use in coatings goes back to the 1920s. The kaolin clays can first be divided into two groups hydrous and anhydrous clays, the latter commonly called calcined clays. Furthermore, hydrous clays can be divided into two distinctive groups, airfloat (dry processed) and waterwashed clays (wet processed).

Airfloat clays (see Fig. 2.1) are primarily used in filler applications and play a very minor role in coating because of their relatively high grit content and low brightness, but can be used in a pre-coat if applied by, e.g., a roll coater or a size press where scratches are not an issue.

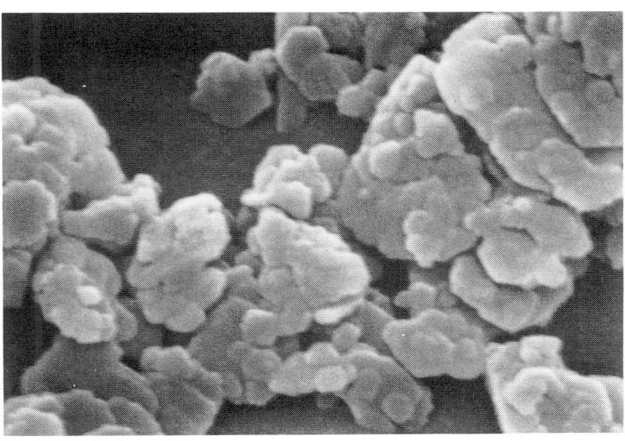

Fig. 2.1. Airfloat clay, 20,000X (courtesy of J. M. Huber Corp.)

Waterwashed clays can further be divided into delaminated (see Fig. 2.2) and regular clays (see Fig. 2.3). Delaminated clays are produced by applying intensive mechanical shear force to a clay slurry, resulting in the clay stacks or booklets being broken down into individual clay platelets or smaller stacks. Delaminated clays can be delivered in coarse or fine particle size. Delaminated clays are, in general, good for coverage and holdout. They produce low sheet gloss but high ink gloss, and are primarily used in coatings for lightweight coated (LWC) papers, where good coverage at low coat weight is important. Rheology at high coating solids can be troublesome.

Regular waterwashed clays are classified by the numbers 1-4, based on percent of pigment less than 2μm with No. 1 being fine (90-94% less than 2 μm) and No. 4 being coarse (65-70% less than 2 μm). There is also a No. 1 fine particle size clay (95-100% less than 2 μm) for maximum sheet gloss development. No. 4, on the other hand, is a coarse clay mostly used in filler applications, but can be used in precoat applications. No. 1 fine and No. 1 and No. 2 clays can be delivered in both standard (approximately 86-88%) and premium (90-91%) brightness form. Delaminated clays can be either coarse or fine, but particle size is in general not reported because the delaminated nature of these clays make particle size measurements less precise, according to the Sedigraph method.

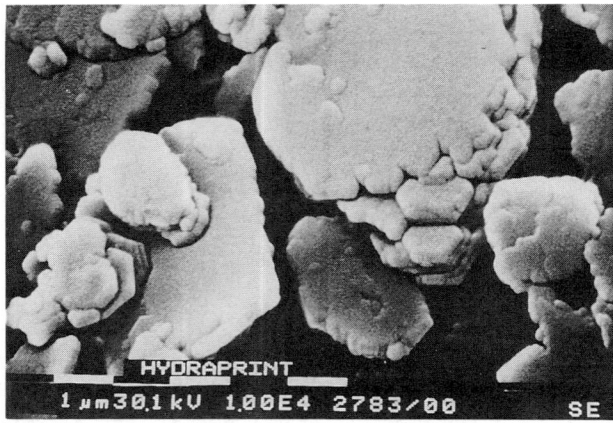

Fig. 2.2. Delaminated clay, 10,000X (courtesy of J. M. Huber Corp.)

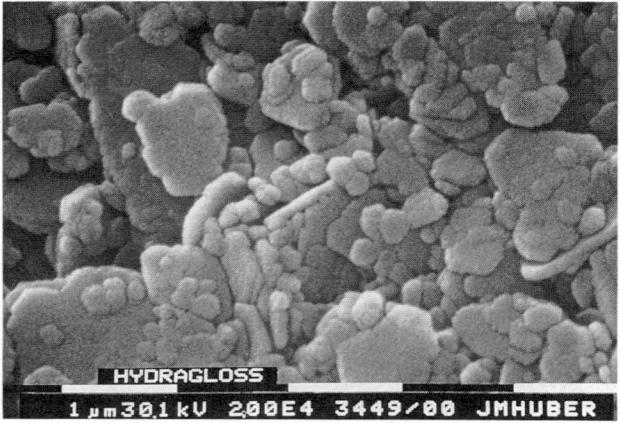

Fig. 2.3. Regular waterwashed clay, 20,000X (courtesy of J. M. Huber Corp.)

Because of its particle size, the No. 1 fine particle size clay has the largest surface area and the highest binder demand, while a No. 4 clay has the smallest surface area and the lowest binder demand. Besides the difference in binder demand, a No. 1 fine clay slurry will have high low-shear viscosity but low high-shear viscosity, while a No. 4 clay slurry or coating will have these properties reversed. No absolute correlation between clay slurry and coating viscosities has been found and rarely would these clays be used as the sole pigments.

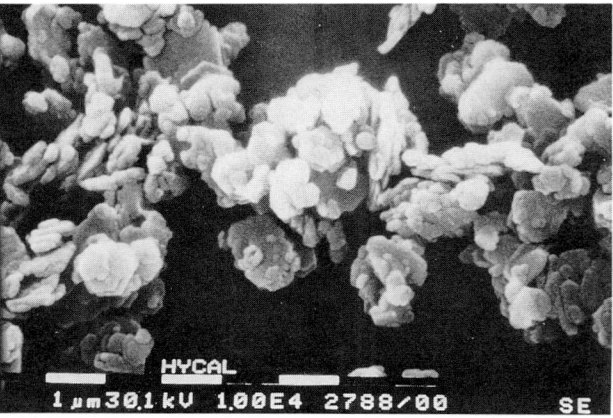

Fig. 2.4. Calcined clay, 10,000X (courtesy of J. M. Huber Corp.)

Finally, calcined clay (see Fig. 2.4) can be coarse or fine and is made from hydrous clay that has been exposed to very high temperatures—approximately 1800°F—in a calciner. The plate-like structure of the clay is changed into a bulky, porous structure resulting in lower density and higher scattering coefficient. Calcined clay is therefore a good TiO_2 extender in high opacity papers, but is also used for porosity and blister resistance in web offset papers and for coverage and compressibility in LWC rotogravure papers. Because of processing costs, calcined clays are significantly more expensive than regular hydrous clays. One drawback with the calcination process is that a certain amount of sintering takes place, resulting in a very abrasive pigment compared to regular hydrous clays.

B. Calcium Carbonate

Calcium carbonate ($CaCO_3$) is a pigment which, for a long time, has been used in Europe, is readily available at low cost, and is now gaining popularity in the United States, especially in mills utilizing alkaline papermaking. Calcium carbonates, in general, have high brightness, oil absorption, and light-scattering coefficient, but low sheet gloss developing ability when compared to, e.g., a No. 1 clay. It, therefore, lends itself well for use in high brightness and opacity dull and matte grades.

Calcium carbonate is produced in dry ground, wet ground (see Fig. 2.5) (from limestone), and precipitated form. They are available in several different particle size ranges. The coarser grades are commonly used in precoat, basecoat, or size press formulation, while the fine ones are suitable for top coat. The particle size—e.g., 90% less than 2µ—and shape of fine, wet ground grades make it possible to use these pigments at high levels and high coating solids without rheology or runnability problems in high-speed blade coaters. Levels of 70-100% $CaCO_3$ are common in Europe. The high solids used—

65-75%—and the use of low-level, all-synthetic binders will, to a great extent, compensate for and even eliminate the gloss depressing properties of the $CaCO_3$.

Fig. 2.5. Wet ground calcium carbonate, 1 μm median size, 5,000X (courtesy of J. M. Huber Corp.)

Precipitated $CaCO_3$ is commonly produced by reacting calcium hydroxide with carbon dioxide, and will produce a pigment of very high brightness and purity. Due to its large surface area, it in general has a higher oil absorption than the ground limestone. Depending upon reaction conditions, particles of different shape and crystalline structure can be produced. Aragonite crystals have acicular shape and are somewhat suitable for coatings, while calcite crystals can have either scalenohedral shape suitable as filler, or prismatic shape (see Fig. 2.6) suitable for coating. A particle size of 0.25-1.0 μm seems to be preferred in coating applications.

C. Titanium Dioxide

Of all coating pigments used, TiO_2 has by far the highest refractive index, resulting in excellent opacifying properties. While all other coating pigments have a refractive index that falls in the 1.50 to 1.60 range, rutile TiO_2 has a refractive index of 2.73, while anatase TiO_2 is somewhat lower at 2.55. TiO_2 is also unsurpassed in "wet" opacity, i.e., a TiO_2 coating will maintain most of its high opacity even after varnishing or waxing. TiO_2 has high brightness and a blue-white shade.

TiO_2 has a small particle size (see Fig. 2.7) and large

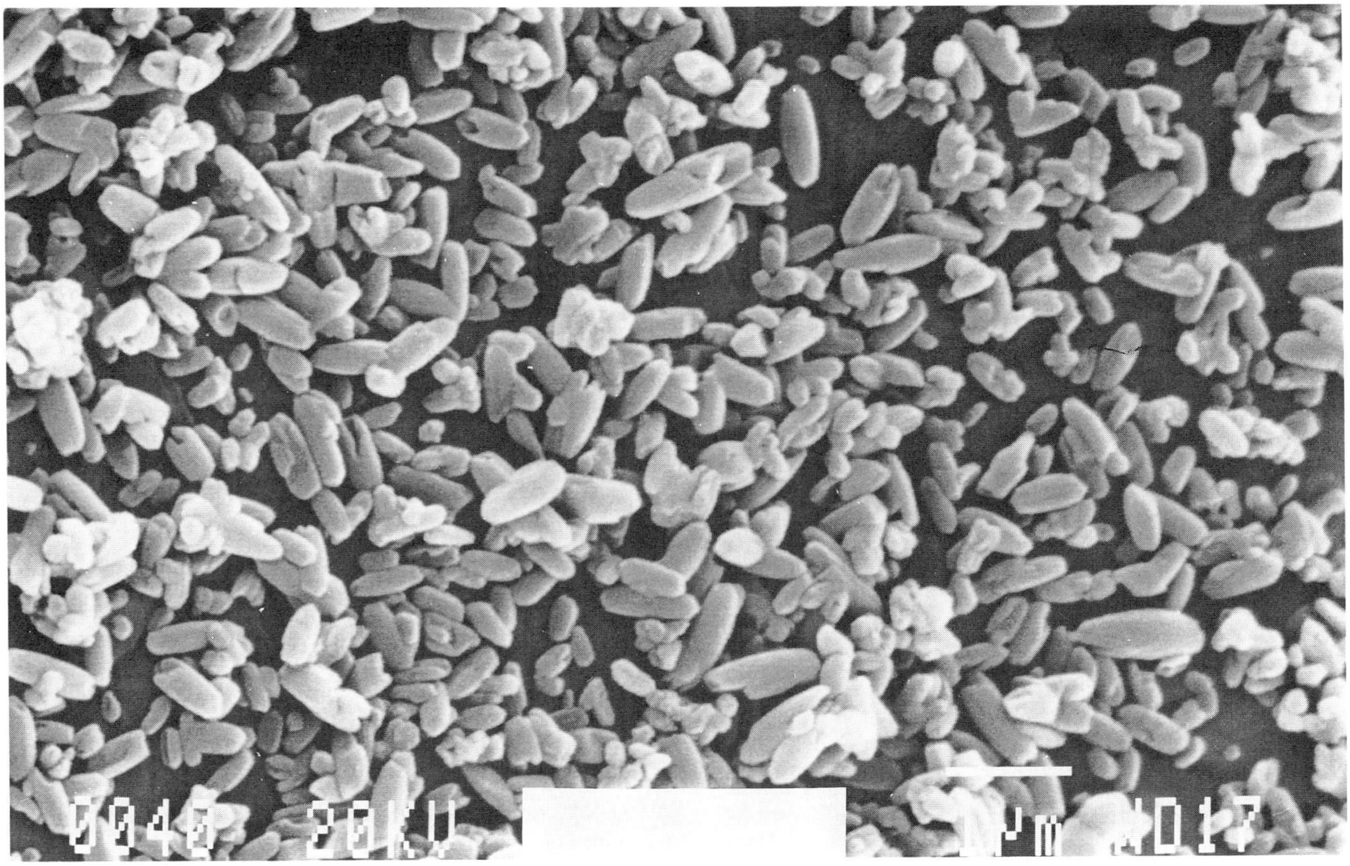

Fig. 2.6. Coating grade precipitated calcium carbonate: calcite crystal, prismatic shape, mean particle size 0.5 μm, 5,400X (courtesy of Pfizer Inc.)

surface area, resulting in a fairly high binder demand. It is also nearly as abrasive as calcined clay. Because of its high cost, TiO_2 is frequently extended by other pigments, e.g., calcined clay and structured pigments, made to approach the opacifying properties of TiO_2. TiO_2 is primarily used in premium grades, like Nos. 1-3, and in coatings for unbleached kraft and recycled board.

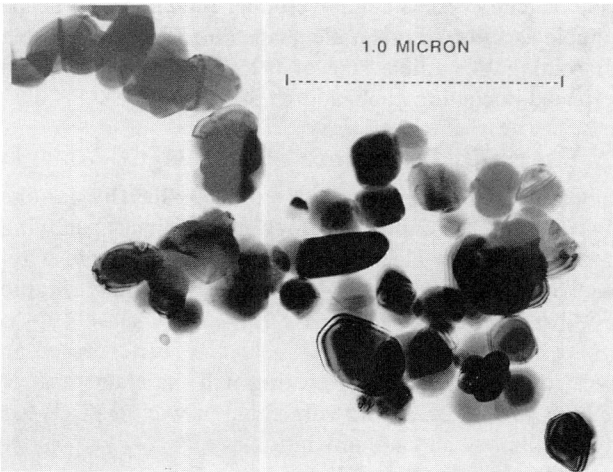

Fig. 2.7. Titanium dioxide, 200,000X (courtesy of DuPont)

Fig. 2.8. Alumina trihydrate, mean particle size 1 μm, 10,000X (courtesy of J. M. Huber Corp.)

D. Alumina Trihydrate

Alumina trihydrate (see Fig. 2.8), or ATH, is a pigment with very high brightness (approaching 100, GE brightness) and blue-white shade. It can be obtained in different particle sizes from coarse to fine, but for coating

application the sub-micron variety is preferred. In the past, fine and ultrafine ATH were precipitated into its fine, final form. Lately, the fine and ultrafine grades are being produced by utilizing a very coarse precipitated product combined with proprietary wet grinding methods. This, plus other processing steps, now makes it possible to produce fine particle size ATH in slurry form. The sub-micron grades have a binder demand similar to that of TiO_2, while the coarser grades are similar to a No. 2 clay in this respect. ATH by itself will generate very little opacity, but will function as an extender for TiO_2 by acting as a spacer between the TiO_2 particles. Besides high brightness, the fine particle size ATH is said to promote sheet gloss and printed gloss. In specialty grades, the fire-retardant properties of ATH can be of high importance.

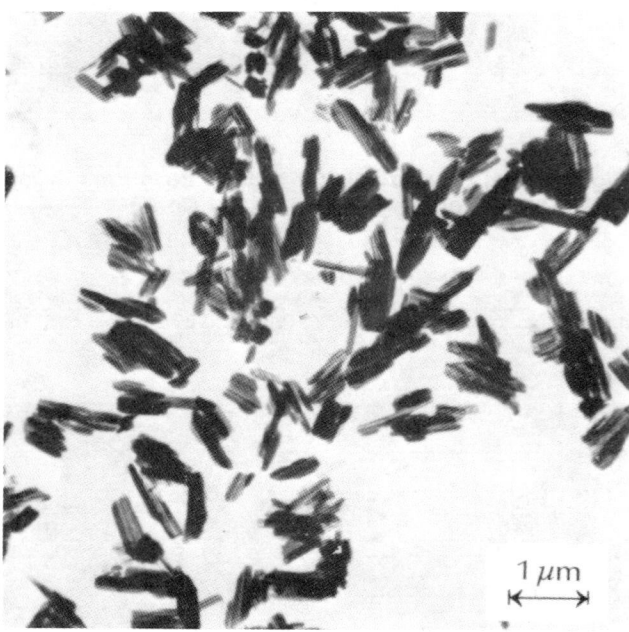

Fig. 2.9. Satin white: needle structure, length: 0.81 μm, thickness: 0.1-0.2 μm (courtesy of Suprasmit USA)

E. Satin White

Satin white (see Fig. 2.9) can be classified as a semi-synthetic pigment and its use goes back a long time, actually to some time prior to 1880. Because of the inherent problems in producing a high-solids pigment slurry, satin white is, or was, most often produced on site by reacting lime with aluminum sulfate. This produced a slurry of 30-32% solids with particles having an acicular shape (needle shape) and small size, approximately 96% below 2 μm and 75% below 1 μm. Today, however, a Dutch company is commercially producing a 50% slurry with good shelf-life stability.

Very little satin white is used today in the United States, but is still quite common in premium coated

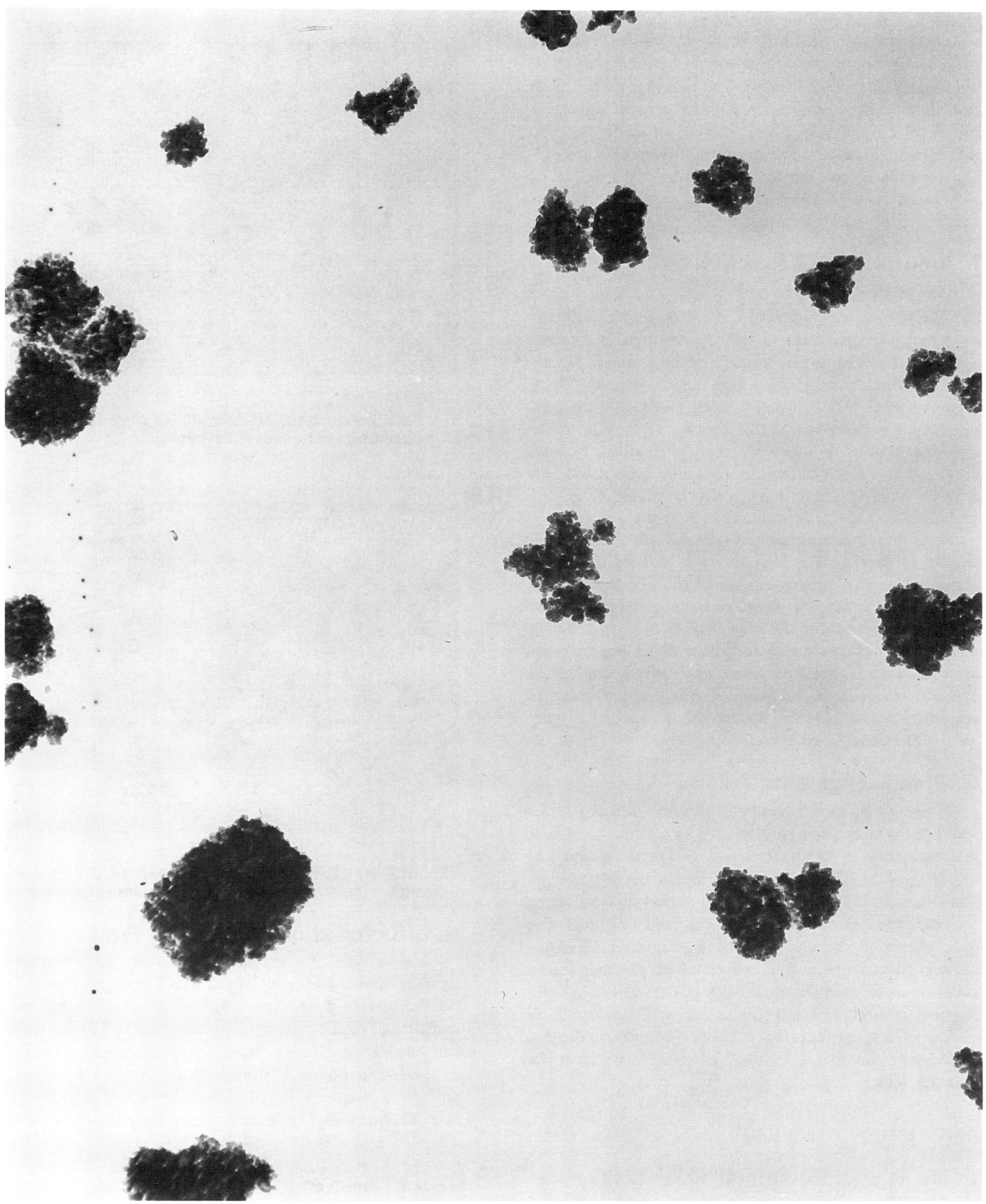

Fig. 2.10. Precipitated silica: mean particle size 1.4 ± 0.25 μm, 10,000X, SAN-SIL® WB 42 (courtesy of PPG Industries Inc.)

grades in Europe. Satin white will produce high void volume, sheet gloss, and brightness, as well as excellent coverage and bulk at low coat weights, but coating solids are usually as low as 50-55%. The immobilization solids of a satin white coating is only a few percent higher, resulting in a highly developed coating structure, but also in a delicate balance between runnability and coating solids. In experienced hands, satin white coatings can be very beneficial to the user, producing among other things excellent multi-color print quality.

F. Amorphous or Precipitated Silica and Silicate

Amorphous or precipitated silica (see Fig. 2.10) are used to a certain extent in today's coatings. In general, these pigments have very high brightness, small particle size, low abrasion, and fairly high binder demands. These pigments will produce bulky, high brightness coatings, but only a nominal amount of opacity. Optical brighteners are said to work well with precipitated silicas because they do not quench ultraviolet light like TiO_2 does. This type of pigment is used primarily in premium grades, such as Nos. 1 and 2 as well as No. 3 to some degree.

Amorphous or precipitated silicate (see Fig. 2.11) is similar to the silica. It has a very high brightness, low abrasion, high oil absorption, and low bulk density. The clusterlike structure of some silicates generate a fairly high light-scattering coefficient and hence good hiding power. Due to particle size and shape, these pigments are more commonly used as fillers but have found use in coatings where high solids and low high-shear viscosity are not required. Like the silicas, the silicates work very well with optical brighteners.

G. Plastic Pigments

Plastic pigment is a synthetic pigment (see Figs. 2.12 and 2.13) which could be classified as a latex, because it is a dispersion of synthetic polymer particles in water at 48-50% solids. The main criteria is that the particles be non-film-forming and remain more or less discrete under the conditions of application, drying, and finishing. The only plastic pigment to gain broad commercial acceptance is polystyrene, which is a very hard polymer with a glass transition temperature (Tg) of 100-110°C. Plastic pigment is highly thermoplastic and will produce high sheet gloss, especially at high-temperature gloss calendering. Compared to many other pigments, cost per pound is high.

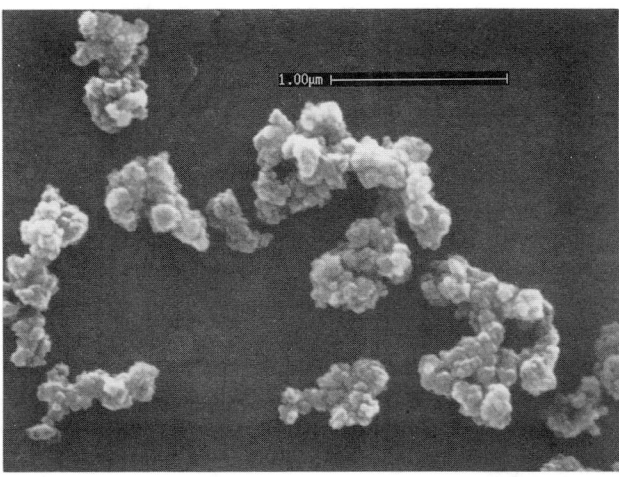

Fig. 2.11. Precipitated silicate, 30,000X (courtesy of J. M. Huber Corp., Chemicals Division)

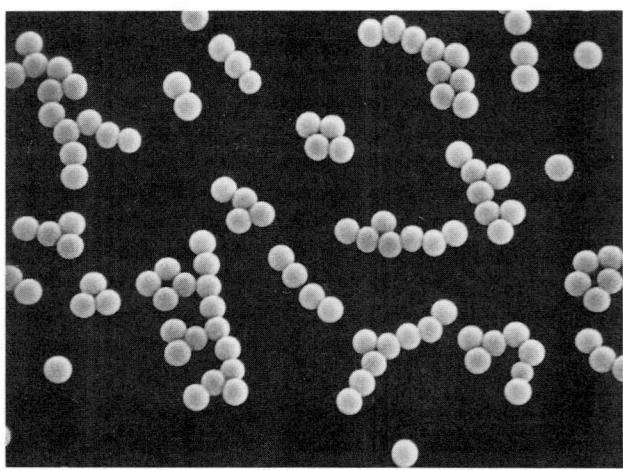

Fig. 2.12. Plastic pigment, monodispersed, 0.5 μm particle size, 10,000X (courtesy of Morton International)

H. Structured or Engineered Pigments

A new type of pigments has recently been brought into the marketplace (see Fig. 2.14). These pigments can be characterized as semi-synthetic pigments with low density and high light-scattering coefficient. The structure can be developed either by a simple chemical flocculation of small pigment particles (e.g., clay) or by a true change in structure, in part due to chemical agglomeration. The most highly structured pigments have very low density—e.g., a bulk density of 10-12 pounds per cubic foot—and high light-scattering coefficient. These pigments are excellent TiO_2 extenders and perform the functions of calcined clays but with a fraction of their abrasiveness, while producing bulk and opacity. Slurries made from these pigments are often of low solids—50% or less—and used at 5-15% of total pigment load. Cost is similar to that of calcined clays.

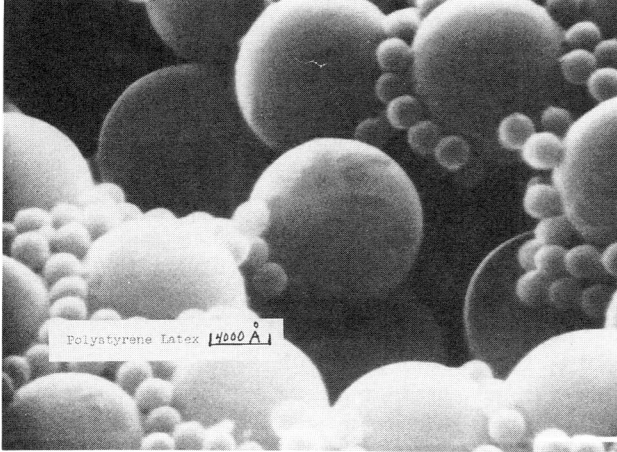

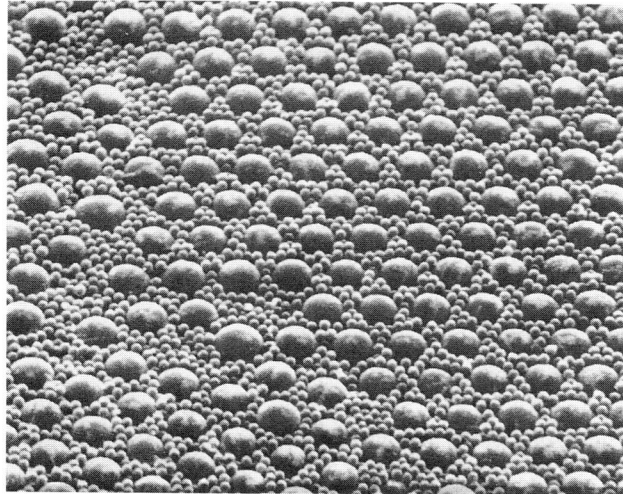

Fig. 2.13. Plastic pigment, polydispersed, approximately 0.9 and 0.2 μm particle size (courtesy of The Dow Chemical Co.)

At the other end of the structure spectrum are the low structure pigments that can be used as the sole pigment in, e.g., LWC coatings for rotogravure grades. Again, these pigments will produce bulk and compressibility, hence good rotogravure print quality. Slurries can be produced at 63-70% solids. Cost is in the same range as hydrous clays.

Binders

Binders can be separated into two groups, natural and synthetic. Of the natural binders used today, starch is by far the largest group, with soy protein being a distant second and casein a very distant third. See TAPPI Mono-

graph No. 17, *Starch and Starch Products in Paper Coatings,* and No. 36, *Protein Binders in Paper and Paperboard Coating.*

Of the synthetic binders, styrene-butadiene latex claims the number one position, polyvinyl acetate (homopolymer) latex being second, vinyl-acrylic (terpolymer) latex third, acrylic latex fourth, and polyvinyl alcohol a very distant fifth. For more details, see TAPPI Monograph No. 37, *Synthetic Binders in Paper Coatings.*

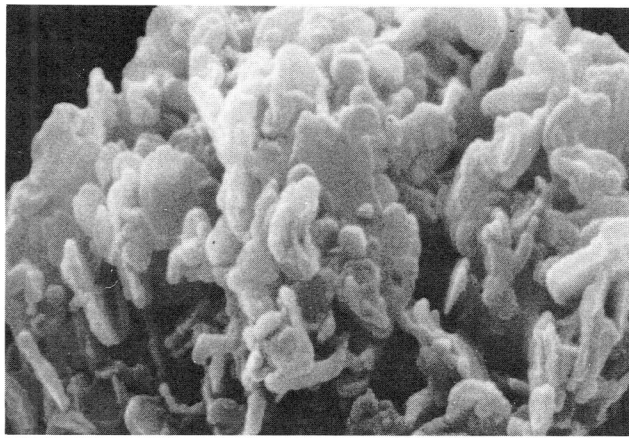

Fig. 2.14. Structured pigment, approximately 30,000X (courtesy of J. M. Huber Corp., Clay Division)

A. Natural Binders

1. Starch

A large number of starch varieties can be found on the market today. The dominant group is made from corn, but a coating starch can actually be made from wheat or potatoes as well, especially in Europe.

Most starches used in coating today are chemically modified to some degree. For high-quality starch coatings, the hydroxethylated corn starch is regarded to be the premium grade in the United States. Oxidized corn starch, and ammonium persulfate (AP) or enzyme converted pearl starch (basically unmodified corn starch) are also common and less expensive than ethylated starch. Most starches used in coatings are jet-cooked, with or without chemical or enzyme pretreatment, to produce a well-dispersed, low-viscosity starch.

All starches are water-sensitive and need to be used in combination with a starch crosslinker, i.e., insolubilizer if used in offset coatings. Coating starches are offered in a wide range of viscosities. Low-viscosity, ethylated starch is primarily used in high starch level or high-solids coatings, while a high-viscosity basic starch, like pearl starch, or lightly oxidized starch, is commonly used in the size press.

2. Soy Protein

Soy protein is used to some extent in today's paper coatings and is frequently used in substantial quantities in board coatings. Soy protein is currently produced in significant quantities by only two suppliers in the United States, but comes in a fairly large variety of viscosities and modified grades. Soy protein is a strong binder with very low thermoplasticity making it useful as a detackifier in high-synthetic board coatings that are gloss calendered.

Soy protein also has the ability to produce structured (i.e., low-density) coatings with good opacity, absorptivity, and gluability. On the other hand, soy protein tends to depress sheet gloss.

Protein, by itself, has quite good water resistance but is most often used in conjunction with an insolubilizer, especially in board coatings, when good wet rub is of importance.

Standard, basic soy proteins are natural binders with some degree of chemical modification, while some new products are more extensively modified and listed as soy polymers by the supplier.

Soy proteins are delivered in dry powder form and have to be made down into solutions of approximately 12-15% solids in water under alkaline conditions and at 140-150°F. Ammonium hydroxide or caustic are used to "cut" the protein, i.e., bring the protein to a true solution. Some very recent developments include special chemically modified soy proteins which are cold-water soluble with no alkali or no increase in temperature needed. These proteins are also available in different viscosities.

3. Casein

Casein is a true, natural binder produced from milk. Primary producers are located in New Zealand and Poland, making the material costly and availability at times in question. Casein is a strong, water-resistant binder made down in a very similar way to protein makedown. Casein comes in only one viscosity, which can vary, depending upon the quality of the raw milk. Casein produces high-viscosity coatings unless solids or casein level is kept low. Casein is known for producing severe pigment shock when being added to a pigment slurry, unless special precautions are taken.

Casein was *the* binder in the '30s and early '40s, but is today replaced by soy protein or starch. There is still some casein used, however, in specialty coatings like cast coatings.

B. Synthetic Binders

1. Styrene-butadiene Latex

Styrene-butadiene (S/B) latex, also called SBR latex, is the dominant synthetic binder used in paper coatings today. It was developed during World War II to replace natural rubber.

S/B latex is a tough, strong binder with a distinctive odor from residual styrene monomer. It is produced via polymerization of styrene and butadiene monomers in a pressurized reactor. Technically, S/B latex is a dispersion of spherical copolymer particles in water with surfactants added for stability (see Figs. 2.15 and 2.16). For simplicity, only the S/B latex is being shown. All synthetic latices will look similar under the SEM.

Depending upon the S/B ratio, these copolymers can be produced with different glass transition temperatures (Tg) resulting in soft or hard polymers. A high styrene level (e.g., 70%) gives a high Tg (hard) resulting in an open, brittle, easy glossing latex with less than optimum binding strength. A low styrene level (e.g., 50-55) gives a low Tg (soft) resulting in a soft, flexible, and tacky latex film with very high elongation but again with less than optimum binding strength.

Disregarding other functional monomers used during production, an S/B latex with an S/B ratio of 65:35 is a good compromise, resulting in high strength, good gloss, flexibility, and low tack.

A common S/B latex particle size is approximately 0.15 µm, but can range from 0.08 µm to 0.30 µm. In general, a small particle size means high binding strength, while a large particle size results in lower binding strength, if everything else in the recipe is equal. Due to oxidation of the butadiene portion, S/B latex coatings tend to yellow under exposure to UV and sunlight.

2. Polyvinyl Acetate Latex

Polyvinyl acetate latex (PVAc) has historically been a low-cost synthetic binder, frequently used in board top coatings due to its good gluability and lack of odor. PVAc homopolymers are hard binders, all with a Tg of approximately 30°C, produced via polymerization of vinyl acetate monomer in a reactor under atmospheric pressure.

PVAc latices have good UV light resistance, low odor, good chemical stability, and a non-tacky film. They are known to produce porous coatings with good gluability and blister resistance. On the other hand, PVAc latex films are water- and alcohol-sensitive and in general 10-15% weaker than an average S/B latex film.

PVAc latex is often used in top coatings for solid bleached board and in blends with S/B latex for improved blister resistance and ink absorption. Sheet gloss and ink gloss are in general lower with a PVAc homopolymer when compared to an average S/B latex with other coating components being kept the same. PVAc latices are often used in combination with a natural binder such as starch or protein.

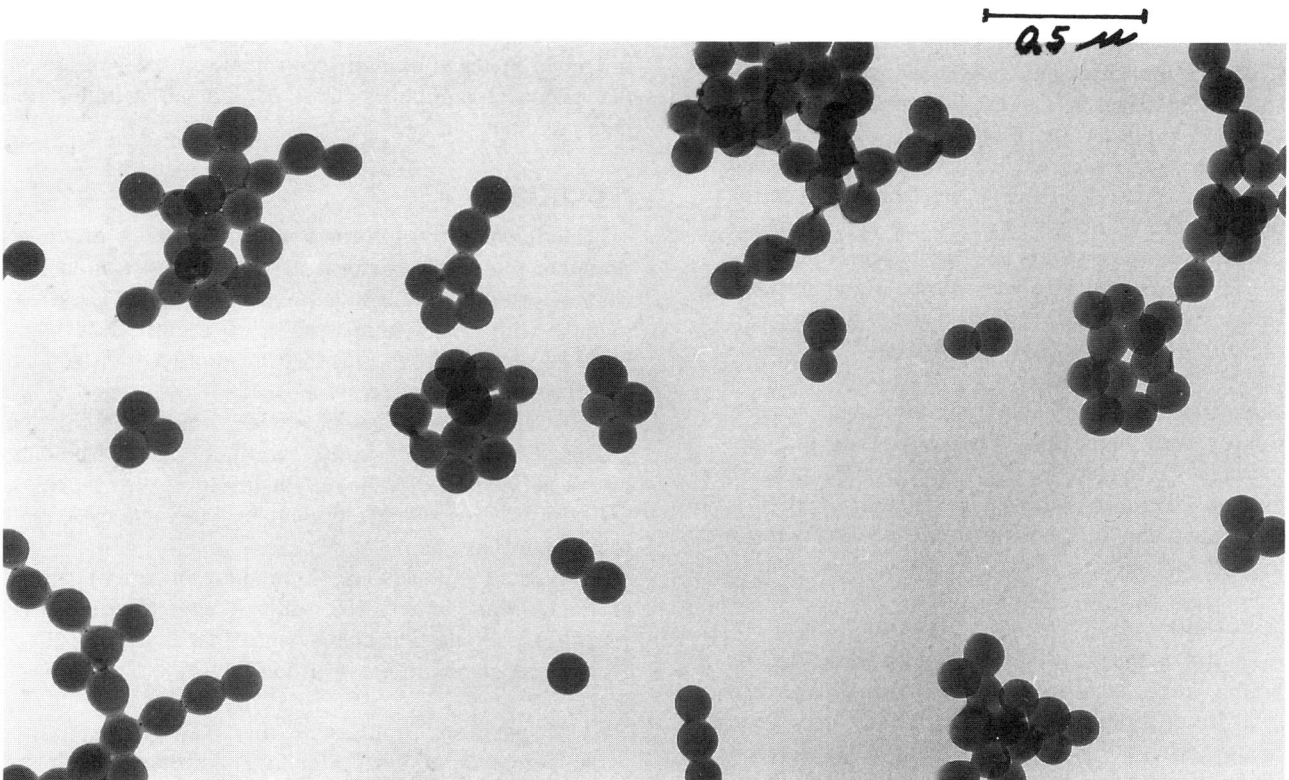

Fig. 2.15. Styrene-butadiene latex, monodispersed (courtesy of The Dow Chemical Co.)

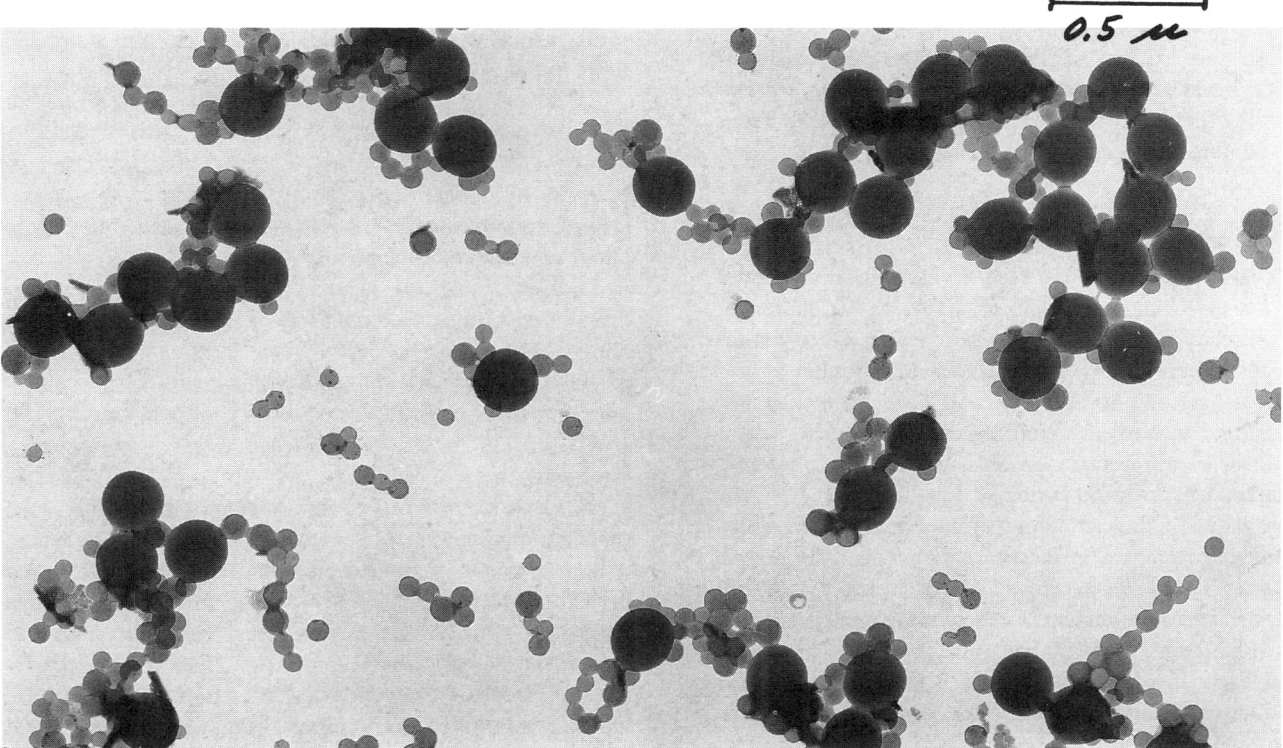

Fig. 2.16. Styrene-butadiene latex, polydispersed (courtesy of The Dow Chemical Co.)

3. Vinyl Acrylic Latex

Vinyl acrylic latices are co-polymers or ter-polymers of vinyl acetate monomer plus one or two acrylic monomers, such as butyl-acrylate, methyl-acrylate, ethyl-acrylate, or methyl-methacrylate. Butyl-acrylate is often preferred because it will produce a very soft polymer, and only a small amount—10-15%—is needed to reduce the Tg of 30°C for the PVAc homopolymer to approximately 15°C for such a vinylacrylic latex. Higher levels of butyl-acrylate will reduce the Tg even further. Ethyl-acrylate, on the other hand, will reduce the Tg much less per unit added, resulting in higher raw material costs to reach a specific Tg.

Because of the high cost of acrylic monomers, a vinyl acrylic latex will invariably cost more than a PVAc homopolymer. Compared to a PVAc, its properties are more like an S/B latex in regard to strength, flexibility, and water resistance, while maintaining good UV resistance and most of the openness of the PVAc.

4. Acrylic Latex

The acrylic latices have long been regarded as the premier latex in the latex industry. This may, to a certain extent, be due to high cost but also to their very good properties. Acrylic latices are strong and flexible like S/B latices and will produce good sheet gloss and ink gloss. They have very low odor, very good light fastness, i.e., UV light resistance and excellent shear stability. Acrylic latices used to be the choice for premium grades of coated paper where non-yellowing and brightness was of paramount importance. In the United States, there is still some acrylic latex used in these situations, but to a great extent they have been replaced by S/B latex or high acrylic content vinyl acrylic latex, primarily due to their high cost.

5. Polyvinyl Alcohol

Polyvinyl alcohol, or PVA, is a very strong synthetic binder delivered in powder form. PVA has to be made into solution by dispersing the powder in cold water, heating to approximately 190°F, and holding under agitation for at least 20 minutes. It will then form a clear, colorless, and viscous solution usually at 10-15% solids. PVA comes in different viscosities, as well as in hydrolyzed or superhydrolyzed form.

PVA is an excellent film former resulting in very good barrier properties in the presence of oil and grease. The film is water sensitive, and a crosslinker should be used if the PVA-coated surface will be exposed to wet rub or offset printing. PVA has very good aging properties and bacteria resistance. The PVA solution is known to produce a pigment shock, like casein does, when added to a pigment slurry.

PVA usage in coating is small in the United States but more common in Europe, where low levels (due to viscosity build) are used in combination with latex. Usage in United States is primarily confined to the size press or the calender waterboxes. Cost per pound is high compared to other synthetic binders.

Additives

Coating additives form a large, complex group of products that each perform a specific function in the coating color or in the finished coating layer. They will be dealt with briefly in this section. For more details, see TAPPI Monograph No. 25, *Paper Coating Additives,* on which some of this material is based.

Coating additives can be divided into the following groups: dispersants, viscosity modifier or water-holding agents, lubricants, crosslinkers or insolubilizers, biocides, pH adjusters, repellents, optical brighteners, dyes, and foam control agents.

The above products are all common in paper coatings of today, while others like antistat agents, conductive polymers, cationic flocculants, antioxidants, and plasticizers are less common and will not be covered in this section.

1. Dispersants, as the name indicates, are used to disperse pigments or keep pigments dispersed. Both organic—e.g., sodium polyacrylate—and inorganic products—e.g., tetra sodium polyphosphate (TSPP) or sodium hexameta phospate—are used. Organic dispersants are known to be much more stable than phosphate dispersants, especially at elevated temperature and extended time, but also more expensive.

2. Viscosity modifiers are most often used in all- or high-synthetic binder systems to control Brookfield viscosity and water holding. Common types are carboxymethyl cellulose (CMC), hydroxyethyl cellulose (HEC), sodium alginate, high viscosity starch, and acrylic alkali responsive thickeners.

3. Lubricants are used to control dusting in super-calendered grades as well as in grades where extensive slitting and cutting are done. The most common one is probably calcium stearate, but other ones like polyethylene emulsions, ammonium stearate, polyethylene glycol, wax emulsion, sulfonated oils, and esters of unsaturated fatty acids are also used.

4. Crosslinkers or insolubilizers are frequently used in starch coatings for offset and in protein coatings where a high wet rub resistance is important. Melamine and urea formaldehyde resins were very common up until a few years ago, but have been nearly eliminated due to formaldehyde emission restrictions. Other formaldehyde-free or nearly formaldehyde-free products like cyclic amide condensates, ammonium zirconium carbonate, and buffered or modified glyoxal resins have now captured the market. Regular 40% glyoxal has been used extensively in the past and is still used to some extent today. It is an effective crosslinker for starch, protein, and polyvinylalcohol, giving a rapid cure. However, it will increase the viscosity of starch coatings significantly, and loses efficiency above pH 8.5.

5. Biocides or biostats are practically always used in coatings containing natural binders, but quite often in all-synthetic coatings as well. Formaldehyde was frequently used in the past, providing inexpensive, effective protection from bacteria attack.

This material is no longer used due to the emission restrictions. Mercury compounds were also effective biocides but are no longer permitted for use. Common products are aldehydes, organo sulphur compounds such as thiones, isothiazolins, phenates, chlorophenates, and inorganic boron compounds.

6. pH control can use a number of additives, but the most common ones today are ammonium hydroxide (ammonia) and sodium hydroxide (caustic). Ammonia is receiving the attention of United States EPA because the vapors are irritating to eyes, nose, and throat, and its use could be restricted. It is a fugitive alkali and therefore preferred over caustic in situations where maximum water resistance is important.

Soda ash, potassium hydroxide, and amino-methyl-propanol (AMP) can also be used; and, in high-level calcium carbonate coatings, sufficient pH increase may be obtained from the carbonate itself.

7. Repellents and release agents are used in coatings for release paper, and the most common ones are silicone compounds, chromium complexes, and fluorochemicals.

8. Optical brighteners are often used in premium paper grades, such as No. 1 enamel. They work on the principle of fluorescence and make the coated paper look very white to the human eye. Some common types are stilbene derivates, benzophenone derivatives, and dibutyldithio carbamate.

9. Dyes are common in practically all paper coatings and are used to give the paper a desired shade. Insoluble, or pigmented dyes, can be inorganic or organic. Organic pigments are, in general, more expensive, have higher brightness, and higher tinctorial strength than inorganic pigments. On the other hand, inorganic pigments usually have better hiding power and stability.

Some common inorganic pigments are Chrome Yellow, Cadmium Red, Prussian Blue, Molybdate Orange, and Iron Oxide pigments. Some organic pigments are Hansa Yellow, Hansa Orange, Benzidine Yellow, Para Red, Lithol Red, and Benzidine Orange.

Soluble dyes can be acid, basic, or direct dyes. Acid dyes are generally the sodium, potassium, or ammonium salts of the corresponding color acids such as the sodium salt of a sulfonic acid derivative. Acid dyes have good solubility in water and are bright. They have fair to good light fastness and fair to poor bleed fastness.

Basic dyes are usually the salts, such as chlorides, hydrochlorides, sulfates, and oxalates, of the color base. They are acid in nature and sensitive to free alkali, often making them incompatible with alkaline coatings containing casein or protein.

Direct dyes are the salts of dye acids, similar to acid dyes but more complex and less soluble. They have good affinity for cellulose and are stable under alkaline conditions. The direct dyes also have a high affinity for clay.

10. Foam control agents can be separated into antifoaming agents and defoaming agents. An antifoaming agent is not necessarily a good defoamer, while a defoamer can be a good antifoaming agent. The difference between the two types are often minimal, but the terms are still used in the paper industry today.

Foam control agents can vary from insoluble to fully soluble. Emulsion of insoluble materials are common and are regarded as separate entities. The insoluble materials are usually the most efficient foam control agents, but may have damaging side effects, such as fish eyes or voids. Coating drawdowns on glass is a quick and simple method for checking on these side effects. Some of the more common foam control agents are: silicone emulsions (very powerful but "dangerous" problem with fish eyes), emulsified fatty acid and hydrocarbons, fatty acid and esters in mineral oil, silica derivatives, ethyl hexanol and fatty acid, and alcohol derivatives in mineral oil base. Regular ethyl and methyl alcohols are inexpensive defoamers with immediate effect but short duration.

Chapter 2, Section II

Preparation of Starch for Pigmented Coatings

Douglas K. Stinebaugh, National Starch and Chemical Co.
Lawrence A. Gaspar, National Starch and Chemical Co.
LeRoy E. Deters, Grain Processing Corp.
Larry E. Fitt, Corn Products, A unit of CPC International Inc.

Introduction

Starch is a complex natural polymer packaged within a granular structure. When gelatinized or pasted, it loses its granular structure, becomes water soluble, and develops adhesive properties. Starches can either be batch- or jet-cooked, and the effectiveness of the cooking process determines whether the paste reaches its potential in terms of performance. Starches can be modified to provide improved properties. A conversion reaction on the starch polymer will enable the papermaker to formulate the coating at higher solids levels. A derivatization reaction can provide improved strength and physical properties to the sheet as well as stabilize amylose containing starches against retrogradation.

Introduction to Starch

Starch is a carbohydrate storage product found in all plants containing chlorophyll. Within plants, the starch is stored in the form of granules. Each granule is made up of a number of concentric layers originating at the hilum, or nucleus. As the granule matures, it assumes the size and shape characteristic of the parent plant.

Because starch is a major constituent in a wide variety of common vegetables, it is an abundant renewable resource. Source plants include corn, potato, rice, tapioca, wheat, sorghum, and others. Figure 2.17 shows three different types of starch and illustrates the difference in granule shapes and sizes. Tables 2.1 and 2.2 list typical characteristics of several starch types. An experienced analyst can usually determine the source plant of a starch sample, provided that the material is still in its granular structure. Once the granules are dispersed into a starch paste, however, it is extremely difficult to determine the source plant:

Fig. 2.17. a, b, and c. Micrographs showing the size and shape of different native starches (750X and 1500X).

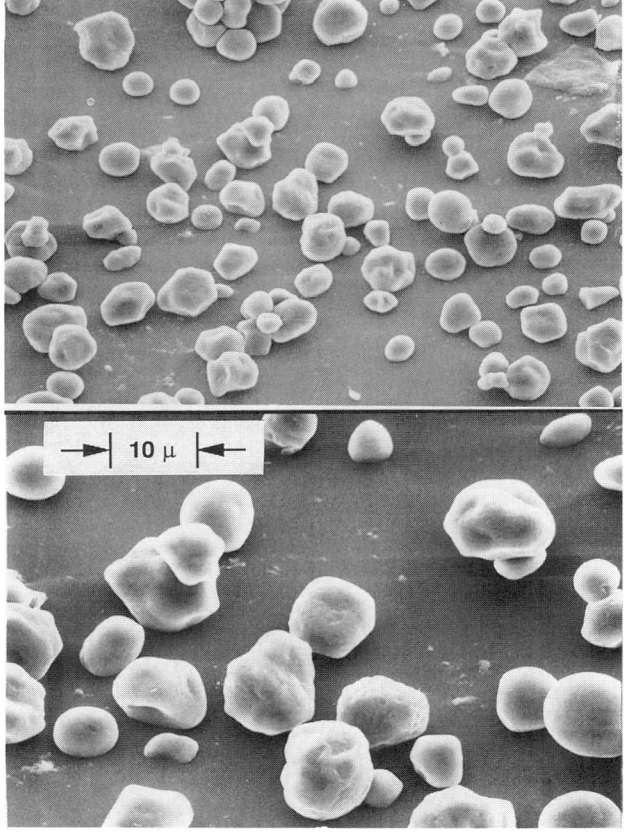

(a) corn

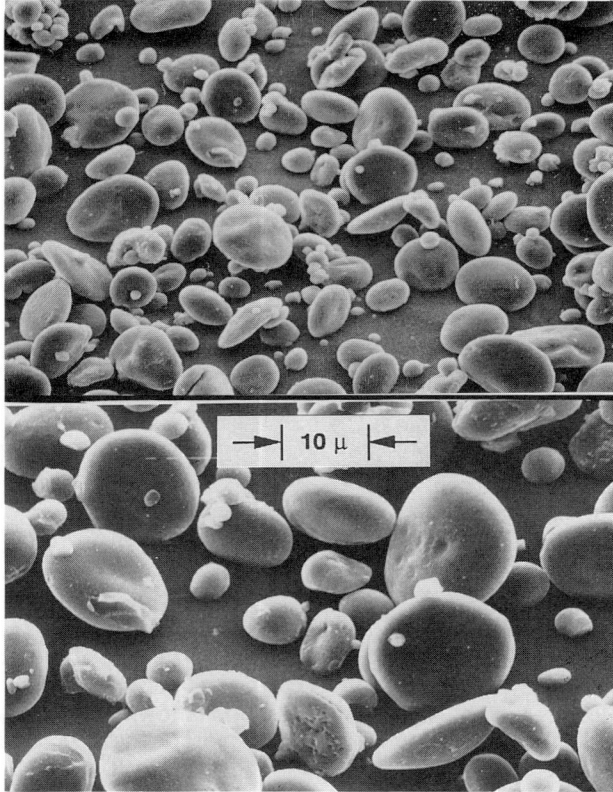

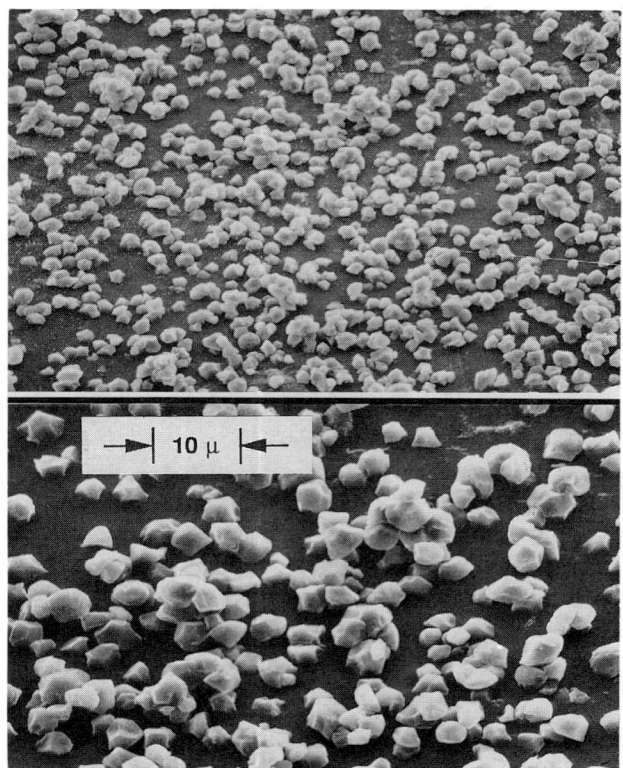

(b) wheat

(c) rice

Table 2.1. Amylose content of starches (11)

Starch	Percent Amylose
Acorn	24
Apple	19
Arrowroot	20.5
Banana	16
Barley	22
Barley, Waxy	0
Easter Lily	34
Elm, sapwood	21.5
Iris Tuber	27
Corn (Zea Mays)	28
Hybrid Amylomaze Class V	52
Hybrid Amylomaze Class VII	70-75
Hybrid Waxy Maize	0.8
Oat	27
Pea, Smooth	35
Pea, Wrinkled	66
Manioc	15.7
Parsnip	11
Potato	20
Rice	18.5
Rice, Waxy	0
Sago	25.8
Sorghum, Waxy	0
Sweet Potato	17.8
Tapioca	16.7
Wheat	26

In the paper industry, starch is used primarily as an adhesive. It can, however, provide other advantages. During the paper formation process, it is used for internal strength but may also provide retention. In surface treatment applications, it is used to reinforce fiber to fiber bonding at the surface, but also provides surface sizing and sheet stiffness. In coatings, it not only bonds pigment particles to one another and anchors the coating to the base sheet, but it also imparts holdout and stiffness to the sheet and can improve the rheology of the coating color.

Commercially available starches are most commonly sold in the form of a white, dry powder. The moisture often ranges from several percent below 12% by weight on corn-based grades to roughly 18% on potato-based grades. The various grades are available in bags ranging from 25 lb to 2000 lb and bulk quantities. Starch powders are best conveyed by pneumatic systems with dust recovery. Starch dust is an explosive material, but is not a difficult material to handle and use, provided the process has been designed with the proper safety systems.

Starch Technology

Starch is composed of two high-molecular-weight polymers, amylose and amylopectin, composed of anhydroglucose units connected at the first and fourth car-

Table 2.2. Characteristics of different starches (14)

Starch	Granule Size[a] (Microns)		Granule Shape	Amylose (%)	Gelatinization Temperature	
					5% Solution (°F)	33.3% Solution (°F)
Corn	4-26	(15)	Round, polygonal	29	176	156
Waxy Corn	5-25	(15)	Round, polygonal	0-6	165	162
Potato	15-100	(33)	Oval	23	147	142
Tapioca	5-36	(20)	Truncated, round, oval	18	145	136
Sago	15-65		Oval, truncated	27	165	160
Wheat	2-38	(20-22)	Oval, round	25	170	140
Rice	3-9	(5)	Polygonal	17	178	167

[a] 1 micron = 0.001 millimeter = 0.0000394 inches. Average size given in parentheses.

bon atoms. As shown in Fig. 2.18, the anhydroglucose unit has three hydroxyl groups. Due to their polar nature, hydroxyl groups on different polymer chains associate with one another through hydrogen bonding. The key to effectively utilizing starch as a binder is to fully gelatinize and hydrate the starch. This process frees the hydroxyl groups from hydrogen bonding and allows bond formation with other materials, such as pigments, that are present in the coating process.

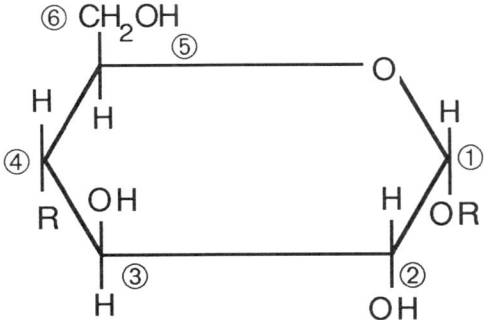

Amylose α 1-4

Amylopectin α 1-6, α 1-4

Fig. 2.18. The anhydroglucose unit, the backbone of the starch polymeric chain

Regardless of the botanical source, starch in the raw, uncooked, state has little value as a coating ingredient.

To increase the usefulness of any starch for coatings, it must first be gelatinized. Starch from different plant sources gelatinizes at different temperatures. Table 2.2 lists the gelatinization temperatures of several different starches. Other terms for the gelatinization process include pasting, cooking, and dispersing. Briefly, gelatinization is a process in which starch granules lose their boundaries and the polymer chains become sufficiently mobile to form a highly viscous aqueous liquid. When a starch slurry is heated beyond the gelatinization temperature or subjected to a certain amount of mechanical shear, the granules burst, large molecules break, and the viscosity of polymer material declines. The viscosity of this material reaches a maximum when the gelatinization temperature is reached as the granules are swollen due to the absorption of water. The viscosity then drops as granules burst and intramolecular bonding is broken and continues to drop until nearly all of the granules are dispersed in the water phase. Figure 2.19 shows the viscosity profile of the gelatinization process.

A feature called the maltese cross is a characteristic of starch granules from a number of plant sources. This feature can be observed microscopically using polarized light. As shown in Fig. 2.20, the maltese cross appears to divide the granule into quadrants. The cross appears due to the systematical arrangement of the macromolecules within the granule. The cross disappears when the granular structure is destroyed in the gelatinization process. This feature is often used as a troubleshooting technique in determining whether the starch slurry has been fully dispersed into a starch paste.

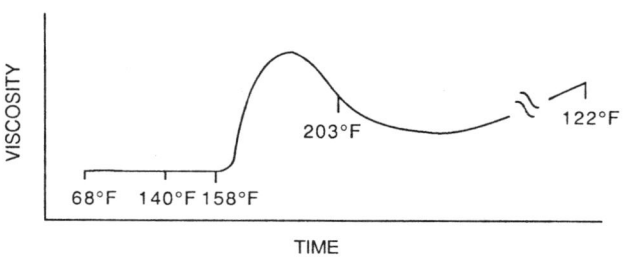

Fig. 2.19. Viscosity profile of the starch gelatinization process

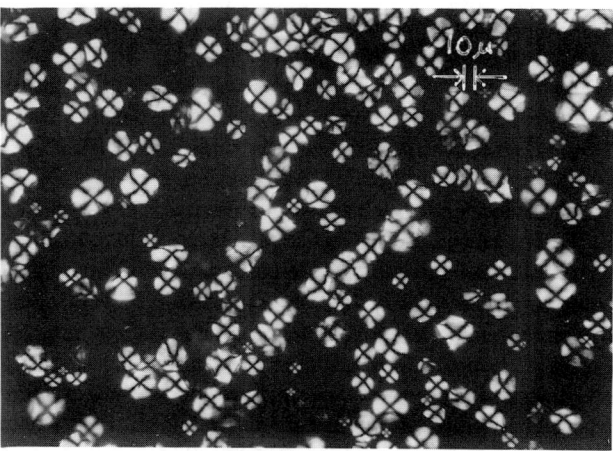

Fig. 2.20. The maltese cross exhibited by starch granules in polarized light

Amylose and amylopectin, the two polymers commonly found in starch, have similarities and differences. Chemically, they are similar in that they both consist of anhydroglucose units connected at the first and fourth carbons, but differ in that amylopectin also has branches that connect the first and sixth carbon units. Physically they are similar in that they are packaged in the same granule, when both are present, and must be gelatinized to be effective, but differ in that amylose takes the form of linear chains and amylopectin takes the form of branched chains. Figure 2.21 illustrates micro views of the $\propto 1 \rightarrow 4$ and $\propto 1 \rightarrow 6$ carbon connections as well as macro views of the straight and branched chains.

These two polymers do have two significant differences, however. The first is in the molecular weight. Amylose is much lower in molecular weight having a degree of polymerization of about 1000 anhydroglucose units compared to approximately 100,000 units for amylopectin. The second difference is the degree of retrogradation. Amylose, being essentially unbranched, is prone to retrogradation.

Retrogradation is the reorganization of cooked and dispersed starch molecules back to a somewhat ordered form. Retrogradation can occur via two different methods: crystallizing and congealing. The crystallizing process occurs when the linear molecules of amylose are allowed to reassociate, align, and subsequently form spherulites. Operating conditions that favor this are a temperature between 160°F and 190°F, a pH less than 7.5, and a presence of a known amylose-complexing agent. Amylose complexation is illustrated in Fig. 2.23. Steric hindrance prevents branched fractions, or amylopectin chains, from crystallizing. One branch point in every 20-25 anhydroglucose units is enough to inhibit the process. Congealing takes place when amylose and amylopectin chains associate, or entangle, and entrap water in a three-dimensional network. The effect of congealing on paste viscosity is illustrated in Fig. 2.22. Operating temperatures below 140°F favor congealing.

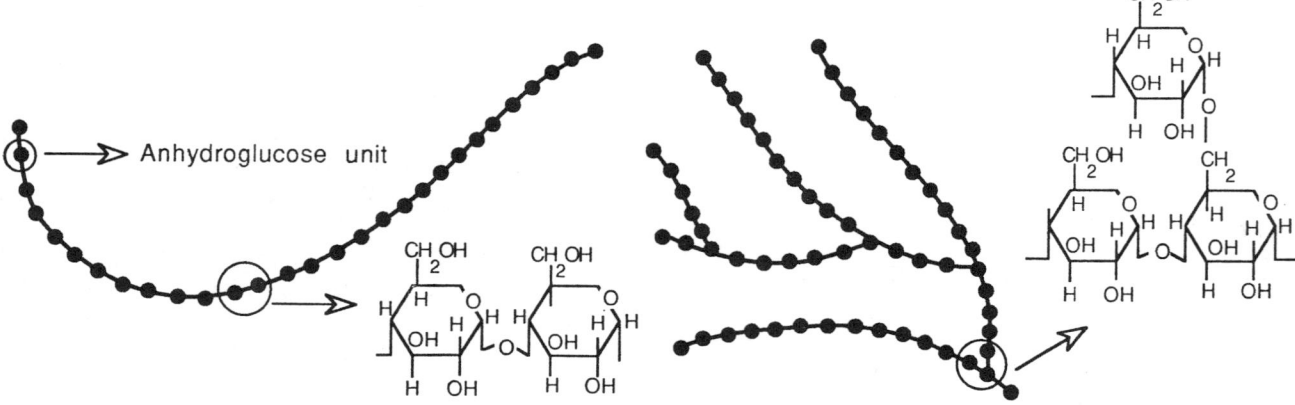

Fig. 2.21. Comparison of amylose and amylopectin molecular structures

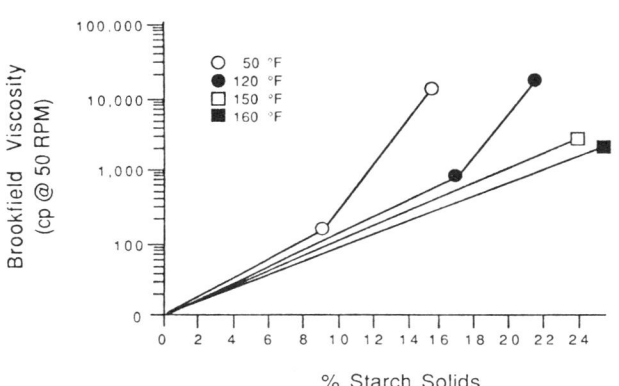

Fig. 2.22. The effect of retrogradation on cooked starch viscosity

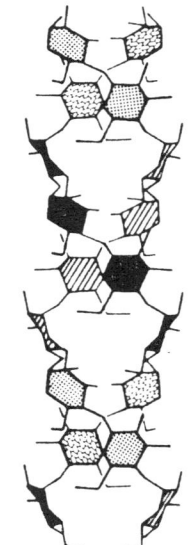

Fig. 2.23. The amylose double helix

The amount of amylose and amylopectin present in the starch is determined by the type of plant source. Table 2.1 lists the relative amounts of amylose and amylopectin in common plant sources. Genetic efforts have resulted in a broad variation available within certain species. As expected, waxy corn starch provides a major benefit over regular corn starch. It is not susceptible to retrogradation, due to the absence of amylose. Much work has also been done on the development of high-amylose corn hybrids. Although high-amylose starch has interesting properties, such as good film forming, the high propensity for retrogradation prevents much application to paper coating technology.

Development of Adhesive Character

As starch is gelatinized, the adhesive character is developed. The objective of starch pasting is to separate the double helixes and dissociate the hydrogen bonds between the hydroxyl groups to maximize the amount of free hydroxyls. Once freed, it is equally important to keep the groups separated so that they are available to form bonds with other molecules such as the pigments in the coatings and the fibers of the basesheet. The adhesive character of starch is irreversibly lost if the hydroxyl groups are allowed to reassociate and retrograde.

Starch Modification

Starches used in the coating area of the paper industry are mostly modified starches. The two types of modification carried out on the starch molecule include conversion, or molecular chain scission, and chemical derivatization. Conversion allows the formulation of coatings at higher solids. Chemical derivatization is done to improve the stability of the cooked starch paste and the coating color slurry and to provide benefits in the coated sheet such as strength, ink holdout, and smoothness. Both types of modification will change the gelatinization temperature of the starch. Most all of the starches used in the coating area are converted, and many of them are chemically derivatized as well.

Conversion, or scission, is the process of reducing the length of the starch molecular chain. Figure 2.24 illustrates this type of modification. Several chemical processes have been developed to accomplish this. Most common are the use of enzymes, acids, and oxidizing agents, but mechanical shear can also be used. The end result is similar in all cases in that the molecular weight of the starch molecule and the viscosity of the starch paste are reduced. This allows the formulator to make starch pastes and coating colors at higher solids levels.

Chemical derivatization of starch molecules takes place when a chemical group is substituted onto the starch molecule. Most derivatized starches are manufactured by suppliers. Several derivatization methods have been developed and each produces a product that is characteristic of the type of chemical linkage or chemical group. In each case, derivatives are attached to both the amylose and amylopectin chains. With enough substitution, the straight amylose chains take on the appearance of branched chains. The end result is a stable molecule with enough steric hindrance to inhibit retrogradation. Many of the derivatives will also contribute to strength through the addition of hydroxyl groups for hydrogen bonding and even charged groups for ionic bonding.

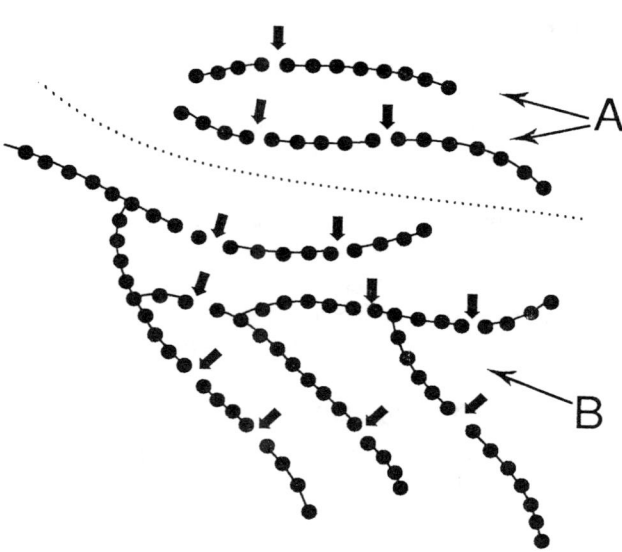

Fig. 2.24. Schematic showing the conversion reaction on the starch polymer chain

Starch Cooking Processes

Today, two different processes are most commonly used to gelatinize starch slurries. One process is batch cooking, and the other process is jet cooking. Both methods require that a starch slurry is first prepared at the desired solids level. Both methods are adequate, but jet cooking provides a more thoroughly dispersed starch paste.

Before starch can be gelatinized, it must first be slurried by sifting it into a measured amount of water with agitation. Violent mixing is not needed, as air could be entrained in the slurry and cause foaming problems. Only enough agitation is needed to keep the granules suspended in the water. If the agitation is inadequate or if the motor stops, the starch will settle to the bottom of the tank. If this occurs, a period of moderate agitation is needed to resuspend the granules. This phenomena is sedimentation and should not be confused with retrogradation.

Sedimentation can occur in pipes and pumps as well as tanks. At some starch concentration, starch slurries can become dilatant and freeze agitators and pumps. This concentration is slightly dependent upon the derivatization and drying history of the starch, but generally occurs near 42% slurry solids. Care should be exercised that flow rates are high enough and not interrupted long enough for the slurry to settle to this concentration.

The cooking process can be done simply on the starch slurry, or it can be done with the presence of the coating pigment. The latter is thought to provide superior disper-

sion, but is not as thermally efficient because the entire pigment slip must be heated and cooled.

Batch cooking is performed by heating an agitated tank containing the starch slurry beyond the gelatinization temperature. The two most common heating sources include steam via an external jacket and steam via direct injection into the starch slurry. Figure 2.25 shows a typical batch cooking tank with direct steam injection nozzles in the tank. The external jacket process is the original process, but significantly less efficient as heat transfer is a problem. Because of the heat transfer problem, the direct steam injection method was introduced. It is important to note that the direct steam injection method results in a diluted starch cook as steam condenses and remains.

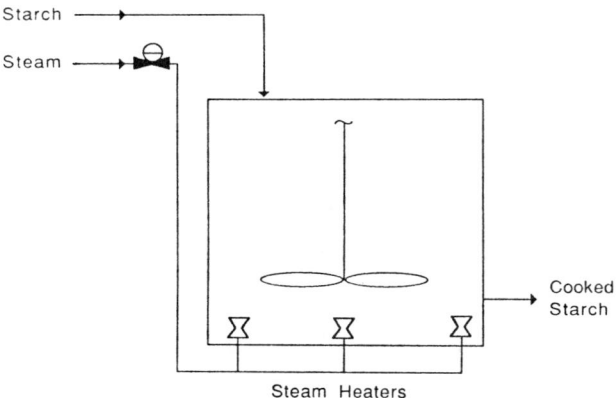

Fig. 2.25. Schematic of a batch cook tank with steam nozzles

Jet cooking is performed by pumping a starch slurry into a specially designed jet-cooking nozzle where live steam is introduced to the slurry. A back pressure valve on the downstream side of the nozzle maintains a pressure condition in the nozzle. Figure 2.26 shows the nozzle configuration. In the nozzle, the slurry temperature is elevated beyond the gelatinization point as the steam is introduced and the intensive mixing action created by the slurry nozzle and the venturi aids in thoroughly dispersing the starch granules. The result is a paste with the maximum amount of free hydroxyl groups and the maximum potential for adhesive strength. As with the live steam batch-cooking process, dilution of the starch occurs during the process. Jet cooking is the method of choice for a continuous preparation process.

Operational Considerations

Starch is a nutrient and is thereby subject to microbiological attack. Dry starch is stable and resists attack, but can become susceptible to attack when exposed to water via a crack in the storage silo, for example. Starch in the slurry and paste form is highly vulnerable. Microbial action occurs much like conversion in that the scission occurs on the polymer chain. These reactions however are not well controlled, and the end result is that the starch polymers are reduced to spoiled sugars with little or no adhesive value. Additionally, the microbes can spread to other parts of the mill if not attended to immediately.

Prevention is the key to minimizing the potential for spoilage. All components of the handling system including the silos, tanks, and pipes should be designed to allow complete drainage and cleaning with minimal effort. It is equally important to routinely inspect and clean the system to ensure trouble-free operation and maximum benefit from the starch adhesive. Provisions for boiling out the system with a chemical cleaner should be made. Cooker systems should complete a flush cycle before idling. To further prevent microbial action, the use of a biocide is often recommended. Biocides are chemicals designed to retard the growth of organisms and are not a replacement for good cleaning procedures. It is also good practice to rotate or alternate the use of several good biocides to prevent the microbes from becoming immune to the beneficial action of the chemical.

In-mill Starch Modification Processes

In-mill processes almost always include a gelatinization step, and many also include a conversion reaction on the starch molecule. This gives control of the starch paste viscosity to the mill and allows them to control the viscosity as needed to achieve the desired coating slurry viscosity. The most common in-mill methods of conversion include: (a) enzyme conversion, both batch and continuous; (b) continuous thermochemical conversion; (c) continuous thermomechanical conversion; and (d) continuous thermal conversion.

It is possible to apply these modification methods to both native starches and supplier-modified starches. The following is a summary of each method.

Enzymes can be used to enzymatically convert starch molecules. Enzymes, which are a part of all living organisms, are polymers of various amino acids which, depending on the sequence, can participate in numerous and varied chemical reactions. Alpha amylase, an enzyme common to many biological sources, is of particular interest because of its ability to convert, or hydrolyze, starch. Alpha amylases have little if any reaction with granular starch. After gelatinization, however, starch becomes susceptible to enzyme attack. The action pattern of alpha amylases is considered random, not selective, which means that linkages between the anhydroglucose units are broken randomly. If conditions are properly controlled, the enzyme will produce little if any reducing sugars during a brief reaction, but reducing sugars are formed during longer reactions. Reducing sugars have little value as an adhesive.

Starch quality is important for the control of enzyme conversion reactions. First, it is important that the starch paste is uniform and clean. Second, it is important that the system pH is compatible with the enzyme being employed. Third, the starch paste must contain a certain amount of calcium ions for maximum reaction effectiveness. Fortunately, most suppliers manufacture products with the proper buffers and calcium levels for popular enzymes and a wide range of water supplies.

Enzyme conversion can be done on either a batch or a continuous basis. In batch enzyme reactions, it could be important to use an enzyme that is less thermally stable so that it can be inactivated using moderate temperatures for short periods of time at the end of the reaction. In continuous systems, alpha amylases that are more thermally stable may be superior because the functionality

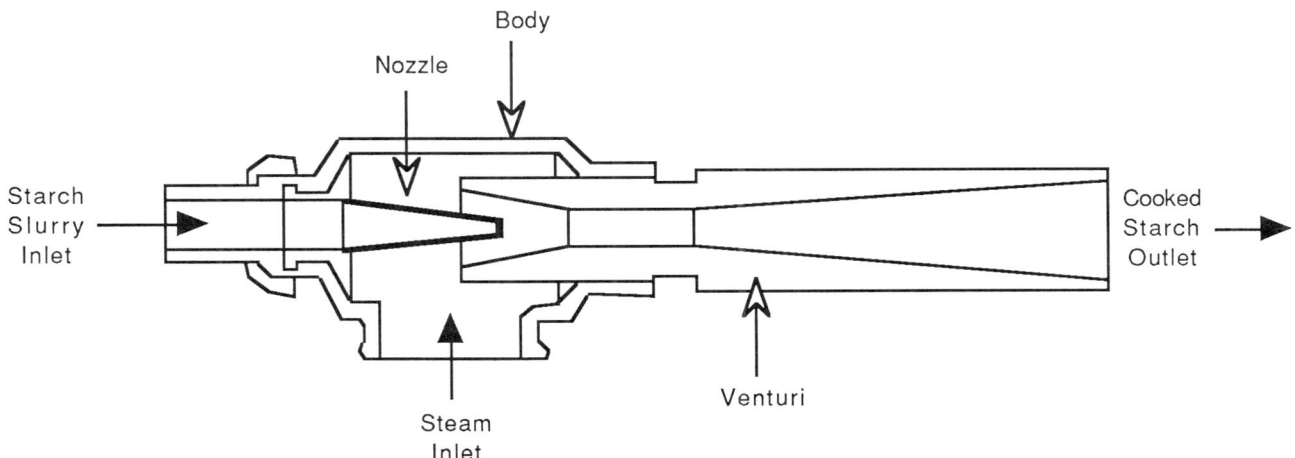

Fig. 2.26. Schematic of the nozzle used in jet cooking

may suffer little, if at all, during the pasting step in the jet heater.

Batch enzyme conversion operations can be based on either time or viscosity end points. In either case, a mixture of starch, water and enzyme is heated to a specified conversion temperature such as 160-180°F. When the defined time period or the desired viscosity is reached, the reaction must be terminated by thermally inactivating the enzyme. This is done by raising the temperature to 205°F and holding the level for 15-45 minutes. The process is flexible in that only minor changes in either enzyme dose or reaction parameters yield a wide range of starch paste viscosities.

A continuous enzyme conversion process is illustrated in Fig. 2.27. The starch slurry is prepared at 20-35% solids and, after thorough mixing, is coarse filtered and blended with 0.02-0.1% enzyme in solution. The slurry is jet cooked under pressure and pasted at approximately 200°F. The paste returns to atmospheric pressure in the agitated holding tank. The paste exhibits plug flow in the holding tank so that the paste in the bottom of the tank near the outlet is the most highly converted. Either the degree of conversion or the residence time is controlled by controlling the holding tank level. The converted paste is then pumped to a second jet cooker where the temperature is raised to 265-270°F under pressure. The primary purpose of this second heat treatment, which lasts only a matter of minutes (including holding time), is to inactivate the enzyme and to prevent further enzymatic hydrol-

ysis. A secondary purpose is to thoroughly disperse the paste to ensure uniformity.

The continuous thermochemical process is different from the continuous enzyme process in that an oxidant is used rather than an enzyme and the holding time is very short. Chemically, this reaction performs a chain scission on the starch polymers just as with the enzymatic conversion, but also attaches carbonyl and carboxyl groups to the opened molecular ring. Figure 2.28 shows the continuous thermochemical conversion process. An oxidant solution, either peroxide or ammonium persulfate, is added to a starch slurry via a pump or pneumatic system (0.05-0.3% oxidant on starch) and pumped to a jet cooker. In the jet cooker, only enough steam is added to achieve complete condensation under pressure at a temperature of 290-300°F. (Failure to desuperheat the steam used for cooking the starch may cause problems.) A holding coil or retention device is used to lengthen the reaction time and allow the desired degree of hydrolysis to take place. Prior to the steam flash chamber, caustic is added to raise the pH of the paste and inhibit amylose retrogradation. Control of the paste storage tank temperature is also needed to prevent retrogradation. In some cases, calcium stearate is added, at levels below 1%, into the slurry preparation tank to reduce the degree of retrogradation. This is done even though calcium stearate impedes the primary reaction and requires the addition of more oxidizing reagent. Fatty acid derivatives have been suggested as viscosity stabilizers (4).

The thermomechanical conversion process differs

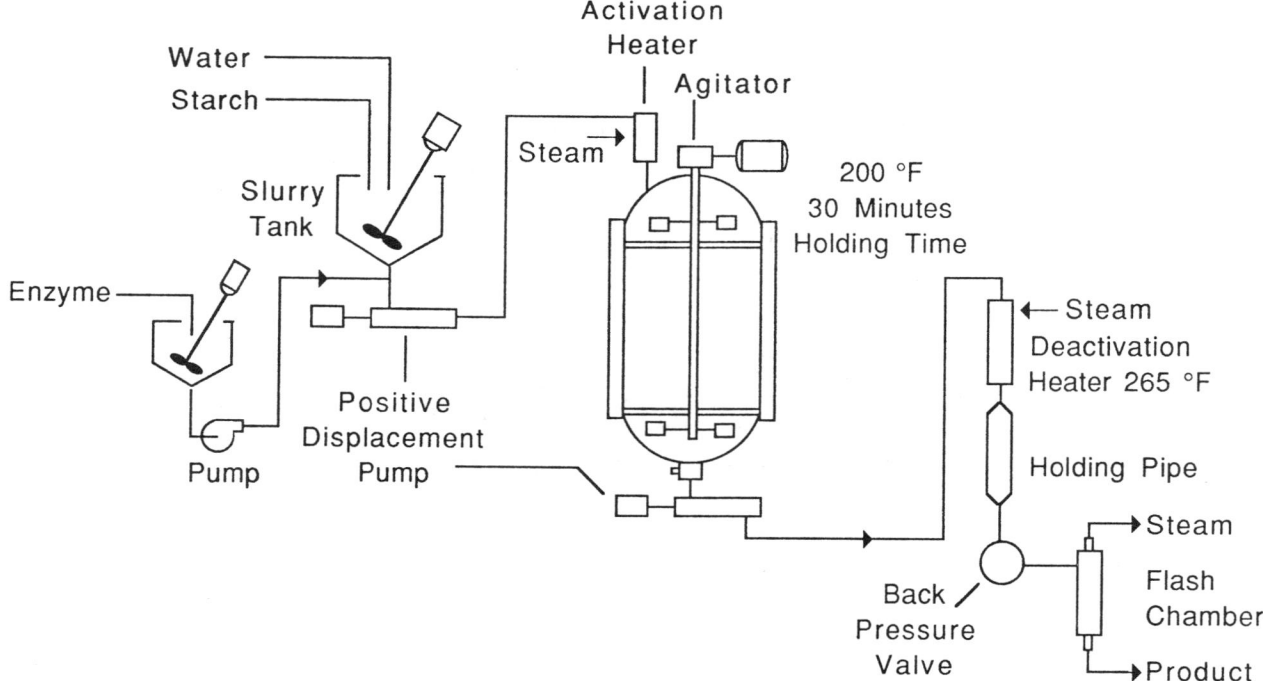

Fig. 2.27. Flow diagram of the continuous enzyme process

from the previous two processes in that it utilizes shear to convert the starch. This process is illustrated in Fig. 2.29. This process does not utilize chemicals, but rather mechanical shear created by the steam, to convert the starch. As with the other in-mill processes, the starch slurry is made down to the desired solids and pumped to the jet cooker. The process requires that excess steam be applied

in the nozzle to reach starch slurry temperatures in excess of 250°F and provide the mechanical shear. With the combination of temperature and shear, the thermomechanical effect is achieved. The converted starch then moves on to the flash chamber where water is added to cool the paste and, if needed, adjust the solids level.

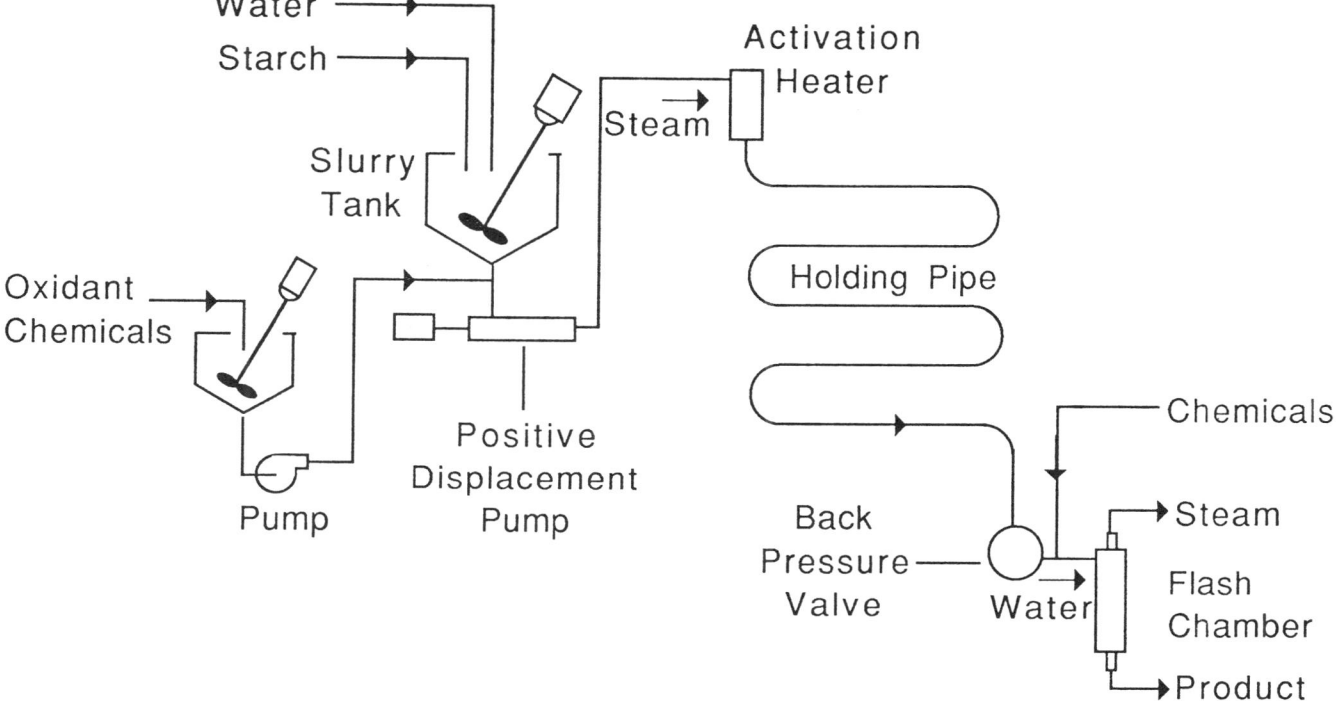

Fig. 2.28. Flow diagram of the continuous thermochemical process

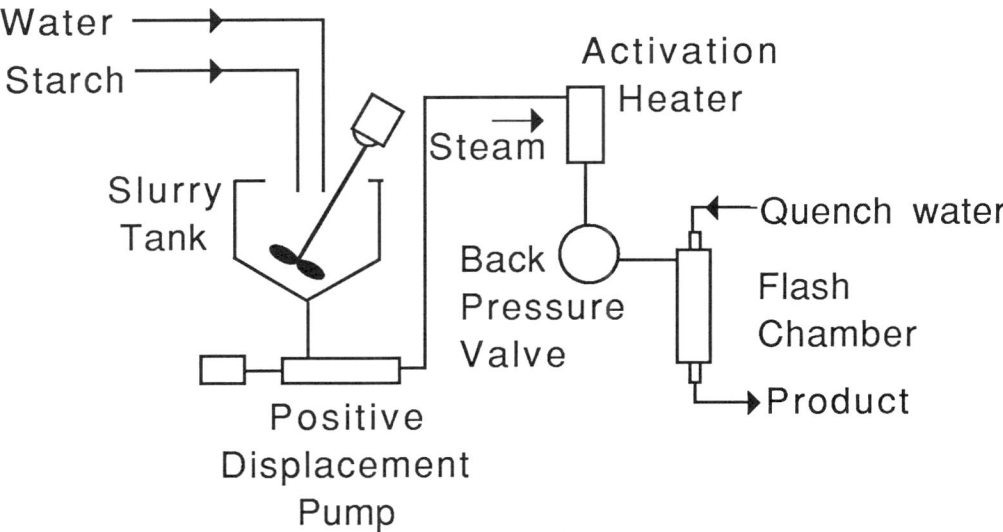

Fig. 2.29. Flow diagram of the continuous thermomechanical process

Supplied Modified Starches

Supplied modified starches can be defined as starches that the mill need only gelatinize for use. The mill may, if it chooses to, further convert these starches by one of the processes described previously. The simple gelatinization process could be carried out through batch cooking or jet cooking. When jet cooked, it is done without excess steam. This is called thermal conversion.

Common types of modified starches include acid-thinned, oxidized, hydroxyethylated, hydroxypropylated, and cationic. The first two mentioned are conversion-type reactions, and the latter three are chemical derivatizations. One of the conversion types is utilized in most every modified starch, and one of the latter three is often used in addition. A brief summary of each of these starches follows.

Acid-thinned starches are chemically similar to the enzyme converted starch molecules. These starches are prepared by subjecting a granular starch slurry to acidic conditions and temperatures slightly below the gelatinization point. The reaction is stopped when the desired viscosity is reached by neutralizing the slurry. The starch is then dewatered, dried, and packaged for shipment. This starch type is susceptible to retrogradation.

Oxidized starches are similar to the thermochemical starches previously discussed except that these starches can be substituted to a higher degree. This reaction is carried out by adding an oxidizing agent, such as peroxide or alkaline sodium hypochlorite, to the granular starch slurry. The reaction can be stopped by neutralizing the oxidant when the desired viscosity is reached. These starches provide increased strength and better film-forming properties than acid-thinned or in-mill modified starches. Highly oxidized starches retard retrogradation and can have increased dispersive powers which may or may not be detrimental to the papermaking process when returned to the furnish through the broke system.

Hydroxyethylated and hydroxypropylated starches are normally converted in addition to being derivatized. These starch ether derivatives are produced by reacting a granular starch slurry with either ethylene oxide or propylene oxide under alkaline conditions. The two carbon ethyl group or three carbon propyl group is attached to the starch molecule by an ether linkage. These starch ethers are stable under a wide range of conditions and are compatible with most any chemical used in the paper coating and papermaking process. These starches provide significant strength improvements due to a higher degree of hydrogen bonding and have improved film-forming properties over in-mill converted starches.

Cationic starch treatments are normally done in conjunction with the acid conversion or oxidation reactions in preparing these starches for coating applications. These starches are produced when cationic reagents are added to granular starch slurries under alkaline conditions. The end result is a cationic derivative attached to the starch molecule. Like the starch ethers, these starches are less susceptible to retrogradation. Cationic starches through additional ionic forces reportedly provide strength and holdout advantages superior to the other types of starches. The majority of the cationic starch remains attached to the pigment and fiber upon repulping of coated broke. Makedown of the coating color is critical due to ionic interaction. Adequate agitation power is necessary during coating makedown to handle the initial high viscosity.

Bibliography

Resources

1. Banks, W. and Greenwood, C.T., *Starch and Its Components,* John Wiley and Sons, New York, 1975.

2. Banks, W. and Muir, D.D., *Structure and Chemistry of the Starch Granule,* The Biochemistry of Plants (J. Preiss, Ed.), Academic Press, New York, 1980.

3. Dixon, M. and Webb, E.C., *Enzymes,* Academic Press, New York, 1979.

4. Harvey, R.D. and Welling, L.J., Tappi 59(12):(1976).

5. Radley, J.A., *Examination and Analysis of Starch and Starch Products,* Applied Science Publishers, London, 1976.

6. Radley, J.A., *Starch and Its Derivatives, 4th Edn.,* Chapman and Hall, London, 1968.

7. Rutenberg, M.W., *Modified Starches, Water-soluble Resins* (R.L. Davidson, Ed.), 2nd Edn., Reinold Book, New York, 1968.

8. Rutenberg, M.W., *Starch and Its Modifications, Handbook of Water-soluble Gums and Resins* (R.L. Davidson, Ed.), McGraw-Hill, New York, 1980.

9. Whistler, R.L. and Paschall, E.F., *Starch: Chemistry and Technology, Vol. 1,* Academic Press, New York, 1965.

10. Whistler, R.L. and Paschall, E.F., *Starch: Chemistry and Technology, Vol. 2,* Academic Press, New York, 1967.

11. Whistler, R.L., BeMiller, J.N., and Paschall, E.F., *Starch: Chemistry and Technology, 2nd Edn.,* Academic Press, New York, 1984.

12. Wurzburg, O.B., *Handbook of Food Additives* (T.E. Furia, Ed.), Chemical Rubber Publishing, Cleveland, 1972.

13. *Starch and Starch Products in Paper Coating* (R.L. Kearny and H.W. Maurer, Eds.), TAPPI PRESS, Atlanta, 1990.

14. "Story of Starches," National Starch Products, Inc., 1953 (National Starch and Chemical Co.).

Chapter 2, Section III

The Preparation of Soy Polymer, Isolated Soy Protein, and Casein Coatings

Dale R. Dill, Protein Technologies International
C. E. Coco, Protein Technologies International

Introduction

This chapter discusses the preparation and use of soy polymers, isolated soy protein, and casein as cobinders or coating modifiers in pigmented coatings. Both isolated soy protein and casein are essentially natural products—perhaps slightly modified by hydrolysis, while soy polymers are chemically engineered products specifically designed for paper and paperboard coatings. Technology recently developed in this area has allowed proteinaceous materials to be greatly expanded in breadth of application and to the most modern coating equipment worldwide.

These three proteinaceous materials are incorporated into pigmented coatings by one of two general procedures. The proteinaceous coating modifier is either predissolved into an aqueous solution and mixed with the other aqueous ingredients or added dry to the pigment slurry and solubilized in the presence of the pigment (dry cut). While all of these materials can be handled by these two generic methods, there are significant differences between the classes of materials in how they are handled. The choice of methods is selected based on mixing and dissolving equipment available, the particular product used, and the solids level needed in the final coating. The tendency for "protein shock," a thickening and flocculation of pigment, may be affected by the method of preparation and certainly by the product chosen. The advent of soy polymers with semi-bulk and bulk handling, continuous processing units, significantly controlled interaction with the pigment, and "no cooking required" products has substantially improved the utility of this group of materials.

Preparation of Solution

The cooking (when necessary) of these proteinaceous materials requires heat and temperatures in the range of 40°C to 70°C. Because of the moderate temperatures involved, hot water and direct or indirect steam can be used. Alkali, preferably ammonium hydroxide, is used to solubilize these normally water-insoluble materials. The general procedure is to slurry the selected material in an appropriate amount of water to 10-25% solids depending on product, being sure to allow for dilution if live steam is used. The temperature is adjusted to approximately 60°C.

The time needed for dispersing depends on the class of material selected—soy polymers the fastest, then isolated soy proteins, and casein the slowest. The soy polymers disperse readily in a minute or two, isolated soy proteins may require 5-10 minutes, and normally 30 minutes is used for the slightly more hydrophobic casein. An appropriate amount of alkali, 5-20%, 26° Be, ammonium hydroxide or 2-5% sodium hydroxide based on proteinaceous solids, is added to obtain a final pH of 8.5-9.0. Caustic (sodium hydroxide) can be successfully used for rotogravure coatings where the coated paper is never contacted by aqueous solutions during the printing process. To achieve the expected performance from a coated substrate to be offset printed, ammonium hydroxide, a fugitive alkali, and a crosslinker are best to achieve the decreased level of wet rub and wet pick required during printing. After adequate dispersion, temperature, and alkali are present, the material is cooked for approximately 30 minutes to achieve final solubilization. Casein is frequently processed for 45 min to 1 hour. Normally the materials in solution are not held at elevated temperatures (over 70°C) or for prolonged periods of time (over 4 hours) to minimize any changes in composition.

These materials normally supplied in a dry form only need to be solubilized, unlike starch which requires higher temperatures at least briefly to effect chemical, physical, or both changes. The rate of solubilization is dictated by the temperature and level of alkali. While ambient temperatures could be used, the time required would be unreasonably long for some products. Higher temperatures (80-90°C) will increase the rate of solubilization, but if maintained for very long (15 minutes) especially at a higher pH of 10-11, product deterioration may occur. The more soluble soy polymers can be solubilized at pH 8.5 or above, while casein normally requires pH 9.5-10.0. It is these considerations that lead to the selection of approximately 60°C and pH 8.5-9.0 as typical conditions for solubilization.

No general statement can be made regarding the concentration which can be successfully prepared. Some higher molecular weight unmodified materials are limited by the viscosity of the final solution to 10% solids. Other products in commercial use are being prepared at 25% solids.

The type of agitation and horsepower requirements will obviously depend on the concentration, molecular weight, final viscosity, and temperature of the final solution. Unlike the preparation of the pigment coating, the requirements in this procedure are to achieve an adequate dispersion of the dry product and uniform composition of the final product. High shear is not necessary although it presents no problem as long as foam is either not generated or is controlled.

While stainless steel equipment is preferred by some users, in practice, it is not necessary and not widely used.

The development of soy polymers which are easier to disperse and solubilize has led to the simultaneous development of automatic process units which allow the soy polymer to be continuously cooked and automatically controlled by a patented process unit. This has greatly reduced concerns about the handling of ammonium hydroxide since it is piped directly to the unit in a closed system. Also, large volumes of the soy polymer material may be processed very quickly and with minimal labor. While continuous methods have been investigated for casein and isolated soy protein, no commercial units are known to be in use.

Casein and isolated soy protein are supplied in 50-100-lb bags. The bags are slit and the material is dumped directly into the mixing tank. Soy polymers are supplied in 50-lb and 2500-lb semi-bulk bags and bulk. Along with the semi-bulk bags, continuous semi-bulk conveying systems have been installed to allow handling to be done by forklifts and automatic systems. Full bulk-handling systems are also in commercial use.

proteinaceous materials are biodegradable, as are other organic materials like starch. Standard good industrial hygiene normally prevents spoilage. In situations where solutions are to be held for long periods of time, suitable biocides can be used to prevent spoilage. However, there is no substitute for good housekeeping and periodic cleaning of equipment with hot water or a caustic wash. These soy polymer materials are processed at high temperature during manufacture, and spoilage normally only occurs after inoculation. Also, casein and some isolated soy proteins may necessitate cooking at high temperatures, and then cooling to prevent hydrolysis and to destroy naturally occurring enzymes.

The exact details for solution preparation of each material should be obtained from the supplier.

Preparation of Coating Color

No particular equipment can or need be specified for preparation of the final coating since a wide variety of almost every conceivable type of equipment is currently used. Generally, the pigment dispersion is optimally prepared and the proteinaceous solution is added to the pigment slurry in a controlled fashion. Finally, the latex binder is added, followed by the additives to be used. In some cases, the pigment dispersion is added to the proteinaceous solution, but, depending on the rate of addition and the level of agitation and shear, this may or may not present difficulties in handling or improvements in the coating. While low-speed, low-shear agitation is often used by necessity (what is available) for blending or dispersing the pigment and for blending the components of the formulation, in recent years more and more high-speed, high-shear agitation is used to obtain coatings with optimum properties. With the increased capability for agitation and mixing, the components can be blended in minimum time to uniform consistency at the highest feasible solids at a sheared viscosity for easy application. Mixers in service include Lightning mixers, Cowles dissolvers, Abbe mills, Kady mills, and heavy-duty sigma blade or dough mixers. Generally, any mixer with sufficient agitation can be used.

Addition Of Prepared Coating Modifier

Historically, proteinaceous materials were limited in utility and suffered in reputation because of spoilage (now not a problem) and too great an interaction with the pigment. A high level of pigment and protein interaction can lead to high viscosity build and flocculation or protein shock, which will limit performance and runnability. To quote a section from this publication's predecessor, Monograph 28, which is still true today, "When working with coatings having above 50% solids content, considerable trouble with protein shock may be encountered when the casein or isolated soy protein solution is first added to the pigment slurry. When the first 1 or 2% of casein or isolated soy protein is added, there is a considerable increase in viscosity and some pigment agglomeration." It is this unfortunate aspect of casein and isolated soy protein that had created a negative image for these materials that held so much promise. The "protein shock" aspect caused make-down problems, and the pigment agglomeration either limited the level of the proteinaceous material or the practical formulation solids. In many cases, this also limited their use in high-speed modern coaters. Fortunately, this general problem no longer exists with soy polymers.

Proteins are amphoteric polymers containing both anionic and cationic charges. While the balance of charge on the polymer is anionic, it is the high level of cationic charge that causes the severe interaction of casein and isolated soy protein with pigments, particularly clays of all types. Yet it is the interaction with the pigment that provides many of the beneficial properties of proteinaceous binders. If the interaction could be controlled, their utility would be greatly expanded. That is one of the major changes in the chemically engineered soy polymers. The balance of charge on the polymer is changed to make the soy polymer more anionic and less cationic to control interaction with the pigment. There remains ample cationic charge to provide the beneficial properties. Several different products are available to provide the level of pigment interaction needed, the rheology for trouble-free mixing, and excellent runnability on air knife, long- or short-dwell blade, roll, or other types of coater. With soy polymers, less care is needed during mixing, higher levels can be used in the formulation, higher solids formulations can be used, and less "protein shock" is encountered.

Dry Cutting

If adequate shear exists in the mixing equipment, the proteinaceous material does not necessarily need to be predissolved before adding to the pigment, but rather can be added dry to the pigment slurry. This procedure is known as dry cutting, and is easier with chemically modified soy polymers than the isolated soy protein or casein. In most cases, the pigment is preslurried, extra water is added for dissolving the proteinaceous material, and then the alkali needed to dissolve the protein is added. Finally this soy polymer is added dry. While a variety of temperatures and time are necessary depending on the product used, generally 30-60°C for 5-30 min is needed. This method works well when properly applied, but requires more care than simply adding the prepared solution. This method is used when solution preparation facilities are not available, higher total solids are desired, or "no cooking required" products are used.

Typical Formulations

The wide variety of applicator types as well as coating formulation ingredients prevents one from providing specific typical formulations. The following general comments can be made, however:

- In higher solids rod and blade formulations, 3-4 parts of proteinaceous materials provide the appropriate benefits of structuring, improved runnability, binder migration control, and water retention.
- In lower solids air knife or ultra-lightweight coatings, 6-8 parts of proteinaceous materials are typically used for the similar benefits.
- Ammoniacal salts are most frequently used for offset coatings so that adequate wet pick and wet rub are achieved, but sodium salts are commercially used.
- Ammoniacal or sodium salts are used for rotogravure coatings where water sensitivity is not an issue.
- Soy polymers are more forgiving and generally can be used at higher levels, if needed, than the more cationic isolated soy proteins or casein, and generally can be run at higher solids and faster speeds.

The supplier of the proteinaceous material of interest can advise the potential user of the specifics for products and provide technical service support to allow the product to be used to the best advantage.

Coating Temperature

Normally, during off-machine coating, the coating is applied at ambient temperatures or the temperature at which the coating happens to reach the applicator from the coating kitchen. With on-machine coaters, particularly where the coating is in direct contact with the hot paper or paperboard and then recirculated, the temperature of the coating may range from 90°F to 130°F depending on recirculation rate, basis weight, and season of the year. In most operations, no attempt is made to maintain the temperature of the coating at any given level, but it reaches an equilibrium value that remains relatively constant for any given time of the year.

Screening Coating

The coating color would typically be processed through some type of screening operation before entering the coating supply tank. Commonly used screen sizes are 80, 100, and 200 mesh. Some mills screen each ingredient going into the coating with a scalping screen of slightly coarser mesh and then screen the final coating.

Summary

The information presented in this section is obviously general in nature and should be used only as a guideline. Variations in all other components of the coating formulation—pigments, latex, insolubilizer, alkali, pH, and certainly solids content—will have a significant effect on the rheology and preparation of the coating. While coatings historically made with isolated soy proteins and casein potentially caused coating color make-down problems, the more recent soy polymers are relatively trouble-free. Careful rheological studies with the selected coating formulation can determine the proper make-down sequence, the solids to be run on a particular coater, and the type of make-down equipment required. This technology has developed to the point that coating colors with appropriate pseudoplastic—thixotropic rheology of high performance can easily be prepared.

Information on the benefits and application of these materials can be found in sections of the TAPPI seminars on binders, air knife coating, and blade coating. Instead of the demise of proteinaceous cobinders as might have been predicted several years ago, they are a growing, vital part of the coating formulation possibilities.

Chapter 2, Section IV

Mixing and Dispersion in Coating Operations

G. Gordon Bugg, Mead, Mahrt Mill

Introduction

Mixing is a unit operation common to the paper coatings industry, and in fact is one of the primary operations carried out during the preparation, blending, and storage of coating materials and coating color.

In this discussion, we will define the categories of mixing as they relate to coating operations. We will also review some basic calculations and relationships that may be used as guides for the design and sizing of mixing equipment.

Mixing operations may be classed as agitation, mixing, or dispersion. These categories are for convenience and, as you will see, tend to blur into each other.

We will define agitation as the process of setting up flow patterns and turbulence in a tank so as to prevent the settling, sedimentation, or stratification of the material in the vessel. Shear is relatively low, and the goal is the development and sustainment of a desired flow velocity and flow pattern within the tank.

Mixing is the blending of multiple materials which may be miscible or immiscible with each other, to the point where the concentration of each ingredient is constant throughout the volume of interest. This "volume of interest" is usually a tank. Mixing may involve gross aggregate reduction, although the amount of shear imparted to the mixture is by definition not sufficient to form fine emulsions or colloidal dispersions.

Looking at a mixing operation, velocity gradients within the fluid (both localized and general) are higher than those developed for "pure" agitation, and the impellers are usually designed for shear as well as nominal agitation. Note, however, once the material in the vessel is homogeneous, the mixing equipment might be used as an agitator to maintain the uniformity of the material until the mixture is released from the vessel.

The third class is referred to as dispersion. Dispersion in this context has two components—mechanical and chemical. From the mechanical standpoint, dispersion is characterized by high shear with the objective of breaking down droplets or aggregates to form an emulsion or dispersion, depending on the miscibility of the materials involved.

On the chemical side, dispersion refers to the use of chemical dispersants which may be incorporated into the coating to minimize the tendency of the coating components to agglomerate once they have been mechanically dispersed. These materials, acting on the surfaces of the coating particles, serve to reduce or destroy the interparticle forces which cause the building of clusters or flocs.

The degree of mechanical dispersion, particularly of pigments, is usually determined by measuring the percentage of material in the finished product that exceeds a certain particle size standard, or by plotting the particle size distribution and comparing that to a desired profile.

The System

Mixing systems, to include agitation, mixing and mechanical dispersion, consist of three factors: (a) the material(s) being processed, (b) the vessel or tank, and (c) the mixing equipment.

Agitation

The goal of agitation is to keep a given material in suspension by setting up specific mass flow patterns and turbulence in a vessel.

Agitation systems consist of a vessel or tank that contains the materials being processed, and a means of imparting movement to the fluid. Fluid movement is usually accomplished by a shaft-mounted impeller; however, there are a number of other options available. These include devices such as eductors which operate by injecting a stream of relatively high-velocity material into the vessel, entraining surrounding fluid, and then creating and maintaining the desired flow pattern by directing the discharge plume.

In coating systems, the shaft-mounted impeller system is the most widespread.

The Impeller

The purpose of the impeller is to generate motion in the surrounding fluid through the transfer of momentum. Depending on the design of the impeller, the principal motion may be axial (along the shaft) or radial (perpendicular to the shaft). In many tank and impeller combinations, there is also a rotational or tangential element that comes into play.

A typical impeller designed to generate radial flow is shown in Fig. 2.30. An axial flow design is shown in Fig. 2.31. Note that the design shown in Fig. 2.31 will also generate a radial component due to the open blade ends. This could be reduced by the addition of a circumferential ring to reduce spillover.

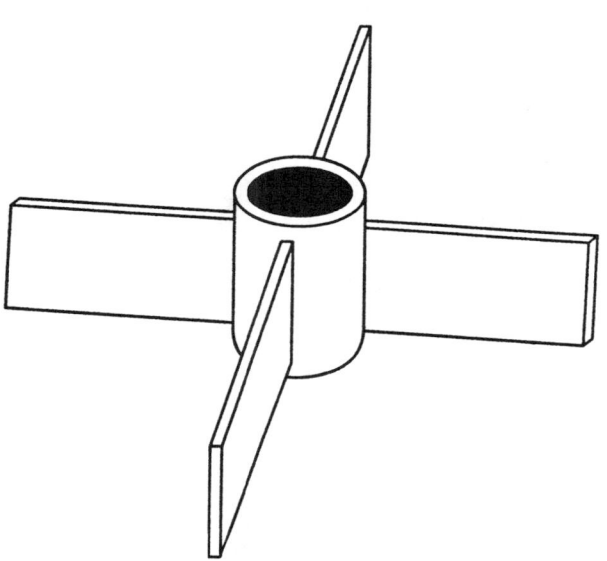

Fig. 2.30. Radial flow, straight-blade turbine

In the coating industry, three impeller designs dominate. These are the marine-type propeller, the straight-blade turbine (Fig. 2.30), and the pitched-blade (Fig. 2.31) or high-efficiency turbine. Many specialty designs exist and are used successfully in many areas. However, the majority of agitation applications are fulfilled using one of these three impellers.

The Tank

The tank, pit, reactor, or chest design plays a big part in how well a particular agitation scheme performs.

Tanks may have either an angular (square or rectangular) section or circular section. Cylindrical tanks may be oriented vertically or horizontally, with a wide range of height-to-diameter ratios. Rectangular vessels also exhibit a wide range of height-to-width ratios.

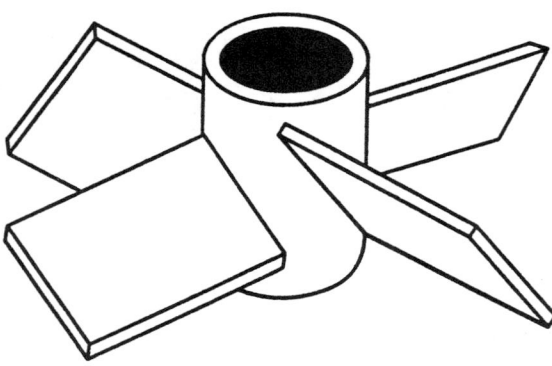

Fig. 2.31. Axial flow, pitched-blade turbine

In coating operations, the vertical, cylindrical tank with a top-entering agitator is the most common.

Selection of the tank design and material of construction is based on numerous factors including cost but, from the standpoint of agitation, the following paragraphs offer guidelines that should be considered:

In viscous systems (viscosities greater than 5000 cps), unbaffled circular tanks are preferred. The tank walls, reacting to the viscous drag of the fluid, serve the same purpose as baffles, reducing unwanted rotational flow. In very viscous systems, the presence of baffles, internal heating coils, or other fittings will degrade flow patterns to the point where stagnation may occur. Rectangular tanks are not recommended for viscous service as the corners of the tank act as baffles, reducing material movement, possibly below the minimum required for circulation.

Most coating components and coatings occupy a lower viscosity range. For these materials, a different set of guidelines comes into play. An unbaffled vertical cylindrical tank containing a low-viscosity fluid under agitation using the types of impellers discussed, will have a tendency to develop rotational flow patterns (Fig. 2.32) in addition to the desired vertical patterns. If a rotational pattern becomes strong enough, it is referred to as vortexing (Fig. 2.33), and may reach the point where air (or other surrounding gas) is drawn down into the impeller and entrained in the fluid.

Side   Top

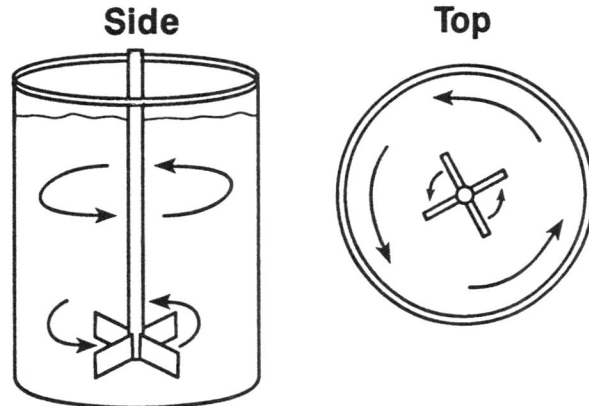

Fig. 2.32. Rotational flow patterns

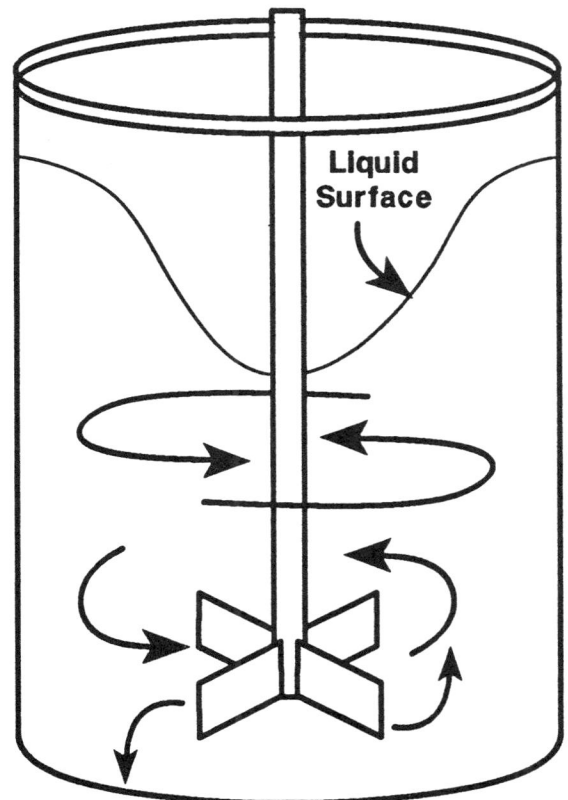

Fig. 2.33. Vortexing

Adding baffles to the walls of the tank (Fig. 2.34) neutralizes rotational flow and enhances vertical currents.

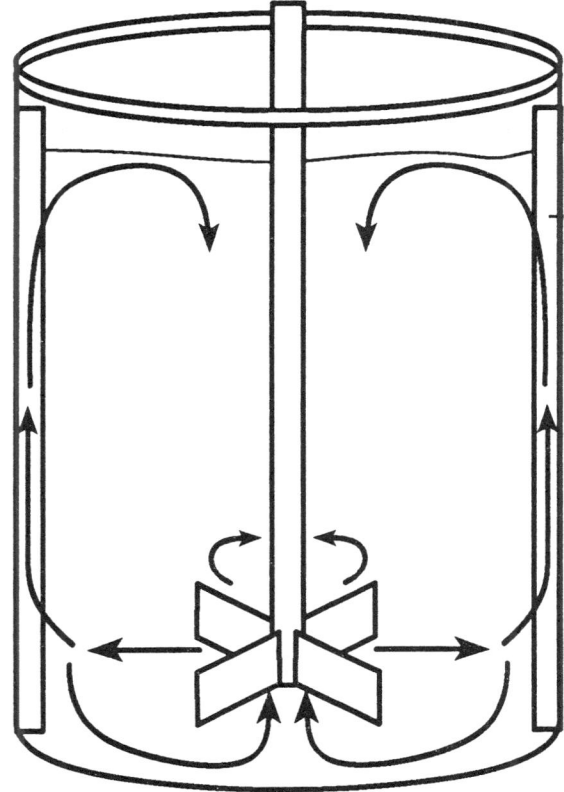

Fig. 2.34. Baffles produce vertical currents

System Design

When designing an agitation system, the following proportions may serve as starting points:

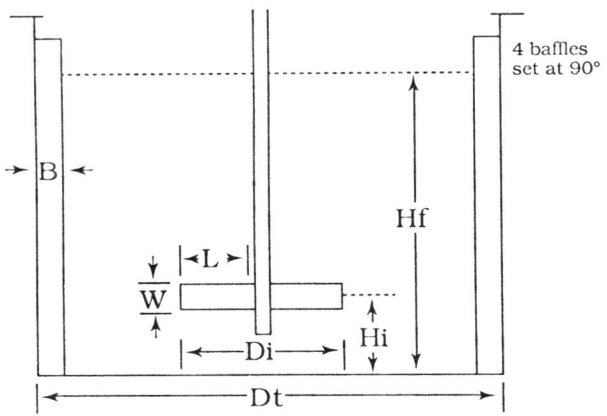

Fig. 2.35. Agitation system measurements

B = Baffle width
Dt = Tank diameter

Hf = Height of fluid
Di = Impeller diameter
L = Impeller length
Hi = Distance of impeller above tank floor
W = Impeller width

Note: Turbine-type impeller and four baffles set at 90 degrees.

Typical relationships: Hf is the average working level of the material in the tank.

Hf ≥ Dt
W = (1/5) Di
Hi = Di
B = (1/12) Dt
Di = (1/3) Dt

Multiple impellers may be mounted on the central shaft for larger tanks; however, these systems will have a greater tendency to entrain air depending on the amount of fluctuation in the tank level.

If multiple sets of turbine blades are used, the upper set of blades should be at least one impeller diameter below the average fluid level.

Our discussion thus far has centered on the flow patterns developed by an agitator mounted on the vertical center line of the agitated tank. In many cases, such as when dealing with smaller (3000 gallons or less) vessels or when having only an unbaffled tank available, good agitation may be achieved by using an off-center impeller. Examples of off-center impellers and typical flow patterns that can be generated are shown in Figs. 2.36, 2.37, and 2.38.

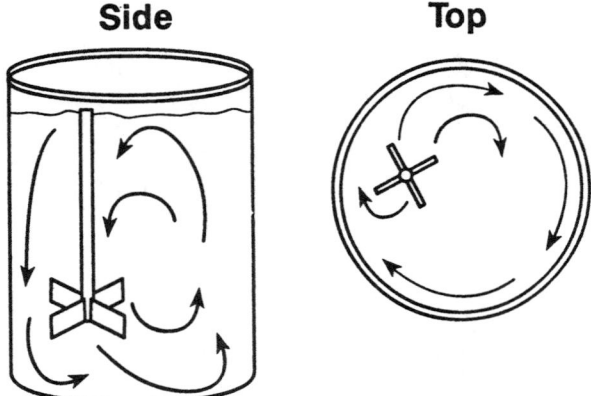

Fig. 2.36. Off-center impeller flow

Please note that these impellers are either marine-type propellers or angled-blade/high-efficiency turbine design. Flat-blade turbines are not generally found in these applications.

Mixing

The objective of mixing is to intimately combine two or more materials to the degree that the concentration of any material is constant throughout the vessel or mixing zone.

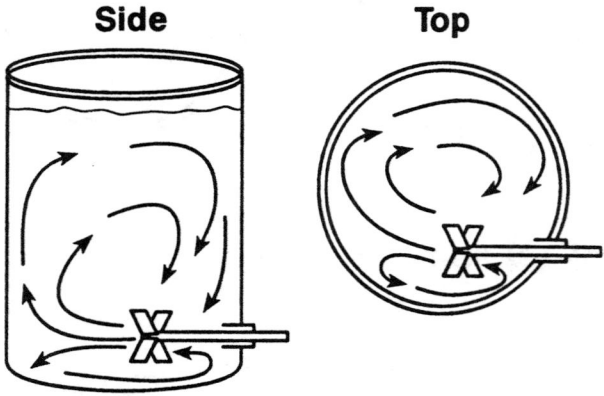

Fig. 2.37. Side impeller flow

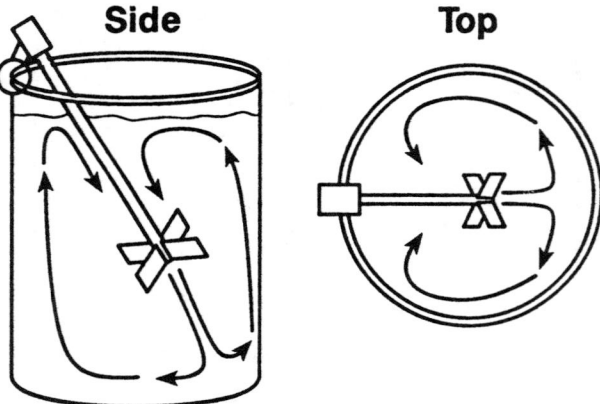

Fig. 2.38. Axial impeller flow

Mixing, as compared to agitation, is characterized by relatively high-velocity gradients in the area around the impeller. This is accomplished by using a smaller impeller than would be used for an agitation scheme, but operating it at a higher speed. The shear developed by the blade tips as they move through the fluid generates the mixing action.

Mixing operations must also be able to generate sufficient fluid circulation in the mixing vessel to ensure that each unit volume of material has been processed through the high-shear zone within a time that is compatible with the requirements of the process.

Indicators of good mixing include such qualities as visual appearance, texture, lack of viscosity variation between multiple samples, minimal solids variation with time, and minimal color changes in samples collected over time.

Mixing vessels are typically smaller than tanks designed only for agitation. Cone bottoms are used in most coating applications and enhance material circulation in the vessel as well as providing a low point for removal of the finished product. More and more mixing tanks may be found with heating or cooling jackets for better process control.

Mixing impellers can be grouped into two main types: open or enclosed rotors. In the open rotor scheme (Fig. 2.39), mixing is accomplished by the high tip speed of the rotor as it moves through the surrounding fluid. The high-velocity material leaving the impeller entrains enough slower-moving fluid to generate circulation in the tank.

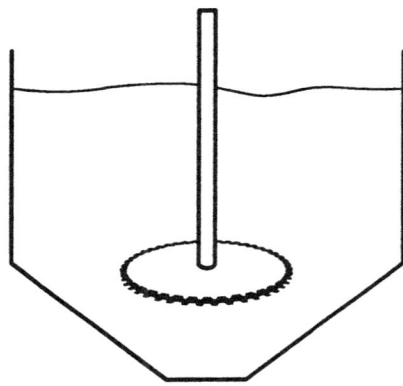

Fig. 2.39. Open rotor

The enclosed rotor design generates additional shear by directing the impeller output through a stationary, slotted ring that surrounds the impeller. While enhancing the mixing action, this method reduces circulation in the tank for a given power input.

Many mixing impeller designs combine regions that develop high-shear, low mass flow with sections that emphasize mass flow through the high-shear region and development of circulation within the vessel.

A third type of mixer should be mentioned at this point: the motionless mixer. In a typical motionless mixer, named "motionless" due to the absence of moving parts, multiple streams of material, in the desired proportions, are fed into the mixer (Fig. 2.40). The mixer repeatedly divides the flow or generates intertwining currents that result in a uniform stream of material exiting the unit. Circulation is accomplished through normal pipe flow, and the mixer serves only to blend the streams. The lack of shear, such as that which occurs with a rotating element, makes these units unsuitable for applications where breaking down flocs or clusters of particles is re-

quired. Residence time is a function of the stream velocity and mixer length and is therefore not as easily controlled as with vessel-type mixers.

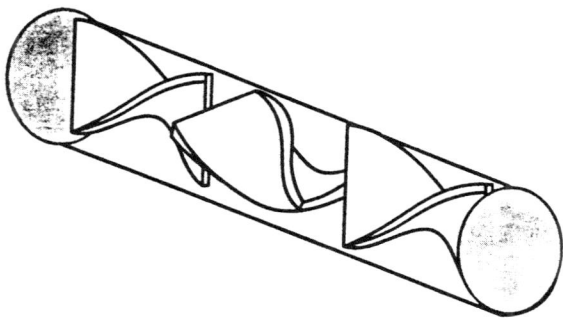

Fig. 2.40. Motionless mixer

Motionless mixers are particularly well suited for processes where homogenous fluids such as cooked starch, dispersants, or dyes are blended in a set proportion with water or similar material. Their small size, low cost, effectiveness, and absence of external power requirements are advantages that can be exploited in the right applications.

Dispersion

Dispersion may be defined as the production of a finely divided material that is evenly distributed throughout a fluid matrix. The main distinction between dispersion and mixing is that mixing does not change the size and surface area of the particles in the medium; whereas in dispersion the particles are physically broken up, increasing the total surface area exposed to the fluid and decreasing the average particle size.

The indicators of good mixing also apply to dispersion operations. In addition, determination of good dispersion includes two factors not previously discussed: (a) particle size reduction, distribution, or both and (b) minimum system viscosity.

Particle Size Reduction

The high shear developed by a dispersion unit serves to break apart flocs, clumps, or aggregates that may have formed in the feed stream. The degree of size reduction may be measured by such means as determining the fraction of the exit stream that is retained on a 325-mesh screen, then comparing that to a similar measurement made on the inlet stream.

Instead of ensuring that a given percentage of the processed stream will be less than a certain size, i.e., 98% of the stream will pass through a 325-mesh screen, it might be more desirable to determine the distribution of particle sizes in a given formulation at a given dispersion unit power. This distribution can then be compared to a desired profile to ensure product uniformity.

System Viscosity

Coating viscosity is typically used to determine optimum coating dispersion. Once the coating or coating component is mechanically dispersed, a chemical dispersant may be added to prevent re-forming of the disrupted agglomerates or flocs.

To determine optimum dispersant dosages, coating viscosity is measured and plotted over a range of dispersant dosages (Fig. 2.41). Most coatings will show an optimum dispersion zone in which the system viscosity is at a minimum for a range of dose rates. Note that adding excessive dispersant can be as harmful as not adding enough.

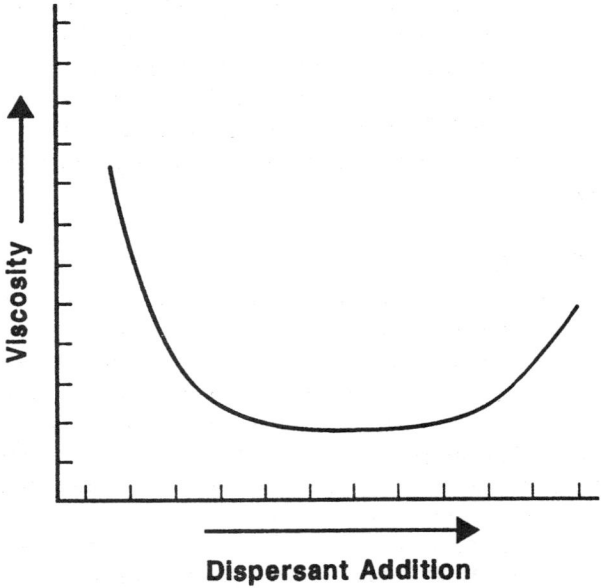

Fig. 2.41. Coating viscosity as a function of dispersant level

Dispersion Systems

Dispersion is characterized by very high shear applied to a given volume of material. Mechanical dispersion, in the context of coating operations, primarily refers to the breaking up of clusters of pigment particles to attain a desired particle size distribution. By this definition, mechanical dispersion is an operation found primarily in the pigment manufacturing process, rather than coating makedown and application.

Mechanical dispersion units are designed to apply a large amount of shear to a small volume of material. An important point to remember is that dispersion has virtually no effect on the liquid medium so, for a given

power, dispersion units are more effective when the liquid portion is held to a minimum.

Mechanical dispersion units may be employed in a number of ways. The unit may be mounted in a tank, in the manner of a mixer, in which case some additional mechanism must be present to ensure adequate circulation in the vessel. The high-shear unit can also be piped into a recirculation loop which gives better control.

Summary of Mixing Operations— Factors to Consider

Overall, mixing operations are not well defined due to the wide range of equipment, operating conditions, process objectives, and ingredient interactions that can take place.

In recent years, a number of computer-based mixing simulation programs have been developed. These simulations allow engineers to evaluate various configurations and designs of mixing vessels and equipment using the requirements of their particular application at a fraction of the cost and time involved in setting up and evaluating a pilot installation.

Use of these programs greatly reduces the risk of installing a mixing system that is incapable of meeting its process goal.

To maximize the probability that the process and system you are planning to use will accomplish their purpose, the following questions should be answered early in the design stage and in as much detail as possible.

A. Define and evaluate the process objectives:

- Desired end use—agitation, mixing, or dispersion
- Desired end quality—i.e., ". . . less than 2% of the initial charge will be caught on a 325-mesh screen after 8 minutes of operation for a given feed"
- Operating conditions to include chemical, mechanical, and thermal effects and process throughput
- Are the objectives feasible as they are now defined?

B. Carefully consider alternative solutions:

- There is usually more than one right answer.
- Factors such as cost, safety, and maintenance may be used to differentiate alternatives at this point.
- Is there existing trial or experimental work to validate the choice that was made?
- Is there a similar system already in operation that can be visited to help validate the selection? Or are there systems that can be visited to help eliminate some of the choices?

Keep in mind that unless your process is to be a duplicate of an existing system with a known history, vendor information and theoretical and technical literature should be used as guides, not absolute truths. Experience has shown that as the number of ill-defined assumptions rises, so does the probability that Murphy's Law will override all other design principles.

Chapter 2, Section V

Continuous Coating Make-down System

Richard H. Bublitz, Minnesota Corn Processors
Editor, Robert V. Hershey, Potlatch Corp.

Introduction

In the early 19th century, the invention of the paper machine by the Fourdrinier brothers introduced the first continuous system to the paper industry. The invention of the on-machine coater by Peter Massy, and the vast market for coated papers created by Life Magazine in the mid-1930s, began the need for the preparation of large amounts of aqueous clay coatings. It was not until the 1960s that there developed a real need for a total "systems" approach to coated paper production.

Several factors were responsible for the demand for a continuous coating make-down system. All of the components of the coating were being prepared in individual batches. The batches lacked uniformity because of inconsistency in preparation and mixing times. There was a large labor force required to prepare the coatings, and it took several hours to produce a single batch. This system, because of the length of time required to prepare a coating batch, dictated that large amounts of coating be prepared before the coater needed it. Keeping large amounts of prepared coating in storage created such problems as significant sewer losses due to over-production of coating and incomplete orders due to too little coating to complete the run. The space for the required number of tanks was large and expensive. The existing coating kitchens were difficult to keep clean, and residues from the process were an effluent problem.

During the 1960s, color kitchens were developed where ingredients were readily available to be metered into the coating mixer from bulk storage. This greatly reduced the amount of time required to prepare a batch of coating.

The ideal continuous coating make-down system is in reality a blending system where all ingredients are accurately blended together in the proper order to avoid flocculation or other undesirable reactions between the various components. The result under such conditions is continuous preparation of final coating color with uniform properties, i.e. solids, pH, and high- and low-shear viscosity. Proper preparation, mixing, dispersion, and storage of the various ingredients is covered in the chapters dealing specifically with the various pigments, binders, etc. In practice, the system described above often results in a final coating color with lower than desired solids. Dry pigment must be added to overcome this problem. In this situation, the blender becomes a combination blender/mixer/disperser.

Batch vs. Continuous Coating Preparation

A batch system implies coating is prepared one tankful or one batch at a time, with storage tanks accumulating the necessary volume for production to begin. A continuous preparation system, as the word implies, continually makes coating as close to the usage rate at the coater as practical.

It was in the 1970s that many technological breakthroughs were made that allowed a "systems" approach to coating preparation. Highly accurate flow measuring and controlling instrumentation was developed. On-machine quality-measuring devices were installed. Computers became both practical and affordable for process control. Continuous starch binder preparation by both chemical and enzyme processes were perfected and accepted widely by the paper mills. These pieces completed the puzzle, and practical continuous coating preparation became a reality.

Both proprietary and commercial continuous coating make-down systems are currently being used in the major coated paper-producing countries of the world. It is not possible to cover all the systems used due to the variety of approaches taken and the lack of specific information on the proprietary operations. However, there are certain requirements common to all systems. These are:

1. A mixing device that will blend all the coating ingredients

2. An accurate measuring system to assure that all ingredients are added in the precise amounts dictated by the coating formula

3. A computer and program that will input recipes, video display status of the system for operator information, allow remote control of the system in a supervisor location, give an alarm, and indicate which equipment has failed and all other necessary control functions

4. When starch is used as all or part of the binder, a preparation system that will provide a viscosity stable (no retrogradation) adhesive to the mixer

5. Necessary in-line screening

6. An automatic or manual method to clean the system

7. Continuous measurement of key coating properties, i.e., viscosity, pH solids, etc.

Mixers and dispersers come in a variety of configurations. Some examples are also shown in the section on mixing, also found in this chapter. The dispersion of dry ingredients and the mixing of the liquid ingredients into finished coating can be done separately or can take place in the same tank. A widely used system currently is based on a mixer-disperser developed in France. It was used originally for batch-prepared coatings. It proved to be an effective coating mixer and is still installed for batch coating preparation (Fig. 2.42). This mixer is jacketed for temperature control, has a center-mounted high-speed agitator and a slow-speed counter-clockwise rotating scraper. In the dome of the mixer is installed spray equipment used to clean the tank. At the top of the mixer is the outlet for the finished coating, which is pumped to storage. The center agitator can be equipped with a variable-speed drive. For batch operations, a mix tank can be mounted on load cells and the ingredients added in order, mixed, and the completed batch pumped to storage. Other dispersers include high-shear mixers using a rotor and stator combination or rotating blades.

For the continuous system, the basic mixing tank remains but with certain modifications. One method to produce coating continuously is to convert the tank to an overflow operation with level controls. The level in the tank is kept steady by the continuous addition of liquid. At the bottom of the mixer are located a series of ports to introduce the required ingredients for the coating recipe. These ports are sized and designed to inject specific ingredients. The order of addition will follow a pattern of starting with the ingredient with the highest to the lowest volume of addition. Care must be taken that the ingredients are compatible. For example, caustic should never be introduced adjacent to the clay port as it can flocculate the clay. All ingredients must be reviewed to assure that problems are not created by order of addition.

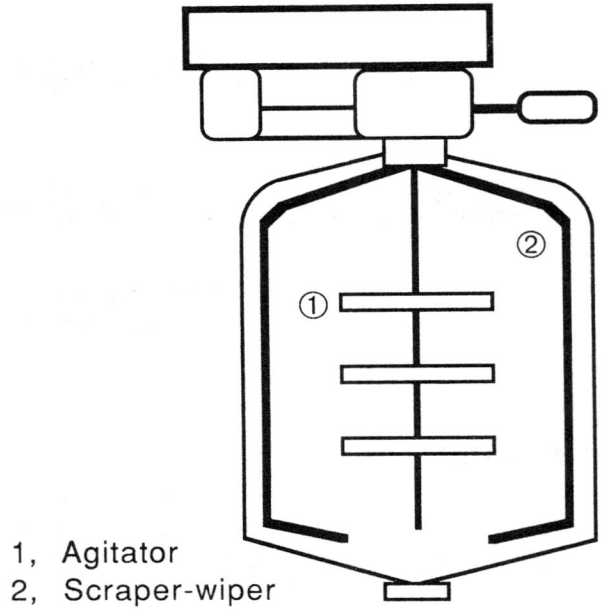

1, Agitator
2, Scraper-wiper

Fig. 2.42. Continuous mixer-disperser (courtesy of the Cellier Corp., 1991)

The size of the mixer will depend upon the volume of coating to be produced per minute and the amount of retention time required. This is the primary design parameter, and all ingredient addition requirements will be derived from this production demand.

System Components

System components involve the following:

1. A process control computer

2. Bulk storage and good accurate solids control for all coating ingredients

3. A mixer and disperser, which can be in the same unit

4. Progressive-cavity positive-displacement pumps with back pressure

5. Calibration tanks

6. Coating storage tanks

7. Proper instrumentation for flow and level control

8. When starch is used in the coating, a continuous starch preparation system.

An Example of an Automatic System Operation

Starting from a clean and empty system, the operator in the control room activates the computer formula for the coating to be produced. The system will start in the automatic mode. The mixer will start, and the pumps will inject each of the ingredients into the tank bottom. The flow rate of addition is based on the recipe. The mixed and dispersed coating will overflow and get pumped to coating storage. The system will run until the system is shut down by a high-level signal from the receiving tank. The mixer-disperser may continue to run in the standby mode. All ingredients should be recirculated in their re-

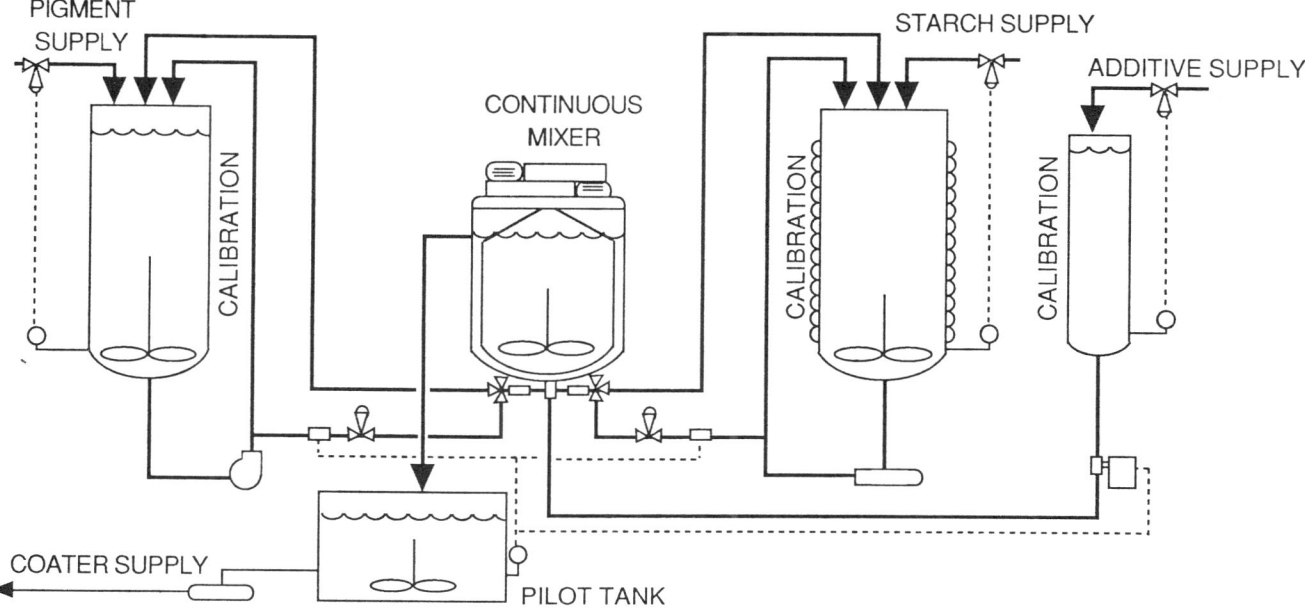

Fig. 2.43. Continuous coating preparation system

Fig. 2.44. Pipe array in a continuous coating preparation system (courtesy of the Cellier Corp., 1991)

spective loop lines to prevent plugging or settling. Some systems also recommend recirculating liquids from the bottom of a bulk storage tank to the top of the same tank. On receiving a low-level signal, the system will again produce coating until the system is satisfied.

Accuracy of addition of the volumes or weight of the ingredients is very important to assure consistent coating performance. Either at programmed intervals or manually by the operator, calibration is done on the flow of each ingredient. The rate of flow is determined by measuring the volume drop or weight loss in a calibration tank (Fig. 2.43) in the supply line for that ingredient or using a mass flow or density meter. The readout is shown on the computer display and the flow can be corrected, if necessary. The system will continue to produce the coating formula until the production run is terminated.

To make a grade change, the continuous addition is stopped and the mixer-disperser contents are pumped to the receiving tank. The new computer coating recipe is activated, and the system starts. To assure that all of the previous coating is out of the system, a ball is introduced into the line to push the residual coating into the receiving tank. There is a basket to catch the ball in the receiving tank.

For shutdown, the system is emptied and an automated cleaning program can be activated. One system in-cludes spray balls in the tank heads that inject cleaning solution into the system, and the pumps circulate the solution. The cleaned system is emptied and put into a standby mode.

Maintenance

This can be a highly automated system of coating color preparation. A properly designed system will have sufficient inventory of coating present to allow for reasonable downtime. Any failure of a component should be announced by an alarm. The operator can determine the fault from a light panel or a terminal display and, possibly, take corrective action. The design must take into account each item to assure that it can be isolated and repaired.

The valve type and nipple lengths used to shut off flows must allow for possible settling or flocculating that could prevent easy restart of the system. It is essential that a supply of spares be maintained as recommended by the component supplier. Critical pumps should be paralleled to minimize downtime. Preventative maintenance should be performed as recommended by the component suppliers.

Fig. 2.45. Mix tanks with floor level access (courtesy of the Cellier Corp., 1991)

Summary

The basic design for a continuous system is quite simple, but can require the highest degree of automation. If your coating needs are complex or simple, the continuous system can provide the ability to produce the quality and uniformity demanded by current paper specifications. However, if the production requirements are small or grade changes are frequent, coating formulation changes are frequent, or coating solids requirements are high, a batch system may be the system to choose.

Chapter 2, Section VI

Screening of Paper
Coatings and Their Ingredients

Dale Midyette, CRS Sirrine Engineers, Inc.
John D. McInnes, JDM Associates, Inc.

Introduction

The concern for quality is ever increasing in today's paper industry. In no area is quality more important than in coating preparation and supply. A coating which is free of contamination is an absolute necessity in producing an acceptable product.

A major tool in achieving high-quality coating is screening of the prepared coating and the coating ingredients. Screening removes oversized particles from the liquid being filtered. These particles may come from several sources including raw material contamination, poor preparation of the coating ingredients in the coating itself, or scale from the walls of tanks and piping.

Screen Performance Considerations

The performance and capacity of equipment designed to filter coating is often difficult to predict because of the complexity of the rheological behavior of most coating. Rheology refers to the study of deformation and flow properties in a liquid. Typically, coating have non-Newtonian flow characteristics. That is, viscosity of coating varies with the amount and duration of stress and strain on the coating. The stress and strain come from flow induced by pumping, agitating, or other disturbances of the coating including filtration and screening.

Coating colors may be made up of 20 or more different ingredients, each having unique rheological properties which should be considered prior to selecting screening units. Still more consideration should be given in determining filter specifications for the finished coating. The coating filter needs to be capable of maintaining efficiency and capacity for varying rheological properties associated with formulation changes, temperature changes, changes in flow rate, and other process variables.

Filter Media

Filter media is the permeable material which allows the clean liquid to pass but retains oversized solids. There are four basic types of filter media used in the coating area. These are woven synthetic fabric, woven stainless

Table 2.3. Effects of wire size on screen characteristics

Mesh Size	Wire Diameter inches (cm)	Opening inches (cm)	% Open Area
100	0.0050 (0.013)	0.0050 (0.013)	25.0
100	0.0045 (0.011)	0.0055 (0.014)	30.3
100	0.0040 (0.010)	0.0060 (0.015)	36.0
100	0.0035 (0.009)	0.0065 (0.017)	42.3
100	0.0030 (0.008)	0.0070 (0.018)	49.0

steel, cylindrical wedge-wire, and perforated or slotted tubes. Each type has different characteristics such as open area, durability, strength, and ease of regeneration.

Table 2.4. U.S. Standard Testing sieve series

Sieve Number	Opening inches (cm)
50	0.0117 (0.030)
60	0.0098 (0.025)
70	0.0083 (0.021)
80	0.0070 (0.018)
100	0.0059 (0.015)
120	0.0049 (0.012)
140	0.0041 (0.010)
170	0.0035 (0.009)
200	0.0029 (0.007)
230	0.0024 (0.006)
300	0.0021 (0.005)
325	0.0017 (0.004)
400	0.0015 (0.004)

Screen cloth, which is woven synthetic fabric or stainless steel, is usually woven in a square pattern. There are the same number of wires in the warp and shute directions (Fig. 2.46a). The warp and shute wires are the same diameter. Other types of weaves are available, such as the plain dutch weave in Fig. 2.46b, where the warp wire diameter is larger; the twilled weave, where the warp and chute wire each pass alternately over and under two wires, and the twilled dutch weave.

The number of openings per linear inch of the screen is referred to as the "mesh." Mesh, along with wire diameter, determines the actual screen opening size which is the important dimension for removing unwanted material. Table 2.3 shows the screen openings for various wire diameters in 100-mesh screen cloth available from one manufacturer.

Table 2.4 shows the U.S. Standard Testing sieve series. This series gives a standard opening for each mesh size. Screen cloth is correlated to the equivalent U.S. Standard Testing sieve series.

Stainless steel screen cloths are supplied in one of two specifications: market grade (MG) and tensile bolting cloth (TBC). The MG cloth has a larger wire diameter for a given opening and, therefore, as can be seen in Table 2.5, the percent open area in less. With a larger wire diameter, the strength and durability of the screen is higher and the life of the screen longer. With a smaller

Table 2.5. Effect of screen type on characteristics

Screen Type	Opening, inches (cm)	Mesh Size	Wire Diameter, inches (cm)	Open Area, %
MG	0.0070 (0.018)	80	0.0055 (0.013)	31.4
BC	0.0071 (0.018)	94	0.0035 (0.009)	45.0
MG	0.0029 (0.007)	200	0.0021 (0.005)	33.6
BC	0.0029 (0.007)	230	0.0014 (0.004)	46.0

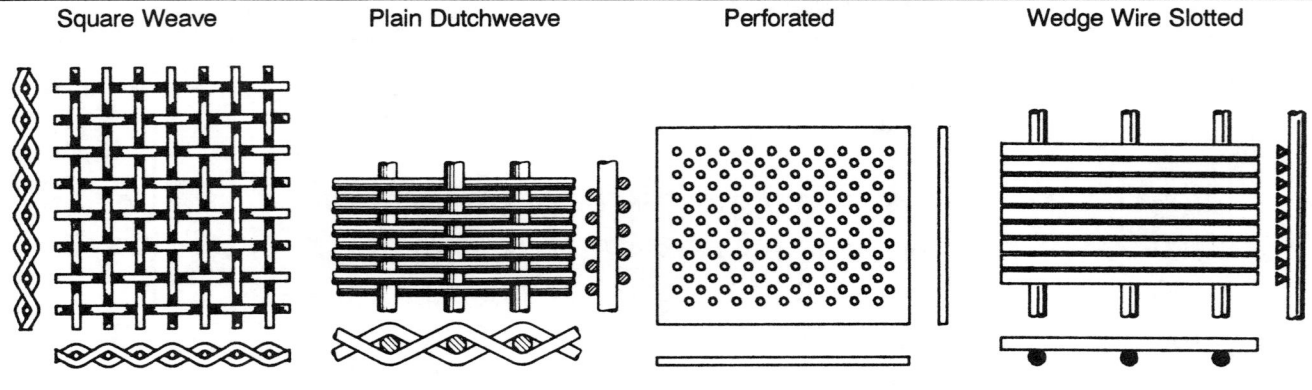

Fig. 2.46. Filter media (courtesy of ProGuard Filtration Systems) a. Square Weave, b. Plain Dutchweave, c. Perforated, and d. Wedge Wire Slotted.

percent open area media, the flow through the screen is restricted and therefore lower. Or, in the case of pressurized screens, the differential pressure across the screen is higher.

A cylindrical wedge wire screen, sometimes called profile wire or slotted, is constructed by helically winding a continuous wedge wire (triangular cross section) around a series of vertical stringers and welding the wire to the stringers (Fig. 2.46c). The wedge wires are separated by an opening which is equal to the element's equivalent mesh rating.

Wedge wire screens are much stronger and more durable than an equivalent woven screen. However, they also have a much lower percent open area. A wedge wire screen with an 80-mesh equivalent and an opening of 0.007 in. (180 micron) would have a maximum open area of only 25%. Compare this to 31.4% and 45% open areas of MG and TBC grades, respectively, of 80-mesh woven screens.

A slotted screen is basically a steel tube or plate with slots cut into it (Fig. 2.46d). These types of screens have the lowest percent open area of all the types discussed. The primary advantage of using slotted screens is that they have smooth surfaces and are more easily cleaned in certain applications.

Perforated material is now available because of high quality, close tolerance, and precession photochemically machined methods recently developed. Standard perforated diameters range from 0.006 in. to 0.020 inches (150-508 micron) and thicknesses of 0.005 in. to 0.014 in. (0.013-0.036 cm). Other hole sizes, materials, and metal thicknesses can be custom-fabricated to meet special requirements.

Table 2.6 shows recommended filtration ranges of coating and common coating ingredients. These ranges can be narrowed down considerably depending on the specific criteria of the filter situation.

Table 2.6. Filtration ranges

Liquid	Filtration Range (mesh equivalent)
Coating Color	30-250
Fresh Water	30-200
Synthetic Binders	40-100
Natural Binders	60-150
Titanium Dioxide	40-200
Calcium Carbonate	30-100

Location of Filters

The locations of filters within the coating operation will vary from mill to mill depending on the mill's philosophy on quality control. Generally, the high emphasis on quality in recent years has led to more filters in more locations than in the past.

Individual additives may be filtered at unloading docks, prior to entry into storage tanks. This filter location helps eliminate contaminants which come from the supplier or an improperly cleaned tank car or truck. Individual additives may also be filtered between storage and mixing. In this case, lower flow rates would decrease the size and cost of the filter over an unloading filter on the same liquid.

Filtration is often done between mixing and storage of the coating. In this location, a filter can eliminate lumps due to improper mixing and poor conversion of organic binders such as starch and protein.

The most critical location for coating filtration is at the coater. Contaminated coating at this point will have a direct effect on the quality of the finished product.

Modern coaters require a large amount of recirculation of the coating. Recirculation rates of 25 units (unit weight or volume) of coating returned to supply tank to every unit of coating applied to the sheet, i.e., 25:1, and higher are common and result in high recirculating flows. There is a high probability coating will pick up paper and felt fibers at the coater head. These and other contaminants need to be removed prior to the coating returning to the supply tank or between the supply tank and the coater. In most cases, the return line to the supply tank is gravity-fed and eliminates pressure filters in this location. Gravity strainers, such as vibrating circular screens, were often used in the past but do not have the capacity to handle the high flow rates of today's coaters in a reasonable space and with media fine enough to actually remove returned debris. Most often pressure filters are used between the supply tank and the coater.

During a web break, there is a good possibility for large pieces of paper to get into the recirculated coating. A danger exists, for these large pieces of paper could plug piping or "blind" the filter unit which is generally designed to remove fine contaminants. A very large wire mesh screen, with a particle retention size of 0.75-1 in. (1.9-2.54 cm), is sometimes used at the supply tank just under the return line from the coater. Some coater manufacturers provide such a screen as a part of their coater.

Materials of Construction

The materials of construction for the various parts of screens and filters used in the paper coating industry should be selected based on sound design considerations. Items which should be considered are durability, chemical or physical reaction between the material of construction and the product to be screened, material costs, and compatibility with standard spare parts at the mill site.

In paper coating, it has become standard to use stainless steel as the material for the screening media and for all parts of the filter or screening housing which actually come in contact with the coating or additive. A 300 Series stainless steel is usually recommended. Portions of the housing, which are not in contact with the liquid, may be made of less expensive materials. Proper coating of these materials, whether galvanized or painted, should be specified.

Where seals are used, the type of material used should be given consideration. Very hot liquids, such as just cooked starch or reactive ones such as alkali, may require seals which are not the manufacturer's standard.

Motors, local control boxes, and other electrical parts should be constructed for the sometime harsh environment of a paper mill. Control panels should be corrosion proof, air tight, and dust tight. Electrical motors should be splash proof or totally enclosed.

Types of Screens

The descriptions which follow are of filtration units commonly used in paper mills. The descriptions are intended to be generic. The mention of any particular type of equipment or equipment option, which may be unique to one vendor, is not intended to promote that vendor, but to provide the reader with a good understanding of the equipment available.

Vibratory Screens

Circular vibratory screens are gravity-type screens where the filtration media is stretched over a circular frame. The liquid to be filtered enters the top of the screen, and gravity pulls the liquid through the media, leaving the oversized solids on the top of the media. Vibration of the screen increases throughput of the liquid to be filtered, inhibits blinding of the screening media, and causes oversized particles to move toward a discharge on the periphery of the unit.

The vibration is caused by a motor which is mounted vertically inside the base of the unit. Two eccentric weights are mounted on the upper and lower ends of the motor shaft. Rotation of the top weight causes the screen to vibrate in the horizontal plane. A heavier top weight would increase this horizontal vibration. Rotation of the bottom weight causes the screen to vibrate in the vertical plane. Again, a heavier bottom weight would increase the vertical vibration. The rotation of each of the weights on relation to each other causes a tangential movement of the screen. This movement can also be increased or decreased by adjusting the angle of lead given to the bottom weight in relation to the top. Using these adjustments, the speed and spiral pattern of the material over the screen cloth can be set by the operator.

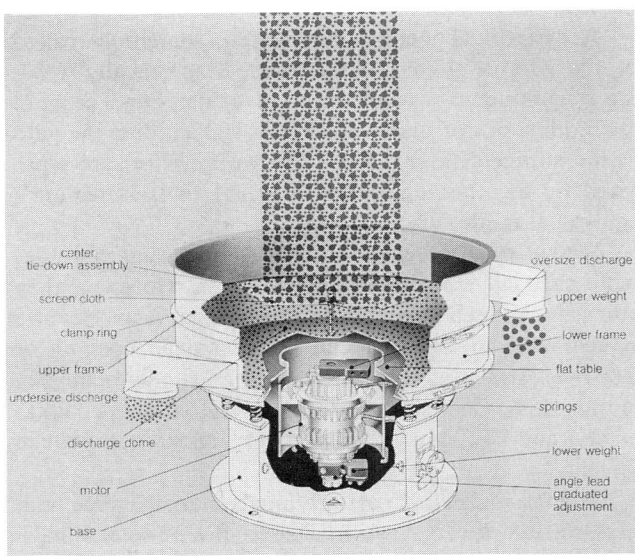

Fig. 2.47. Vibratory screens: circular and rectangular (courtesy of SWECO)

Some mills choose to increase throughput of the screens by running a "head" in the unit. In this case, there will be no discharge for oversized particles or the discharge will be blanked off. The entire screen can then be used, and the liquid level can be run at 34 in. deep. Any solids would have to be manually cleaned out when the unit is taken out of service.

There are several manufacturers of vibrating circular screens. Each manufacturer offers several options including self-cleaning mechanisms.

Certain manufacturers also make square or rectangular vibratory screens. These screens have slopped tables to direct oversized particles to the low end. In some cases, greater flows per area can be attained with rectangular units.

Vibratory screens are used for screening raw materials as well as prepared coating. Their use at the coater is limited due to the high flow rates returned to the coater supply tank and the fact these are nonpressure-type screens.

Basket Filters

Basket or bag filters are the original pressure filters. Liquid enters a basket filter from the top. Oversized particles collect inside the basket and clean filtrate passes through the media. Bags are available for basket filters in any media type.

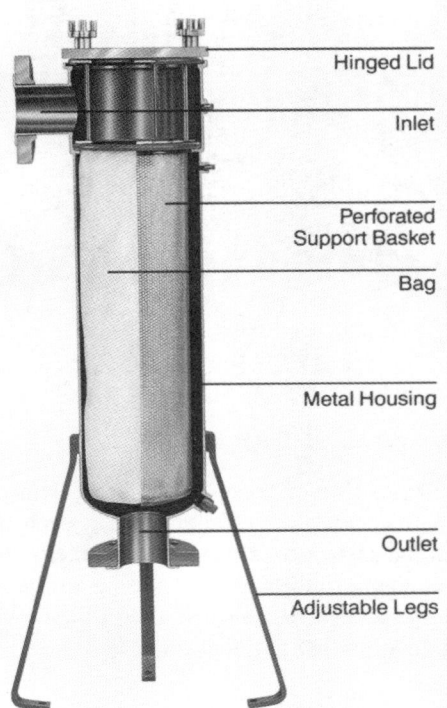

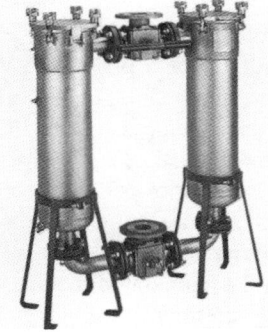

Fig. 2.48. Basket filters: batch (cross section) and continuous service (courtesy of Ronningen-Petter)

Basket filters can provide either continuous or batch

service. In batch operation, the filter has to be moved off-stream and an operator removes the media for cleaning or replacing with a new bag. For continuous flow applications, basket filters are valved together providing the option of running one or both baskets at a time. In either case, basket filters require manual cleaning to regenerate the media.

Basket filters are used to filter smaller coating ingredients such as functional additives, water, or adhesives.

Tubular Pressure Filters

The tubular pressure filter is made up of a tube of filter media, commonly called a filter element, fitted inside a slightly larger-diameter housing tube. The material to be filtered is forced by pressure through the element, and on to the process, leaving the oversized material between the element and the housing. Tubular filters are often manifolded together so that higher flow rates can be handled. These units are referred to as multiplex tubular filters.

Fig. 2.49. Tubular filter: automatic multiplex (courtesy of Ronningen-Petter)

As the oversized material begins to build up and blind the filter element openings, flow through the element is restricted and the pressure differential across the filter increases. It is then necessary to "regenerate" the media by

a. Internal

b. External

Fig. 2.50. Backwashing tubular filters (courtesy of Ronningen-Petter) a. Internal b. External

either external or internal backwashing. External back-washing is performed by using valves to isolate the filter, then introducing water to the discharge side of the filter, and forcing oversized material off the element and into a drain header. Internal backwashing uses the filtrate as the backwashing liquid. Internal backwashing would not normally be found in paper coating applications, other than the filtration of water used in the preparation of coating.

Provisions can be made so that the backwashing sequence is controlled automatically. Automation of tubular filters is very beneficial at the coaters or other places where continuous operation is needed. Automation is not as beneficial at an unloading station. When automation is used, pressure differential gauges or timers initiate the backwash. Actuators open and close the proper valves, isolating one filter element at a time until all the filter elements in the bank are regenerated.

Further provisions can be made so that, prior to automatic backwashing, each tube is purged with air, blowing the contents into a "salvage" header. In this way, much of the good liquid is saved rather than sewered with the contaminates.

There are other options which can be obtained with tubular filters. One of these is clustered elements where three smaller elements are contained in each tube. This increases the surface area per tube. Another option is the elements vibrate, mechanically putting more shear into the coating and delaying any blinding of the media openings due to a forming pigment cake.

Mechanically Cleaned Filters

In-line mechanically cleaned pressure filters have only one single-diameter tubular filter element. This element can be made up of a plate with wedge wire slots, laser cut slots, machined slots, or perforated filter media. Oversized material is removed from blinding the media openings by a mechanical scraper/wiper rotating around the periphery or the interior of the element (Fig. 2.51), by the element itself moving around a stationary doctor, or by a doctor moving up and down inside the element. Oversized material accumulated in the vessel is discarded from the unit by draining, purging, or backwashing the debris from the vessel. Because the filter elements are difficult to remove, it is recommended to have a spare filter installed for times of high contamination to ensure filtration of coating at all times.

Some mechanically cleaned filters have panels and automatic system controls which can be incorporated into the coating system control rooms.

Fig. 2.51. Blades cleaning interior of filter element (courtesy of Cellier Corp.)

Fig. 2.52. a, b, and c. Three ways to install mechanically cleaned filters:

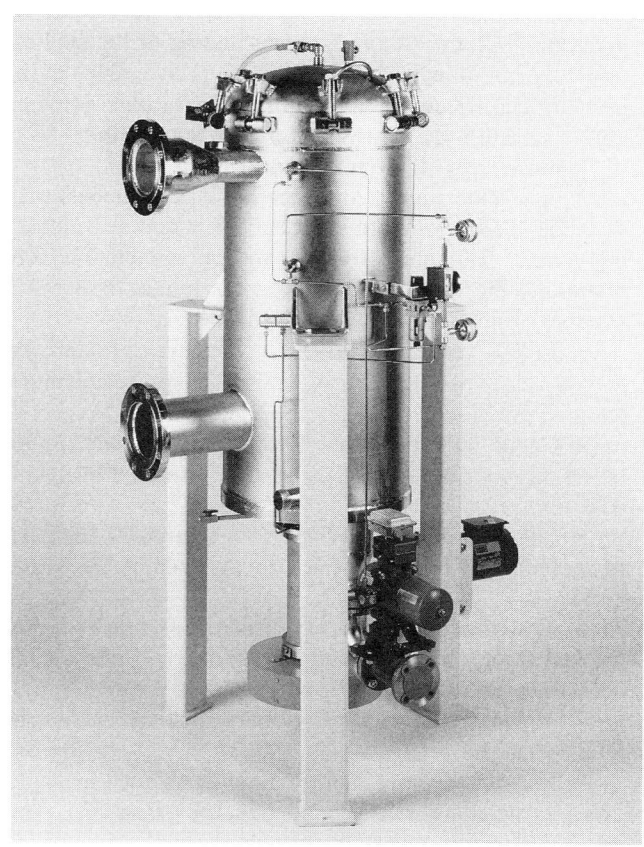

b. on-floor (courtesy of Cellier Corp.)

a. in-floor (courtesy of Jylhäraisio Oy)

c. multiple station unit (courtesy of Ronningen-Petter)

Filter Selection

The selection of the type of filter to use should be made only after a thorough understanding of the application to determine what filter option is available as the best fit. The following are major points to be considered before selecting the best filter for each application:

1. Flow rate: Minimum and maximum

2. Operation time: Will operation be continuous, such as at the coater, or intermittent such as a tank car unloading station? This may affect the media regeneration technique and the amount of automation. It may also affect whether or not an in-line spare is needed.

3. Liquid properties: Viscosity, percent dry solids by weight, amount and size of contaminants, temperature, and line pressures

4. Capital cost restraints: Purchase investment, installation requirements, spare parts costs, use, and availability.

5. Space limitations: What, if any, exists? Remember to include the required space to disassemble the filter unit.

6. Utilities requirements: Electrical power, compressed air, backflush water availability, and cost should be considered.

7. Noise: From motors, vibrating, etc. Can they be tolerated in this area?

8. Pressure differentials: Some coaters such as short-dwell coaters are sensitive to pressure pulsations. This may affect media-regeneration techniques.

9. Manpower: For regenerating the media when not automatic. Also for routine inspections and cleanup.

Summary

Just as filtration and screening techniques in the coated paper industry have evolved over the years, they will continue to change. The industry requires change due to higher coater speeds, higher dry solids coating, new coater types, increased labor and utility costs, and the greater emphasis on quality. It should be expected that filters of the future will operate with more and more automation, as is the trend today. Stronger, more durable filter media with greater percent open area will be developed, as will improved filter housings. Filter media regeneration techniques will continue to be optimized as new techniques are developed.

Chapter 2, Section VII

Pumps and Circulation Systems

Raymond O. Wiener, Simons Eastern Consultants Inc.

Introduction

Coatings are generally non-Newtonian fluids and usually pseudoplastic. However, because coating formulations and solids content vary considerably, a coating could have dilatant properties. At low solids content, the coating could approach a Newtonian flow pattern. These varied properties cause considerable difficulties in determining the flow characteristics of a coating in a transfer or circulating system and the horsepower requirements of the pump to be used.

In addition to the coating flow properties itself, the selection of the proper type of pump, piping, valves, and controls for optimizing the process requirements and serviceability is a study in engineering economics. Coupled with this aspect are operational problems such as power demand of pumps, system maintenance, and flexibility.

Types of Pumps

Pumps of many different manufacturers have been used for coating service. The type of pump considered will depend on the flow volume and discharge pressure needed, the net positive suction head available, and on the flow characteristics, viscosity, solids, abrasiveness, and shear sensitivity of the coating being pumped. The types described below were chosen to represent those most prevalent. There is no implication intended that other types will not perform. Pump descriptions are brief. Additional information should be obtained from the manufacturer.

Progressing Cavity Pump

The progressing cavity pump (Fig. 2.53) is widely used in the coating field for transferring and circulating coatings. The operating principle is based on a single-thread helical rotor turning within a double-thread helical stator. By design, this pump is a positive-displacement unit capable of developing substantially greater discharge pressures than a centrifugal pump. It should be protected against shut-off downstream to prevent damage to the piping system, other equipment, or to the pumping unit itself. To accomplish this, it is sometimes desirable to provide a circulation loop with takeoffs to the operating points with the excess coating returning to the supply tank. In this way, the pump always has a free discharge in the event of shut-off at the operating points. Pressure relief valves, torque limiting couplings, and rupture discs have also been used on pump discharge piping to protect against over pressure.

The capacity of a progressive cavity pump is basically determined by the size of the cavities and revolu-

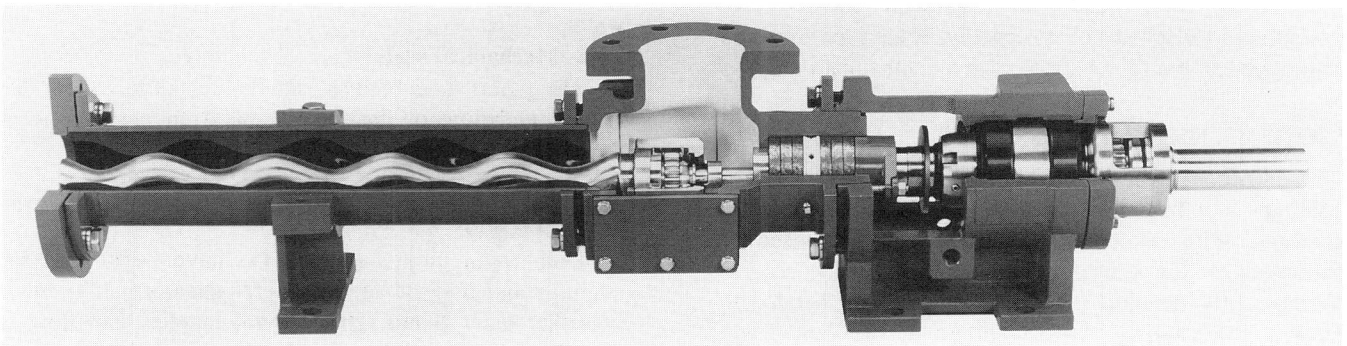

Fig. 2.53. Progressing cavity pump (courtesy Robbins & Meyers, Inc.)

tions per minute. However, the discharge pressure required and the coating viscosity also affect the capacity. This type of pump is run at a substantially lower speed than a centrifugal pump.

These pumps are generally constructed with a chromium-plated stainless or alloy steel rotor and an elastomeric stator usually of natural or synthetic rubber. Flame-sprayed ceramic and metal oxide rotor coatings are also available to increase the wear resistance of the rotor. The materials of construction chosen must be compatible with the abrasiveness, temperature, and corrosion characteristics of the coating. The manufacturer's recommendation for materials of construction as well as pump speed should be obtained to give a unit with reduced maintenance and long life.

While it is not desirable to run any pump dry, it is imperative not to let the progressing cavity pump operate without fluid. Sufficient friction and heat could develop between the rotor and stator in the absence of lubrication to destroy the elastomeric lining. Suitable measures should be taken to stop the pump when it no longer has flow to the suction side. One such method is to install a no-flow switch in the pump suction piping that is interlocked with the pump motor.

Rotary Pumps

Among the kinds of rotary pumps that have been used for transferring and circulating coating are the gear type and lobe type. The gear-type pump operates on the principle of a gear rotating within a gear. The lobe type (Fig. 2.54) operates by means of an impeller with one or more lobes of various designs, mounted on a rotating element turning within the pump casing. Generally, two rotating elements with timing gears to synchronize the action of the impellers are used.

Rotary pumps operate on the positive-displacement principle and must be protected against downstream shutoff. Some models can be provided with an internal bypass arrangement for this purpose.

Like the progressing cavity pumps, the capacity of the rotary pump is determined by the size of the cavities and the pump rotation speed with adjustments for discharge pressure and viscosity of the coating. This type of pump also operates at a lower speed than a centrifugal pump.

Centrifugal Pumps

Centrifugal pumps may be satisfactory for low-solids coatings and in some cases high-solids, low-viscosity coatings. With relatively high-solids and high-viscosity formulations, the centrifugal pump may not be able to economically develop the discharge pressure needed to convey the coating. An advantage of the centrifugal pump is the absence of pressure pulsations.

Fig. 2.54. Lobe type rotary pump (courtesy Tuthill Pump Division, Tuthill Corp.)

The speed of a centrifugal pump handling coating should be chosen to minimize erosion in service. Lower speeds generally decrease maintenance and prolong pump life. However, lower speeds require a larger pump. Each application must be analyzed to arrive at the most economical and serviceable choice.

Conventional ring packing with a lantern-type liquid seal ring in a stuffing box can be used. Water at a pressure to counter balance the pump pressure is connected to the seal ring. It is important that the pressure not be high enough to cause water leakage into the pump. For best results, the seal water pressure must be carefully controlled and a valve in the seal water line, which is interlocked with the pump motor, should close when the pump is not operating.

Because of the danger of coating dilution from seal water with a conventional packing system, other methods of sealing have been used. These methods include:

1. Loosely packed conventional packing with no seal water, which allows a small amount of coating leakage but no dilution
2. Dynamic seal which still may require flushing water
3. Mechanical seals.

The pump manufacturer should be consulted for his recommendations on the correct seal arrangement to meet the process requirements.

Diaphragm Pumps

Diaphragm pumps (Fig. 2.55) have been used for coatings and may be of benefit for shear-sensitive coatings that other pump types cannot handle. The flow is controlled by stroke length or frequency with an air or electric motor. This type of pump gives a pulsating flow. However, pulsation dampeners can be provided.

Pump Drives

Pumps can be motor-driven in a number of ways. The general methods are: (a) direct-connected drives, (b) mechanical gear reduction drives, (c) belt drives, or (d) variable-speed drives.

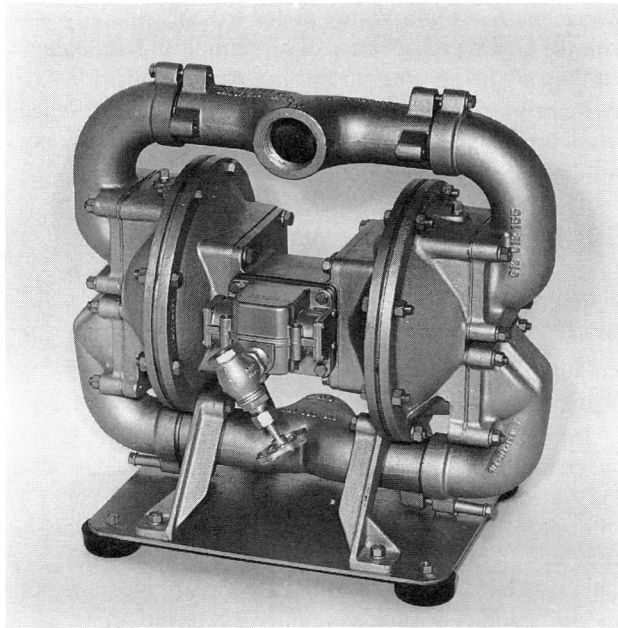

Fig. 2.55. Diaphragm pump (courtesy Warren Rupp, Inc., a Unit of IDEX Corp.)

Direct-connected drives are commonly operated at fixed speeds of 1800, 1200, 900 rpm, etc., corresponding to the common operating speeds in which motors are supplied. This type of drive is suited to constant operating conditions and generally is used only on centrifugal pumps.

Mechanical gear reduction drives provide greater flexibility in pump operating speeds. They can be designed for many fixed speed requirements by proper selection of gear sizes.

Belt drives can be designed for almost any fixed speed need by proper choice of the motor and correct selection of the drive and driven sheaves. Belt drives can easily be changed to another fixed speed by changing sheave sizes. Belt drives are usually the most economical way of meeting the constant speed requirements of the slower positive-displacement-type pumps.

Several different types of variable-speed drives are used to allow pump speeds to be varied throughout the desired range. Among those in general use are mechanical adjustment types, the eddy current coupling, the adjustable-speed direct-current motor controllers, and the adjustable-frequency alternating-current motor controllers. Cost and operating requirements dictate the specific unit to be chosen for a particular case.

Materials of Construction

The piping materials that have been used for coating and their components vary widely and include stainless steel, carbon steel, copper, brass, bronze, polyvinyl chloride (PVC), and fiberglass-reinforced plastic (FRP). The selection depends on the erosion and corrosion resistance needed. Stainless steel is most often the choice. Generally thin wall type 304L stainless can be used for most services in coating plants. If better external and internal corrosion resistance is required, type 316L can be used. Economic comparisons between materials should be made on the installed cost basis.

Valves

The type of valve material used for coating and its components is generally stainless steel. However, cast iron, copper, brass, bronze, and PVC have been used. Process considerations similar to piping apply to valves. It is not essential that the valve material be the same as the piping. Economic considerations as well as process requirements should be used in selecting the valve material.

Among the types of values used are gate, ball, plug, and butterfly. The choice depends on the type of service condition required. For tight shut-off and throttling, ball and plug valves are generally better. Non-lubricated plug valves are preferred over the lubricated type. For space limitations, it may be necessary to use a knife-gate valve. Metal seated knife-gate valves are better for coating service than those with resilient seats. A three-way plug valve may sometimes be used to replace two two-way valves to avoid pocketing.

Pumps

Like valves, it is not essential that the pump be constructed with the same materials as the piping. The choice of the construction depends on economics and the system. Frequently, cast or ductile iron may be satisfactory for the intended service rather than the more expensive stainless steel pumps. The trim on these pumps should be stainless steel, however, because of the better corrosion and erosion resistance. Other types of construction are available for specific purposes. The pump manufacturer should be contacted for the most suitable materials to meet these requirements.

Control Instrumentation

A large number of controls from many different manufacturers are available for coating piping systems. The controls listed below represent the prevalent types in use. It is not intended that other types will not provide adequate control for a particular operation.

Level

Various types of level control elements have been used to control tank level. These include floats, bubble tubes, diaphragm transmitters, capacitance probes, and ultrasonic detectors. The type most often used is the differential pressure, diaphragm transmitter. The transmitter should be mounted as nearly flush to the tank wall as possible and in a position so that it remains submerged in the coating to prevent dry coating buildup on the diaphragm. A special isolation valve, available from at least two valve manufacturers, is often included in the installation so that the transmitter can be repaired without requiring the tank to be empty. Two modes of level control in a tank are used: continuous or intermittent. In the continuous mode, the signal from the transmitter operates a valve in a branch off a circulation line or the pump speed on a direct supply line. The level in the tank remains relatively constant. This mode is most often used where suction head to the pump and coating uniformity are important. For batch-type transfer of coating, the intermittent mode can be used. The level transmitter will supply high and low signals to open and close a valve in a branch off a circulation line or to start and stop the motor on a supply pump.

Temperature

Most types of temperature control have been used. However, temperature well probes are most common.

Pressure

Most types of pressure control have been used. It is important with pressure transmitters used for coatings to employ diaphragm seals for isolation.

Viscosity

On-line viscosity control of coating is not often used because of variation in coating viscosity due to components, coating preparation, and temperature. This property is generally controlled by the laboratory determination of viscosity samples.

Density

Density control of coating has become the most accepted method of controlling coating uniformity. The types used include mass meters, nuclear meters, and U-tube weight beams. The U-tube weight beam generally requires a dedicated pump to supply coating to the instrument at constant flow and pressure. The mass meter measures mass flow rather than volume flow. Density can be measured from the same unit. The nuclear meter is a non-contact instrument.

Flow

The types of flow control that have been used for coating include mass flowmeters, magnetic flowmeters, positive-displacement meters, and Vortex meters.

The magnetic flowmeter measures volume, whereas the mass flowmeter measures mass. The positive-displacement meter and Vortex meter are not generally used with the coating itself because of particle size limitations, but they are being used for coating components.

Positive-displacement pumps with variable-speed drives are also used in certain applications as a flow control device.

System Design

Pumps and Piping

It's good practice with coating piping to strive for a run with as few turns as possible, sloping toward the discharge point. The discharge into tanks should be extended below the liquid level to minimize air entrainment and foam buildup. Pockets in the run should be avoided. The run of pipe should have flanged or quick-coupling connections conveniently spaced so that dismantling for cleaning, if necessary, can easily be done. Drain and water flush connections should be employed on both sides of equipment such as pumps and in the run itself, if needed, so that the entire system can be flushed. Air connections can be used to flush good coating back to the holding tank for reuse. Valves should be located for easy access.

In sizing of the piping, flow velocities should be determined to provide the optimum cost relationship between pipe size and horsepower. In general, reducing the pipe size will increase the pressure drop through the system and result in a higher horsepower demand. In some cases, the higher pressure drop can require a larger, more expensive pump model. Increasing the pipe size will most often reduce the horsepower demand, but increase piping and installation costs and also increase the volume of coating in the pipeline that may be discarded when changing formulations, flushing out, or shutting down the system. With some coating formulations, problems may occur with excessive coating buildup in the pipeline with the lower velocities generated by the larger pipe sizes. Desirable velocities will vary for different types of coatings and different systems, but generally will fall within 2-6 feet per second.

Design of gravity lines is especially critical. The proper slope must be determined to ensure proper flow at suitable velocities. The optimum slope angle will vary with the coating characteristics. As an example, for high-solids coatings it may be necessary to have a minimum slope of one foot in ten foot of run.

The flow characteristics of coatings vary considerably with solids content and formulation. Being mostly non-Newtonian fluids with pseudoplastic properties, coating viscosities generally decrease with an increase in shear rate. The shear rate in a pipe increases with velocity. Ignoring the viscosity changes of the coating through the system can lead to drastically oversized pipe sizes, pumps, and motors.

The correct head loss needed for pipe and pump sizing may be difficult to determine because of the viscosity variations that occur with different flow rates and pipe sizes. Experience with an existing system often is beneficial. Penkala and Escarfail (1) have developed a method for calculating pipe flow viscosities using laboratory viscosity measurements. This method has been used to give acceptable approximations for many coatings.

When the pump volume and discharge pressure requirements have been determined, the proper pump selection can be made. The choice depends on abrasiveness, viscosity, pH, specific gravity, solids content, temperature, Net Positive Suction Head Available (NPSHa), volume, discharge pressure, frequency of operation (continuous or intermittent), application, and any unusual properties or services that may be expected. The pump vendor should be given all the above properties and operational requirements so that his best recommendations can be given.

In designing the pump installation, the pump suction piping should be as short as possible and of ample size to assure that the NPSHa is large enough. If the NPSHa does not exceed the Net Positive Suction Head Required (NPSHr), the pump could operate in a "starved" condition and result in reduced capacity, serious vibration, noisy operation, and abnormal wear. Net positive suction head is the energy that forces liquid into the pump. NPSHr is a characteristic of the pump itself and will vary for different pump sizes and models.

Spare pumps should be included in the installation wherever a pump failure could shut down an operation. As an example, the coating supply pump to the coater should be spared so that a pump failure will not shut down the coating operation. Depending on the design, there may be other applications in the coating additive storage and supply, coating preparation, and coating storage and supply areas where it is essential to have spare pumps.

A variable-speed drive should be used wherever the pump is used for metering. Besides the coating preparation area, this could include the supply pump to the coater itself. For an operation where the coater speeds and coating weights vary, it may be necessary to include variable-speed drives to meet the coater requirements of application and recirculation rates.

Control and Instrumentation

The degree of automation and the possibility of future expansion should be part of the control method decision. Information items such as the amount of coating used, the number of coaters being supplied, the number of different formulations run, the quality reproduction range of the coating required, and the manpower cost should enter into the determination of the degree of automation.

For proper control, process variables must be known in order to make the correct instrument choice. These variables include the following:

Level	– Tank material of construction
	– Tank dimensions
	– Open or closed tank
	– Location of instrument
	– Minimum, maximum, and normal control or indicating level
	– Temperature, percent solids, pH, and specific gravity of fluid
Temperature	– Material of construction of tank or pipe
	– Pipe size
	– Minimum, maximum, and normal control or indicating temperature
	– Fluid corrosion properties
Pressure	– Material of construction of pipe
	– Pipe size
	– Minimum, maximum, and normal control or indicating pressure
	– Temperature and corrosion properties of the fluid
Density	– Material of construction of pipe
	– Pipe size
	– Inlet pressure
	– Minimum, maximum, and normal control or indicating flow
	– Minimum, maximum, and normal control or indicating density
	– Temperature and corrosion properties of the fluid
Flow	– Material of construction of pipe
	– Pipe size
	– Inlet pressure
	– Minimum, maximum, and normal control or indicating flow
	– Temperature and corrosion properties of the fluid

The control vendor should be contacted for proper connection size and type needed for installation and also for special location requirements such as vertical or horizontal pipe runs and minimum straight pipe runs needed before and after the instrument.

Entrained air and foam in the coating should be minimized if accurate measurement and control is expected. Generally, proper design of the pumping and piping system will suffice. However, in severe cases deaerators have been used.

Summary

Many types of pumps, pipes, valves, and controls are available for use in coating piping systems. To properly design a system, the requirements of the system must be known. The maximum, minimum, and normal operating conditions must be defined. The degree of flexibility required and provisions for future expansion, if needed, must be decided upon. The controls needed and the degree of automation desired must be outlined. The realistic flow characteristics of the coating and its erosions and corrosion properties must be known.

When all of these fundamental factors have been resolved, selection, sizing, and layout of the components can begin. To arrive at the final design, all of the following should be considered:

1. Material costs
2. Installation costs
3. Spare parts inventories needed
4. Interchangeability of equipment and parts
5. Power demands
6. Serviceability of equipment
7. Operational reliability of components
8. The possible need for installed spares
9. The anticipated maintenance requirements.

Bibliography

Literature Cited

1. Penkala, J.E. and Escarfail, J.P., Tappi 45(5): (1965).

Chapter 2, Section VIII

Flow and Material Measurement in the Coating Process

G. Gordon Bugg, Mead, Mahrt Mill

Introduction

Accurate measurement of the quantities of the various materials that are combined to form the finished coating is critical to maintaining uniformity in any coating operation.

Unless the coating is manually formulated using calibrated buckets and dipsticks, some type of quantity sensing system will be utilized.

This section describes the main types of flow and quantity sensing devices available today. The intent is to provide a "one stop" basic source of information about the types, advantages, and disadvantages of the various sensors in current use.

Our discussion applies to the handling of liquids and slurries as these are the forms that coating components take in most coating makedown systems.

From a quantity measurement standpoint, coating makedown and application systems can be grouped into those systems that use a sensor to measure the flow past a given point and those that do not.

Flow Independent Systems

This type of system incorporates a method of batching a given amount of material that is later added to the coating make-down process. Typically, this method uses a weigh tank into which is measured a desired quantity of material. These systems are operated based on weight or volume.

Weight Basis Flow Independent Systems

In a weight-based system, the material in the weigh tank is sensed by load cells mounted under the tank or by a pressure-type level sensor mounted on the tank (Fig. 2.56). This weight will be the "wet" or "as received" weight. Since most coating formulations are based on dry

pounds, an adjustment factor will be needed to correct the wet weight for variation in the solids of the material.

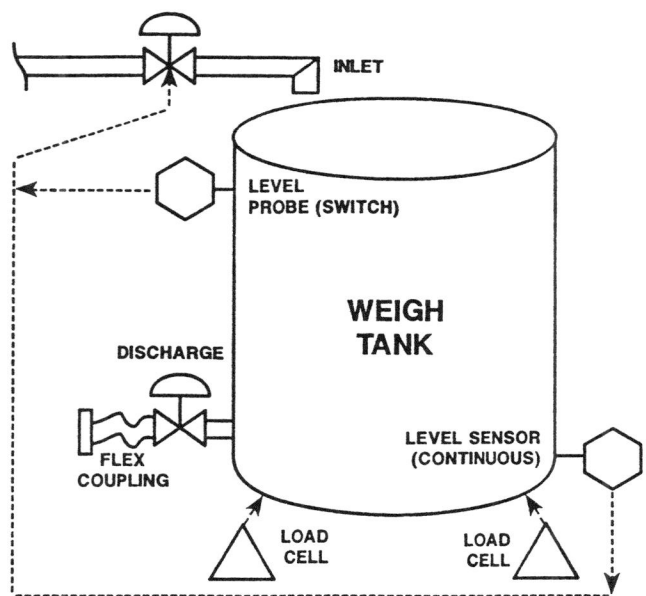

Fig. 2.56. Tank-mounted pressure level sensor

Note: For those not familiar with the terms "wet" and "dry" as they are applied to coating formulation, most coating components are used as liquids or slurries containing a given percentage of active ingredients or solids, the remainder being a carrier or solvent such as water. "Wet" basis refers to the addition of the material to the coating based on the weight of the solid plus the solvent. "Dry" basis indicates that only the solids or active fraction is used in the calculation. The solids percentage normally fluctuates depending on the materials and the manufacturer. Therefore the "wet" weight of material will vary to ensure the proper number of "dry" pounds are added to the formulation.

Volume-based Flow Independent Systems

In systems using the volumetric approach, the weigh tank will incorporate a sensor that measures the absolute level in the tank. This may be done using a top-mounted, sonic-type level transmitter, or by a level switch that will close when the desired level is reached. In either case, once the setpoint is reached, the system sends a signal to the inlet valve causing it to close. The measured quantity is then held for as long as needed before being discharged through the outlet valve.

Flow Independent Systems— Advantages and Disadvantages

In an automated system, the same weigh tank may be used for multiple coating components. Note however: In designing such a system, the intended ranges of each material must be carefully reviewed beforehand to ensure that the system will be able to handle the materials within an acceptable percent error. For example, a weigh system sized to handle 3000 lb of clay slurry should also give good service when weighing out 2000 lb of titanium dioxide, but will be grossly inaccurate if used to weigh out 50 lb of dye. In this case, multiple tanks will be needed. The point is that the desired accuracy must be designed into the system from the beginning on a component by component basis.

Advantages of a weigh-tank system include ease of manual calibration checks, low cost when compared to multiple flow sensor systems, and adaptability to an automated system.

Disadvantages include floor space requirements, a tendency to build up material on the inside of the tank leading to accuracy and housekeeping problems, and increased water requirements (both supply and disposal) for keeping the system clean.

Flow Dependent Systems

The other general class of systems are those incorporating a device to measure the material flow past a given point in a pipe. The majority of coating systems in operation today are of this type. Flow sensors come in many different forms and utilize a variety of measurement techniques.

Direct Flow Measurement

Flow sensors can be divided into two main groups. One group measures flow directly and is referred to as positive displacement flowmeters. In these, the fluid is divided into increments of a given size, moved through the device, and discharged on the other side.

The increments are counted electronically or mechanically, and the total flow is the summation of these increments over a given time. Metering pumps are typical examples of positive-displacement flow measurement devices (Fig. 2.57). Positive displacement flow sensors have the advantage of measuring volumetric flow directly and are widely used to meter coating additives such as dyes and defoamers.

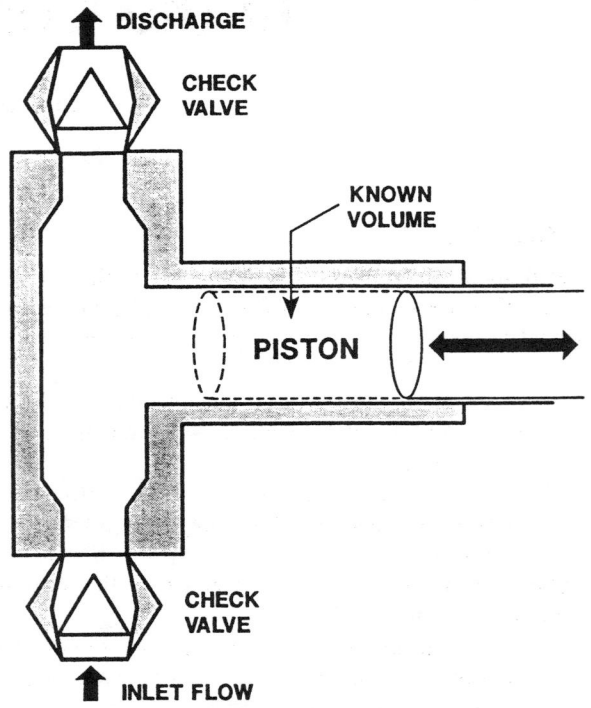

Fig. 2.57. Metering pump

If properly sized and piped, metering pumps are a reliable, accurate means of flow measurement. Some of their disadvantages (using the metering pump example) are high initial cost and relatively high maintenance cost when compared to indirect sensors.

Indirect Flow Measurement

The second group of flow sensors utilize an indirect measurement technique. This discussion will focus on the types of meters found in coating operations. Indirect flow sensors can be generally categorized into one of three types: (a) sensors based on measurement of the fluid velocity, (b) sensors based on measurement of differential pressure, and (c) sensors based on measurement of the mass flow of material.

At this point it is appropriate to discuss the type of fluid flow being measured.

Characterizing Fluid Flow: The Reynolds Number

The performance of flow sensors is greatly influenced by a dimensionless value known as the Reynolds number. This value relates the effects of the pipe diameter, fluid density, viscosity, and flow velocity. Essentially, the Reynolds number is the ratio of dynamic forces of mass flow to the shear stress due to viscosity and is represented by the equation:

$$R = \frac{DVd}{u}$$

Where: R = the Reynolds number
  D = inside pipe diameter, in.
  V = average fluid velocity, in./s
  d = fluid density, $lb/in.^3$
  u = fluid viscosity, cp

At Reynolds values of 2000 and below, the flow is termed laminar, and the fluid flows in smooth layers with the lowest velocities at the pipe walls and the highest velocities at the pipe center.

As the Reynolds number rises, either due to lower viscosity or increased fluid velocity, the smooth flow lines break up into turbulent eddies that flow through the pipe with the same average velocity. This type of flow is called "turbulent" flow.

In coating applications, laminar flow is not desirable. Coating materials containing entrained or suspended solids, such as clay, will be prone to settling out, and the flow sensors described here will not work well. When sizing a system, calculated Reynolds numbers above 4000 (turbulent flow) are desirable.

Three Types of Sensors

A. Fluid Velocity Sensors

Velocity sensors are the most widely used and best understood type of sensors used in the coating area. These devices measure the velocity of a fluid moving past a given point, i.e., the sensor. Since the cross-sectional area of the pipe at that point is known, the volumetric flow can be calculated.

The basic relationship for flow in a full pipe is:

$$Q = A \times V$$

Where: Q = the volumetric fluid flow (example: gal/min)
  A = the cross sectional area of the pipe
    at the point of measurement
  V = average velocity past the
    point of measurement.

In velocity-type sensors, A is known, so only V is needed to calculate the volumetric flow.

The techniques used to determine the fluid velocity vary widely. In coating applications, the main types of velocity sensors in use are magnetic, vortex, turbine, and doppler or sonic meters.

Magnetic flow sensors are the most widely used flow sensor in coating operations at this time. These sensors operate on the principle of Faraday's Law of electromagnetic induction. This law states that a voltage will be induced when a conductor, in this case the fluid, moves through a magnetic field. The magnetic field is created by coils outside of the flow tube. The voltage produced is directly proportional to the velocity of the fluid through the tube (Fig. 2.58).

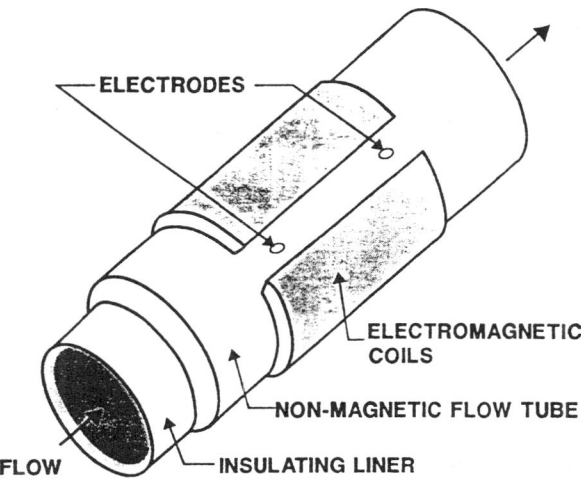

Fig. 2.58. Magnetic flowmeter

These sensors can handle a wide variety of liquids and slurries. There is virtually no pressure drop across the sensor if it is properly sized and these sensors have no obstructions in the flow area that might clog, bind or build up. The wetted areas of these sensors are available in a wide variety of materials. Accuracy is typically ± 0.5% of the calculated flow rate.

At the current stage of electronics development, this type of sensor has the widest usable operating range of any of the sensors we will discuss.

Some disadvantages of magnetic flow sensors include the basic requirement that the fluid be electrically conductive, and a relative insensitivity to entrained air in the process fluid that can result in a loss of accuracy.

Vortex flow sensors make use of a natural phenomenon that occurs when a fluid flows around a non-streamlined object, called a bluff body, fixed perpendicular to the direction of fluid flow (Fig. 2.59). The flow around the body generates a series of vortices that are alternately shed downstream of the bluff body. The shifting vortices create a characteristic pattern called the Von Karmen Vortex Trail. The alternate shifting and shedding of the vortices creates a measurable pressure difference on the bluff body which can be detected. Within a certain range, the frequency of vortex shedding is proportional to the velocity of fluid moving past the bluff body. As the cross-sectional area of the pipe is known, a simple calculation yields the volumetric flow.

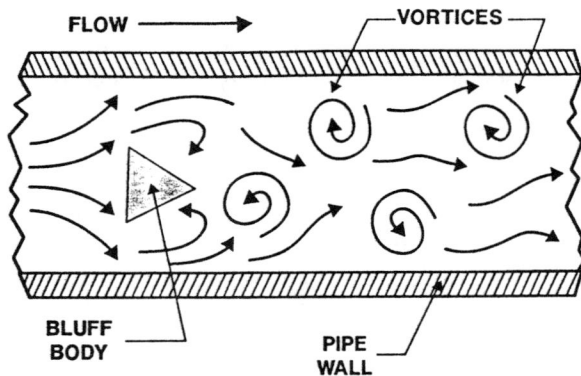

Fig. 2.59. Vortex flowmeter

Vortex sensors are widely available in a variety of materials. They are generally not suitable for coating or many coating components due to their tendency to clog for those models that have ports, and wear due to long-term abrasive effects on the bluff body. Accuracy of these sensors averages ±1% of the calculated flow rate.

Turbine flow sensors incorporate a freely-turning rotor mounted perpendicular to the fluid flow (Fig. 2.60). The rotor spins as the liquid passes through the blades. The rotor spin rate is sensed by a magnetic, optical or mechanical linkage. This rate is a direct function of the fluid velocity from which the volumetric flow rate is calculated. Turbine sensors are best used with clean fluids, as the blades are subject to abrasive wear as well as buildup and blockage. These sensors generate a higher pressure drop in the line as compared to a magnetic flow sensor. Typical accuracy is ±0.25% of the calculated flow rate.

Doppler or transit time flow sensors are very versatile devices that lend themselves to trial and survey work as well as continuous measurement. Typically, these sensors consist of one or two units that are attached to the sides of a straight run of pipe. Doppler devices send out a pulse that is reflected from bubbles or solids in the moving fluid (Fig. 2.61). The returning signal is frequency shifted in proportion to the fluid velocity (Doppler effect). As the original frequency is known, the amount of shift allows the fluid velocity to be calculated, and given

the cross-sectional area of the pipe, the volumetric flow rate can be determined.

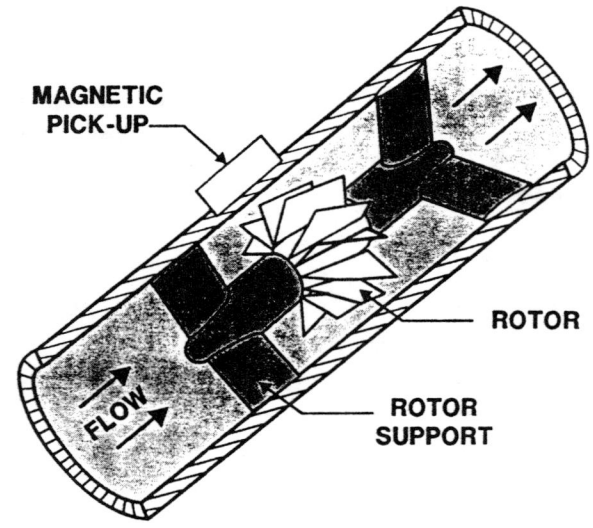

Fig. 2.60. Turbine meter

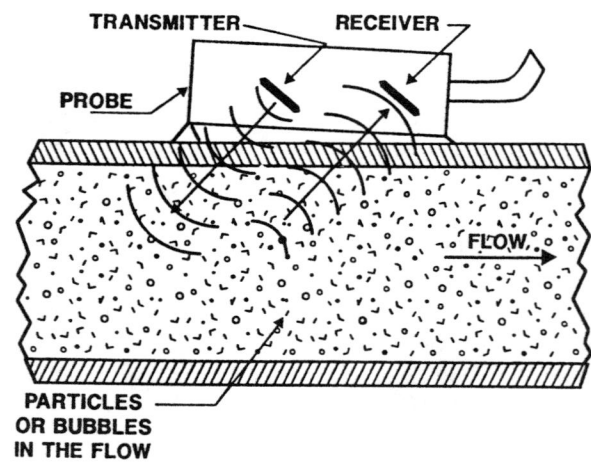

Fig. 2.61. Doppler flowmeter

The transit time flow sensor uses two sensors offset from each other on opposite sides of a straight section of pipe. This device measures the time of travel for sound waves traveling at a 45-degree angle to the direction of fluid flow.

Both types of sensors share the advantages of portability, flexibility in sensor placement, no pressure drop across the sensor, and absence of in-pipe obstructions. The cost is low when compared to a magnetic flow sensor, particularly for a large-diameter pipe. Doppler sensors have the disadvantage of requiring air or solids in the fluid for the sensor to work, precluding their use in clean, gas-free liquids. Accuracy normally runs ±5% of the sensors full-scale reading. Time of transit sensors will work with clean, bubble-free liquids, with typical accuracy of ±5% of the full-scale reading.

B. Differential pressure sensors

Sensors based on differential pressure are the oldest type of sensor of the three types under discussion. The basic principle underlying differential pressure flow sensors is that the measured pressure drop across the sensor is proportional to the square of the flow rate. Since the pressure drop can be measured, the flow rate may be calculated from the square root of the pressure differential.

In coating applications, rotameters and impingement sensors are the most common types of differential pressure sensors in current use. Orifice, flow tube, and Pitot tube sensors also may be found in specific applications.

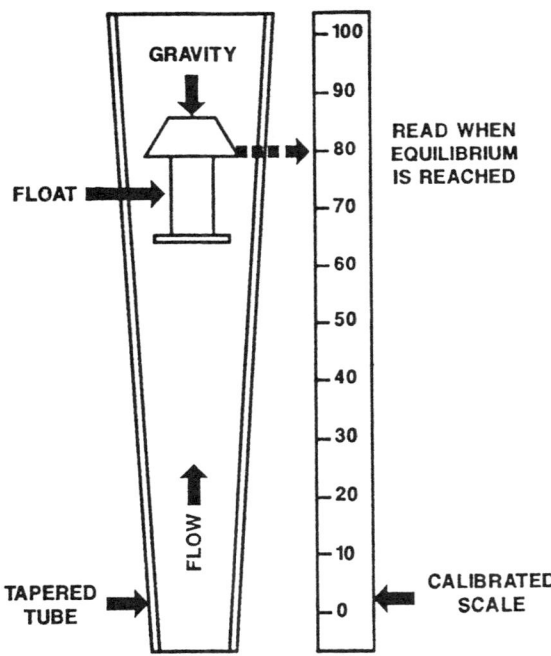

Fig. 2.62. Rotameter

Rotameters, also known as variable area flowmeters, typically consist of a weighted plummet or float in a tapered tube. In operation, as the fluid flows through the tube, the float rises to a point of equilibrium where the pressure exerted by the rising fluid balances the weight of the float (Fig. 2.62). Once calibrated to a particular flow range and material, the sensors are accurate to ±5% of the full-scale reading. The simpler rotameters use a glass tube and flow is read from a scale mounted next to the tube.

Rotameters with magnetically coupled floats can be

connected to a transmitter which will generate a signal that can be incorporated into a control or monitoring system. Technically, rotameters are constant pressure, variable-area devices, although in the context of our discussion their principle is based on a measurement (or balance) of pressure.

Rotameters are primarily used on clean, relatively low viscosity materials such as aqua ammonia, cooked starch, and water. They are subject to clogging, buildup on the tube and float, and the float is prone to abrasive wear depending on the service and material of construction. Glass tubes may break. These sensors must be mounted vertically, as gravity is a necessary part of their operation. If properly mounted and sized, and used on a suitable (clean, low-viscosity, and nonabrasive) fluid, these simple and rugged devices will give years of useful service. They have found wide use as economical visual flow indicators for water, such as those servicing mechanical seals or glands, as well as on-line backups in coating component lines where they are used to verify the accuracy of other types of sensors.

Impingement sensors give a direct flow rate readout for a particular fluid by measuring the force of impact on a target of known area suspended in the fluid stream. These low-cost devices find many uses in low flow applications, particularly mechanical seal water and gland water measurement. As with any type of sensor that uses a flow obstruction, impingement sensors are prone to internal buildup, target damage, and abrasion. Accuracy is typically ±5% of the full-scale reading.

C. Mass flow sensors

In contrast to velocity and pressure differential sensors which operate by determining volumetric flow rates, mass sensors measure the mass flow rate, usually expressed in pounds per minute (lb/min), of the material in the line.

The advantage of mass sensors in coating operations is that the mass flow rate is much more useful when discussing coating and coating components. These devices are not affected by variation in volume due to temperature and pressure. Note, however, that these sensors will register differences in mass flow if the volume varies due to changes in pressure because of the presence of a compressible fluid such as air in the noncompressible liquid medium. Techniques for minimizing this problem will be discussed later.

Mass flow sensors have been used in the paper and coating industry for many years. Due to advances in the associated electronics, the use of mass flow sensors, as compared to the other types of flow sensors, is rapidly growing.

There are a number of variants of this type of device on the market, so an idealized version will be used as an example. In the mass flow sensor, there are one or more curved tubes through which the process fluid flows. These tubes are vibrated at a given frequency by an external driver. As material flows through the tube, the vibration is dampened. The degree of damping and the additional energy input required to maintain the vibration is a function of the fluid density, which is a sensor output. As the fluid moves through a curved portion of the vibrating tube, Coriolis forces cause the tube to elastically distort or twist. The amount of twist is sensed by magnetic pickups mounted by the tube. This twist is proportional to the mass flow rate through the tube (Fig. 2.63).

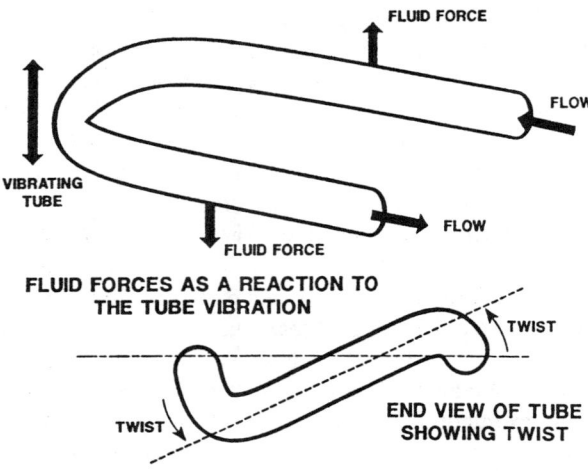

Fig. 2.63. Mass flowmeter

Typical outputs from mass flow sensors are the fluid mass flow rate (lb/min) and density (lb/gal or g/cc). Using these two values, the volumetric flow rate can be calculated if desired.

The availability of mass and density signals from these devices becomes very valuable in terms of knowing and maintaining desired coating solids and coating uniformity. For most primary coating components, particularly pigments and binders, the density is a function of the percent solids. In a properly designed and programmed system, a mass sensor allows you to continuously trend the solids content of the material in the line.

Even those components that do not exhibit the solids-density relationship can benefit from the mass flow sensor's density measurement capability. The specification density of a material can be programmed into the system. If dilution or contamination occurs that causes the density to change, an alarm can be set to alert the operator that a change has taken place.

The main advantage of a mass flow sensor is that they measure mass rather than volume. This eliminates the need for factors or adjustments to convert volumetric

quantities to mass quantities. Also, the mass flow can be combined with the density-solids information to yield a dry basis mass flow value. Typical accuracy of a Coriolis-type mass sensor is ±0.4% of the mass flow rate.

On the downside, mass flow sensors are much more sensitive to mounting errors, external line vibration, and entrained air. They also have a significantly greater pressure drop and are more expensive than their magnetic counterparts.

Many of the potential pitfalls of mass flow sensors can be avoided during the design stage. For example, mounting a mass sensor in a low point in a line will help ensure a full pipe. This also holds true for most of the other sensors we have discussed. Mounting a mass sensor in the bottom of a "U" that is solidly clamped at the top of the "U" will not only help eliminate air pockets, but will also reduce the amount of external line vibration that would otherwise be transmitted through the sensor. When fabricating a system to accept a mass sensor, use a spool piece the exact length of the meter to align the pipes prior to installing the sensor. Any torsion or misalignment will result in degradation of the sensor's performance.

In regards to the problem of entrained air, many locations have found that by taking steps such as ensuring tank agitators are not oversized and putting those agitators on timers rather than letting them run continuously, they can gain a measurable improvement in sensor performance.

Mass sensors operate best when the downstream pressure as "seen" by the sensor remains constant. The opening and closing of downstream valves result in line pressure changes than can cause entrained air bubbles to expand or contract, leading to sensor variability. Installation of a back-pressure valve between the sensor and the first downstream control valve will eliminate this problem. Care must be taken in valve selection, particularly if a positive displacement pump is used. A valve that will maintain a constant back pressure regardless of the flow rate is the best, and valves of this type are widely available.

Chapter 2, Section IX

Raw Material Quality Control

G. Gordon Bugg, Mead, Mahrt Mill

Making a sheet of coated paper or board can be compared to building a brick wall—the individual parts are fitted together to form an integrated whole. A corollary to this is that the components (in the case of the wall, the brick and mortar) must be of uniform quality, otherwise your finished product will contain built-in weaknesses that will lead to premature failure. In the coatings area, we have little (if any) influence over basestock quality, to include variation in fiber supply, the vagaries of the pulp mill and the paper machine, and the general outlook on life of the operators up and down the line.

One area in which coating personnel have a significant amount of influence is in the quality and uniformity of their coating raw materials.

This section outlines a Statistical Quality Control (SQC) program for coating raw materials, and shows how such a program can help take some of the heat off the color plant when finished paper or board quality isn't what it should be. There is a proverb in industry that says (minus the colorful phrases sometimes included), "Last on, first out when things go wrong." A paraphrase that has been experienced many times over many years says "Last on, first blamed when things go wrong." The fact is that coating makedown and application is generally a high-dollar, high-profile operation.

Anything that can be done to show the consistent, high quality of the coating color and build the credibility of such a system can go a long way toward diverting unnecessary attention from the color plant and allowing that attention to focus on other, more suspect, parts of the process.

A statistical quality control program for incoming raw materials is a critical aspect of any program for monitoring color plant operations.

A successful SQC program involves a close, open relationship between the vendor, coating personnel, and the mill testing facility. Typical steps in setting up such a system are:

1. Designate one individual as having personal responsibility for setting up and administering the program. Committees are fine for addressing specific issues, but having one undisputed person in charge is critical.

2. Select the raw materials that are to be included in this phase of the program. If this is the initial round, pick a group of similar materials (i.e., clays, TiO_2, latex, starch) to work on. Keep the list short. Trying to do everything at one time will overtax whoever is running the program. The other reason for a stepwise approach is to allow you to learn from the previous step and improve your procedures as you go along.

3. Coordinate a meeting, either separately or as a group, with the suppliers of that raw material. You can still set up a program on single-sourced materials. In this meeting, review supplier product bulletins, specification sheets, and mill experiences and requirements. The purpose of this meeting is to let the supplier know that you are setting up an SQC program, and to decide exactly which quality variables will be monitored. The key in deciding on which tests to run is that whatever tests you agree on must be able to be run by both the mill and supplier. For example, in the case of starch, you might look at incoming moisture and cooked starch viscosity at a given solids and temperature. For clays, the solids content, viscosity, and pH might be selected.

Also at this meeting, agree upon the reporting interval (which will depend on the frequency of shipments) and report format. Typically, an X-bar chart showing agreed upon specification limits and actual truck/car/lot values is the minimum that should be accepted. A tabular listing of the data, and a moving range chart showing the difference between subsequent values is also helpful.

Once this step is complete, you'll have a steady stream of usable information on the quality and uniformity of the raw materials you receive.

4. Test each load that arrives at the mill using the same testing procedures that were agreed upon with the supplier. If at all possible, have the tests completed before the material is unloaded. This way, substandard ma-

terial can be returned without contaminating your system. As the vendor's charts arrive, overlay your results on theirs. Doing this on a PC-based system gives you a lot of flexibility in generating summaries of the combined data.

5. If you see significant variation between your results and the vendor's, or your test results are different from what you have been averaging, contact the supplier and let him know. This builds internal credibility in your system and lets the vendor know that his product is being actively monitored.

6. Later, when confidence in the supplier has been developed, you can reduce your testing frequency to conserve labor, using spot checks to verify your supplier's results.

Over time, benefits of this type of program include:

A. Identification of the "best" suppliers. This is information which can be used in purchasing/contracting decisions.

B. Rejection of substandard material before it can affect your system

C. A documented and clearly understandable record of the type and quality of the materials in use

D. Building confidence in the integrity and reliability of your suppliers.

And finally, monitoring raw materials quality in this way gives coating personnel a useful tool against that day when somehow things don't go quite right and those fateful words are spoken: "The coating is all fouled up!"

For more information on how SPC and SQC can be applied to coating operations, please refer to Chapter 8 of this book.

Chapter 3, Section I

Blade Coaters

Theodore C. Vanya, Consultant
Editor: E. William Wight, Beloit Corp.

Introduction

The 1945 patent of A. Trist, which specifically involved an oil-phase emulsion and not water-based coatings, is considered as the spark to take paper coating into the high-efficiency, high-speed, high-quality industry it is today. As early as 1906, however, patents were granted to Krouse and Baumann in Dresden for what was the beginning of an inverted-blade coater system.

Two of the more popular means of applying coating to a web of paper are with the applicator roll system and with the short dwell coater. With both systems, either use a bent blade or a rigid, beveled blade to meter the excess coating from the paper web. Blade loading can be accomplished with a blade tip pneumatic tube, pneumatic diaphragms, or a combination of the two for either bent-blade or beveled-blade operation. Mechanical blade loading is specifically designed for operating a beveled blade where grade changes are frequent.

Other coating application systems such as the puddle coater, the BTG Billblade coater, and the fountain coater are described.

The exact descriptions of each manufacturer's coaters are freely available. The purpose of this section is to describe the various coating application systems and the available tools to control the blade in removing the excess coating applied. Most of the time the commonly used name of a coating machine describes the application method such as puddle, roll applicator, fountain, and so on. The term flooded nip is frequently used to describe roll applicator coaters and is a Beloit Corporation trade name.

Flooded Nip and Short Dwell Coaters

The flooded nip coater, also called an applicator roll coater, is designed for the application of a wide range of coat weights either on- or off-machine. The term

"flooded nip" is derived from the nip formed between the applicator roll and the backing roll which is flooded with coating to develop a hydraulic pressure (Fig. 3.1).

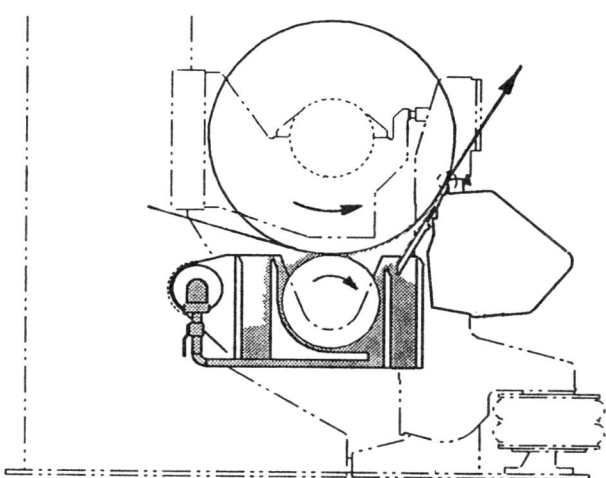

Fig. 3.1. Flooded nip coater (courtesy of Beloit Corp.)

The applicator roll rotates in a pan of coating and applies an excess of coating to the paper web as it passes through the nip formed by the applicator roll and the backing roll which supports the sheet. The hydraulic pressure in the nip eliminates skip coating caused by foam. Any bubbles that form are immediately collapsed. The pressure promotes good adhesion of the coating to the paper web. The excess coating is removed with an inverted trailing blade.

An excess amount of coating is supplied to the pan and is allowed to overflow into the return compartment (Fig. 3.2). Overflow recirculation carries away foreign particles and prevents the coating from dehydrating. The coating is introduced to the pan through a header pipe and a series of small inlet tubes. The color pan is jacketed to accept chill water to promote sweating, making the pan easier to clean. The color pan separates from the applicator roll providing accessibility for cleanup.

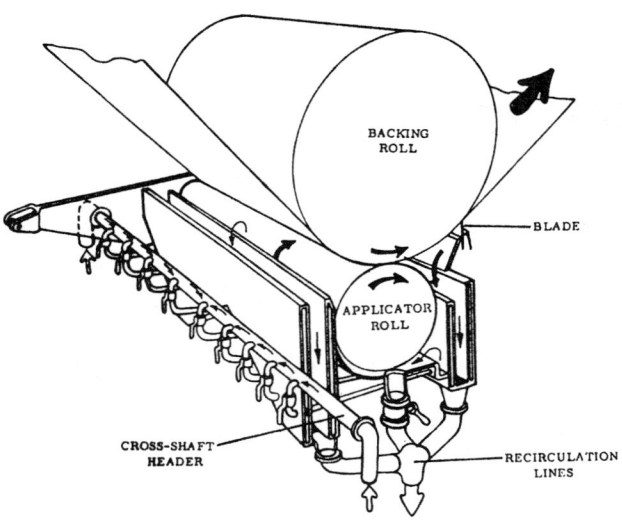

Fig. 3.2. Flooded nip coater flow (courtesy of Beloit Corp.)

Dial indicating micrometer stops with screw jacks allow precise adjustment of the gap between the backing roll and applicator roll. The gap can be adjusted on the run. A wider gap lowers the amount of coating forced into the sheet. This nip gap has a range of operation of 0.005-0.075 in. normally.

The applicator roll speed is adjustable from 10% to 35% of the coating line speed. Both the speed of the applicator roll and the nip gap can be varied to control the amount and characteristics of the coating applied. The applicator roll can be deckled with end wipes for different sheet widths.

Short Dwell Coater

The term "short dwell" refers to the relatively short period of time the coating is in contact with the sheet before the excess is metered off by a trailing blade. The short dwell coater consists of a captive pond just prior to the blade (Fig. 3.3). The pond is approximately 2.5 in. in length and is slightly pressurized to promote adhesion of the coating to the paper web. The excess coating supplied to the sheet creates a backflow of coating similar to the backflow seen with the roll application system.

The excess coating is channeled over a deflector and collected in a return pan before returning to tanks to be screened. Both deflector and return pan are jacketed to accept chill water to promote sweating, making cleanup easier.

Fig. 3.3. Short dwell coater (courtesy of Beloit Corp.)

Coating is supplied to the short dwell coater head using multiple inlets to ensure a uniform coating pressure across the width of the sheet. From the multiple inlets, the coating enters a large cross-machine chamber (Fig. 3.4). The coating must then pass through either one or two orifices before entering a smaller, cross-machine chamber. One orifice is a slot that the coating travels through to enter the cross-machine chamber adjacent to the sheet. The orifices are required to back pressure the coating to obtain a uniform coating pressure against the sheet. The uniform pressure promotes a uniform coat weight profile.

The excess coating supplied to the sheet must pass through a final orifice comprised of a full width gap between the backing roll and a baffle. The size of this gap is usually adjustable to provide the ability to affect the coating pressure against the sheet on the run.

Coating is usually supplied to the short dwell coater head by variable-speed pumps to provide another method of affecting coating pressure against the sheet. Flow rate required is dependent on coating viscosity, baffle height, machine speed, and base sheet properties. The actual operating range is between 0.7 and 1.5 gal/min per inch of width.

The main advantage offered by the short dwell coater over the applicator roll coater is reduced blade pressure required for the same coat weight. The reduced blade pressure is the result of a lower amount of coating penetration into the paper web during application with the short dwell coater. Fewer sheet breaks, especially in the production of groundwood publication grades, has been the result with short dwell operation.

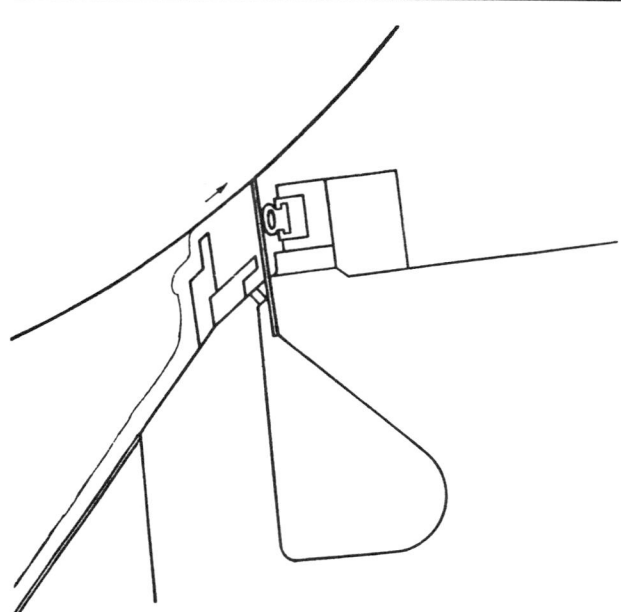

Fig. 3.4. Short dwell coating flow (courtesy of Beloit Corp.)

Coater Head

The coater head strength section and blade clamp arrangement are similar for either roll applicator or short dwell coater. The short dwell coater head must additionally accommodate the coating application system.

The coater head is a rigid box section beam designed for minimum deflection. The blade is clamped by means of a solid clamp jaw and pneumatic tube uniformly along its entire length to ensure a uniform blade load resulting in a more uniform coat weight profile (Fig. 3.5). The blade clamp and blade holder are solid stainless steel.

During a blade change, the clamp opens slightly allowing the blade to be removed and a new one installed without opening the chamber so no coating is lost. On a shutdown or washup, the coating chamber opens wide for cleaning.

Metering of the Coating

After the sheet exits the coating application nip, it comes into contact with the steel metering blade. The blade material is usually blue steel and blade thickness ranges from 0.012 in. to 0.025 in. blade width for beveled-blade operation ranges from 2.75 in. to 3.5 in.

A beveled blade has a pre-honed angle which is precisely ground by the manufacturer. Typical bevel angles range from 30° to 50°.

The blade forms a nip with the backing roll which the sheet must pass through. As the sheet passes through this nip, excess coating is metered off by the blade to the correct amount (Fig. 3.6). The pressure exerted on the

blade determines the force at the blade nip and the amount of coating that will be metered off, thus controlling the coat weight. The excess coating metered off is directed to a return compartment of the coating color pan.

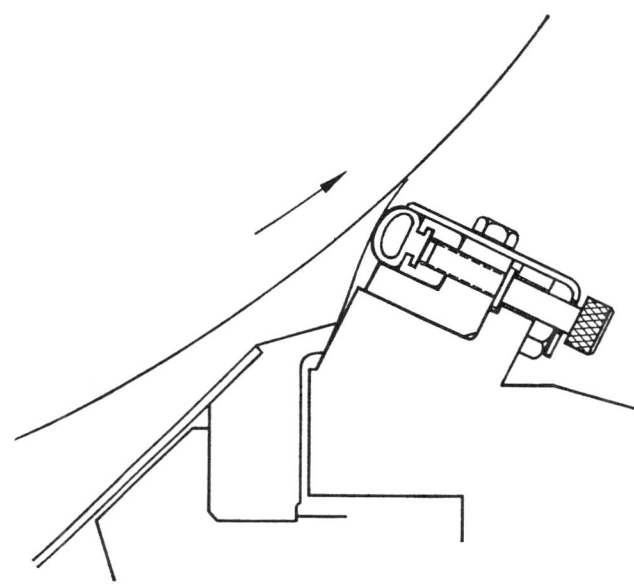

Fig. 3.5. Blade clamp mechanism (courtesy of Beloit Corp.)

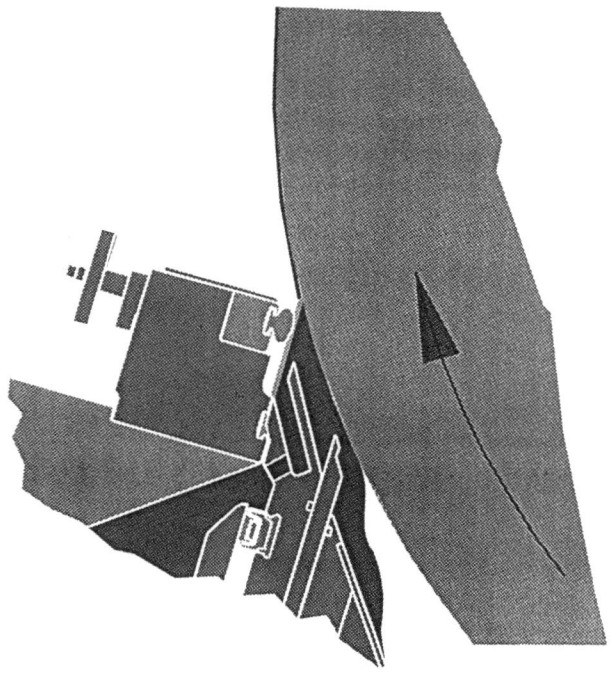

Fig. 3.6. Tube-loading bevel blade (courtesy of Beloit Corp.)

Profiling screws are also provided on some coaters which allow the operator to adjust blade pressure every 4-6 in. to compensate for base paper variations and achieve an even coating layer across the width of the sheet.

An alternate to a beveled blade is to use a bent blade as the metering element. This blade is approximately 0.5 in. wider than the beveled blade, and the sheet runs on the flat of the blade itself (Fig. 3.7). A bent blade may be used in the production of coated board where a high coat weight is required at a relatively slow operating speed, such as 1000 ft/min.

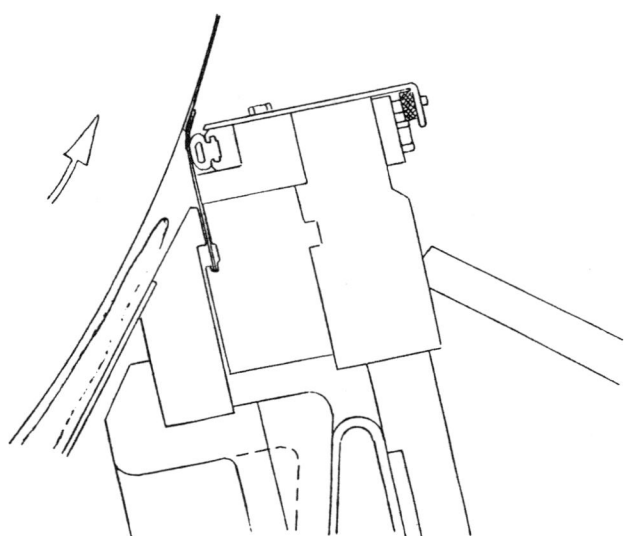

Fig. 3.7. Bent-blade operation (courtesy of Beloit Corp.)

Another reason a mill may operate a bent blade rather than a beveled blade is to reduce coating scratches caused by contamination in the coating. A bent blade is less prone to contamination-caused scratches than a beveled-blade system.

An alternate to the use of a blade to meter the coating is the use of a rotating steel rod. The steel rod is supported in a polymeric holder which is positioned in the blade holder in place of the blade (Fig. 3.8). The rod holder usually includes channels for circulation of water to keep the rod clean. The rod is normally driven by variable-speed motors.

Rods are used in the production of coated board where coating scratches are a problem due to particles lodging under the blade. Rods are also used in coated paper production when contaminants in the paper lodge under the blade and cause coating scratches. In these instances, the coated paper mill will usually change back to a blade after the contamination problem has been solved.

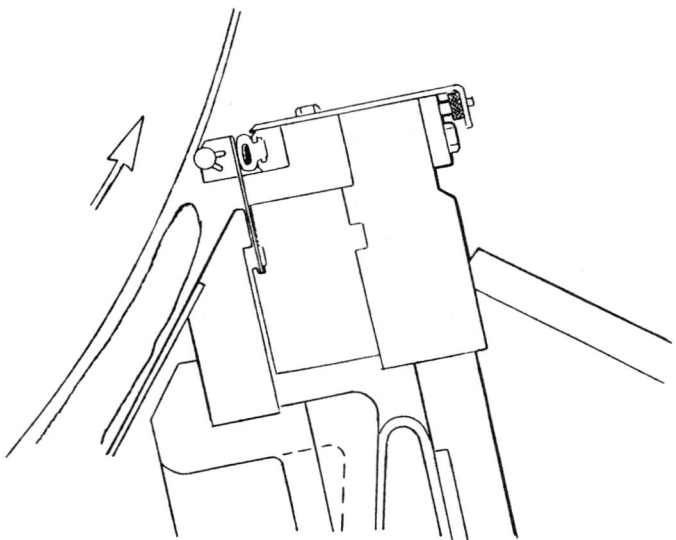

Fig. 3.8. Rod metering (courtesy of Beloit Corp.)

Blade Loading Mechanism

One coater head loading system uses a head positioner cylinder to raise and lower the head to provide clearance for threading, changing the blade, and cleanup (Fig. 3.9). When raised, the blade tip is approximately 0.5 in. from the backing roll.

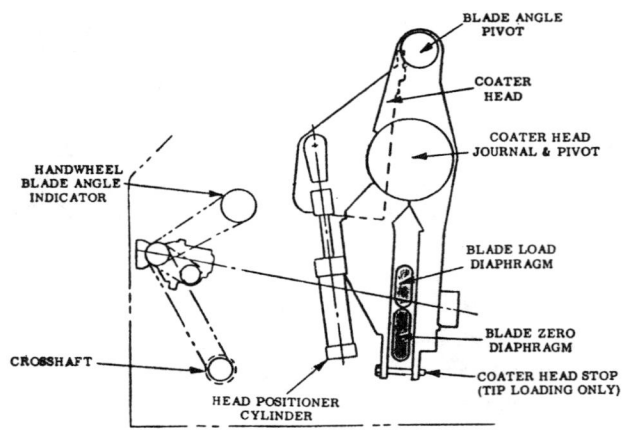

Fig. 3.9. Blade loading and angle adjustment (courtesy of Beloit Corp.)

The blade zero load diaphragm loads the blade against the backing roll at zero blade pressure, thus balancing the coater head so its weight does not affect blade load.

Blade loading is provided by the blade load diaphragms using either a solid bar or a heavy-duty extruded pneumatic tube located near the tip of the blade serving as the fulcrum point. When the air tube at the tip is used with the blade load diaphragms, both pressures are controlled simultaneously to obtain the desired blade loading. The air pressure controller is normally tied into the coat weight scanning gauge, and the air pressure is controlled by a signal from the gauge to automatically maintain a target coat weight. The flooded nip coater is commonly supplied with the above mentioned loading arrangement.

The first short dwell coater heads simply replaced the standard roll applicator coater heads using the same mountings and loading system. This made the conversion to short dwell coaters from roll applicator coaters very simple and required little downtime. In this case, the zero load and blade load diaphragms were usually loaded against a mechanical stop, and the blade loading was accomplished with the pneumatic tube at the blade tip (Fig. 3.6). Automatic coat weight control is possible with this system, too.

Another loading system very popular today is the pivoting frame arrangement (Fig. 3.10). The head is mounted rigidly in pivoting side frames to allow clearance for threading, blade changing, and cleanup. The unit is operated either by large pneumatic diaphragms or hydraulic cylinders to position the blade approximately 1 in. from the backing roll. The blade is then loaded to zero blade pressure by the diaphragms that position the coater head against a stop. The blade loading is accomplished with the pneumatic tube located near the blade tip. By varying pressure in the air tube, blade pressure can be varied and coat weight can be controlled automatically.

Dial-indicating micrometer stops with screw jacks mounted on the front and back pivoting frames are adjusted by a handwheel independently to square the coater head to the backing roll after each roll change.

Screw jacks operated by an air motor pivot the head about the blade tip to set the correct head angle and minimize blade wear-in time. Head angle typically ranges from 25° to 60°. Either a beveled blade, a bent blade, or a rod can be used with any of the above mentioned loading arrangements.

Mechanical Blade Loading

An alternate method of blade loading is with the mechanical blade loading systems. Each equipment manufacturer has systems with similar operating features. They all rely on physically bending the blade around a pivot near the tip to control blade loading (Fig. 3.11). The greater the bend in the blade, the higher the resulting blade loading.

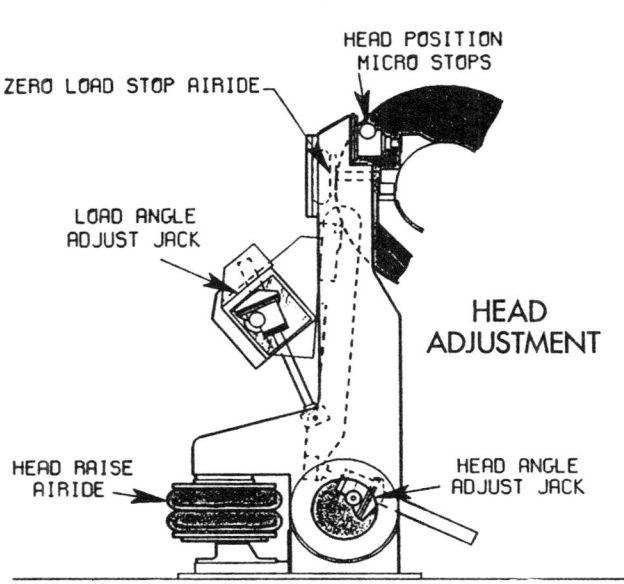

Fig. 3.10. Pivoting frame arrangement (courtesy of Beloit Corp.)

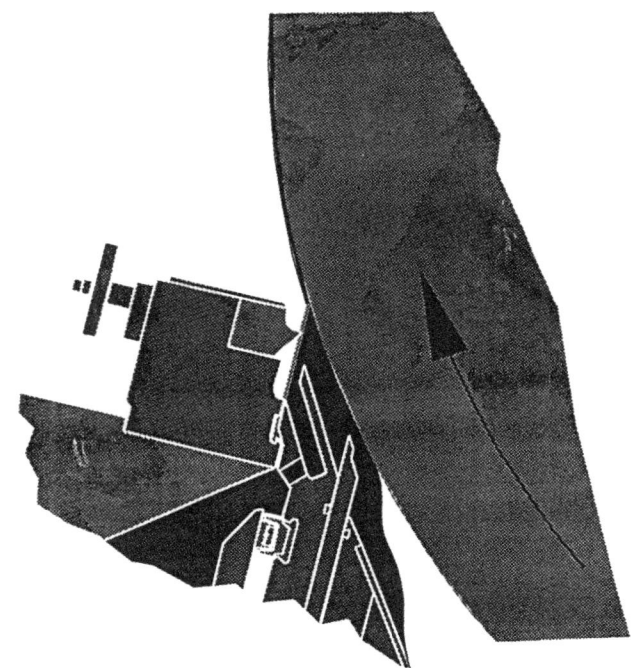

Fig. 3.11. Mechanical blade loading (courtesy of Beloit Corp.)

The unique feature offered by these mechanically loaded systems is the ability to change blade loading without affecting the blade contact angle relative to the web against the backing roll. This is an important feature if several grades, thus several blade loading requirements, are run frequently on the coating line.

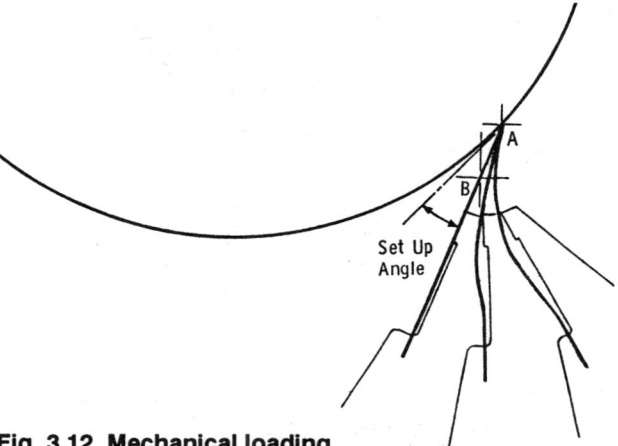

Fig. 3.12. Mechanical loading

With these mechanical loading systems, the blade contact angle is set by pivoting the coater head around the blade tip, point "A" (Fig. 3.12). The blade load is changed by pivoting the coater head around point "B," a specified distance from point "A." As the coater head pivots around point "B," the blade is bent over a solid fulcrum bar located near point "B." The amount the blade is bent determines the blade load and can be varied without changing the blade tip contact angle. This loading arrangement is designed specifically for beveled-blade metering systems to eliminate blade wear-in time

associated with changing loading conditions such as grade changes.

Control of the Blade

These mechanical loading systems also offer benefits for high coat weight applications. In these cases, as coat weight increases and the pressure on the blade is decreased, control of the blade profile becomes more difficult due to lack of sufficient pressure on the blade. This problem is reduced with the mechanically loaded systems using rigid support behind the blade (Fig. 3.13). Pressure on the blade is then applied by varying the position of the pivoting frame which increases or decreases the amount of bending of the blade, thereby, increasing or decreasing pressure on the blade. This system is also tied into the coat weight scanning gauge for automatic coat weight control.

Tip Angle Control

In recent years, much attention has been focused on controlling the "blade tip angle," which is the angle formed by the tangent lines of the backing roll and the blade and is controlled by the position of the blade beam (see Fig. 3.14). Ideally, once the coating station is setup and the coated quality is good, the tip angle should not change. However, measurements of the blade on production coaters has shown that the tip angle does change with changes in blade pressure. Blade pressure will

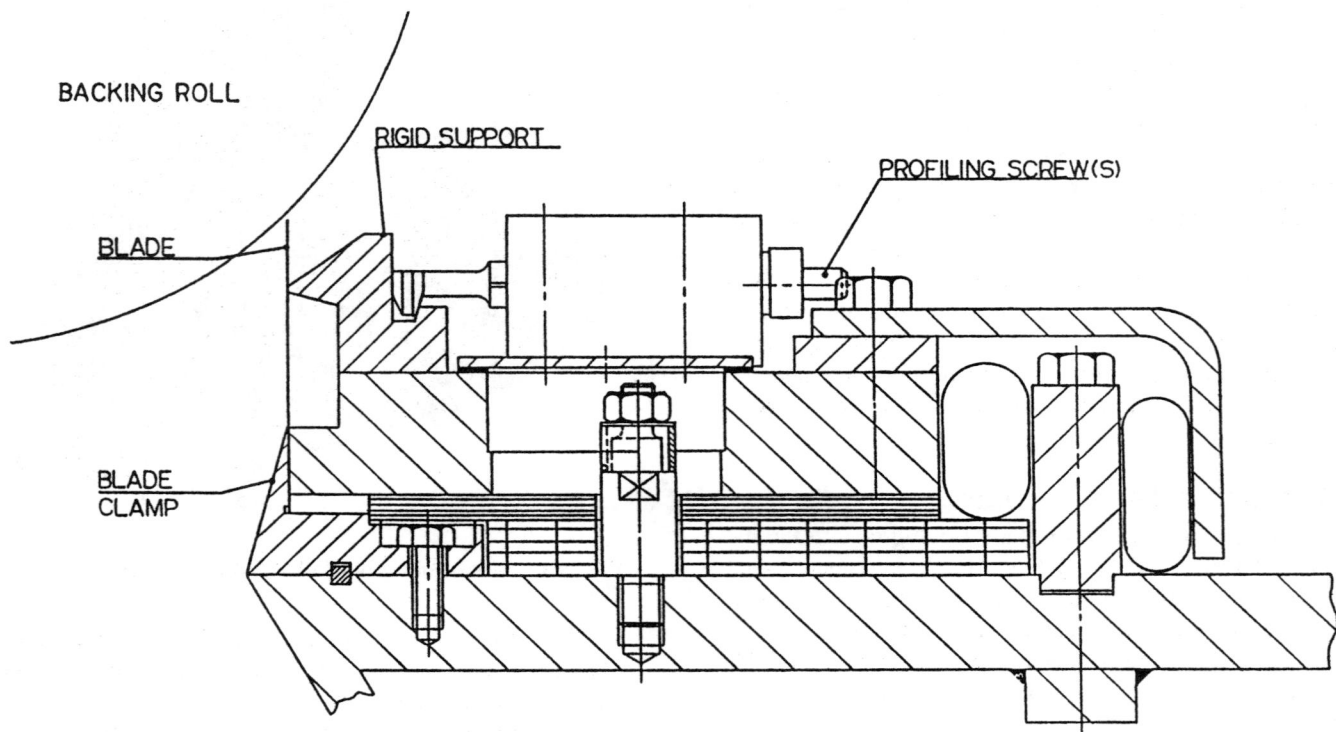

Fig. 3.13. Blade holder with fixed support (courtesy of Valmet Paper Machinery)

change during the production run as the automatic coat weight control system makes pressure adjustments to maintain target coat weight. As a result, problems with coat weight and quality control could occur as illustrated in Fig. 3.15. This phenomena creates an endless circle for the automatic coat weight control system as it will adjust pressure to maintain a target coat weight. However, the change in pressure will then cause a change in tip angle and a subsequent detrimental change in coat weight. The automatic control system must then make another correction and the process repeats itself.

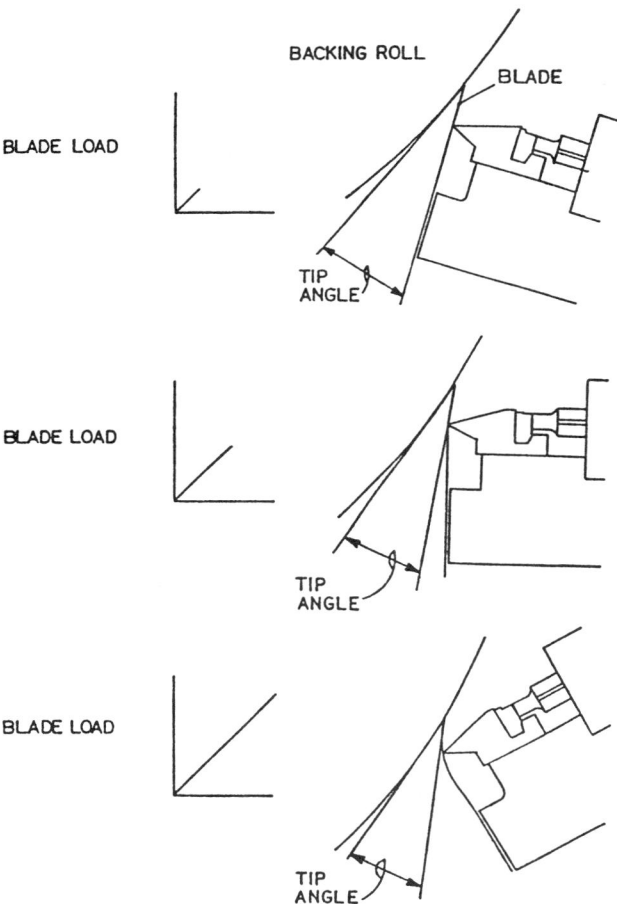

Fig. 3.14. Blade loading and tip angle (courtesy of Valmet Paper Machinery)

As illustrated in the upper right- and lower left-hand corners of Fig. 3.15, changes in tip angle also cause changes in the contact conditions of the blade to the sheet. An increased tip angle causes the blade to ride on its toe. A decreased tip angle causes the blade to ride on its heel. Both conditions create problems with the quality of the surface of the coated layer. In order to control or

maintain the blade tip angle, it is necessary to make a corresponding change in the beam angle anytime the blade loading is changed.

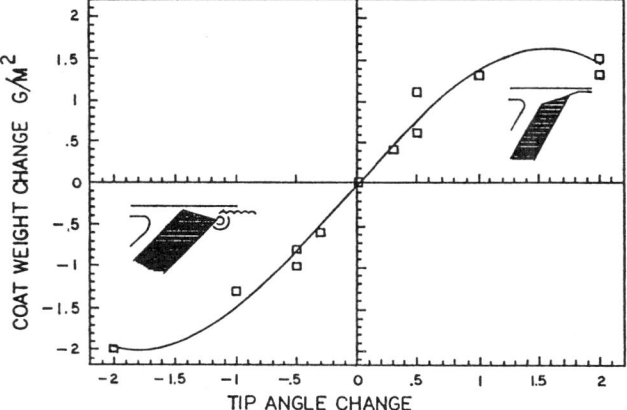

Fig. 3.15. Influence of tip angle changes (courtesy of Valmet Paper Machinery)

Due to the importance of tip angle in controlling coat -weight and coated sheet quality, most new coating stations are now equipped with devices that automatically control the blade tip angle. In the new coating stations, the adjustment of the blade beam in response to a change in blade loading takes place automatically. Automatic tip angle control can be accomplished either mechanically via linkages (Fig. 3.16) or electrically using encoders, stepping motors, and a mini computer (Fig. 3.17).

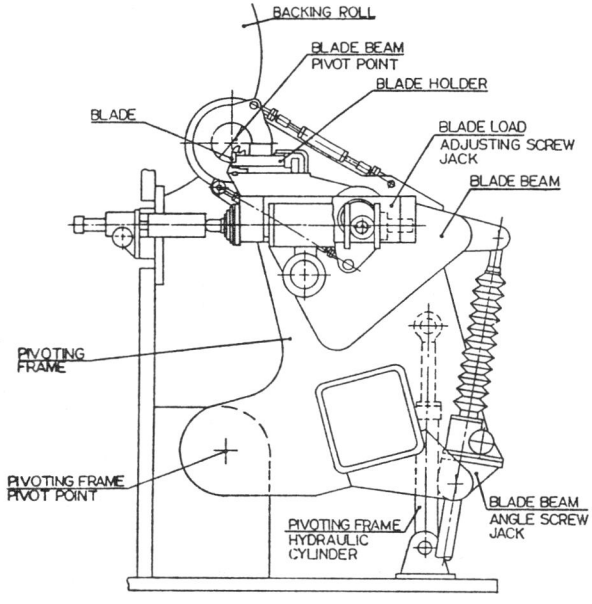

Fig. 3.16. Constant tip angle mechanism (courtesy of Valmet Paper Machinery)

Washup

Washup is performed on the coating station anytime the sheet breaks or the coater is shut down for an extended period of time. During washup, the pivoting frame pivots the blade away from the backing roll and the applicator roll and color pan are also pivoted away. Automatic spray showers activate to wash the backing roll, applicator roll, and color pan (Fig. 3.18). The operator may also use a water hose to clean the entire coating station. Most new coating stations include circulating cool water through the color pan, causing it to sweat, preventing coating color from crusting on it.

Roll and Blade Maintenance

The blade is changed usually every shift or sheet break on most coaters for coated paper. The backing roll and applicator roll changes are generally every one to two months, but longer lives have been recorded. Blade, backing roll, and applicator roll life is, of course, determined by the particular mill production conditions.

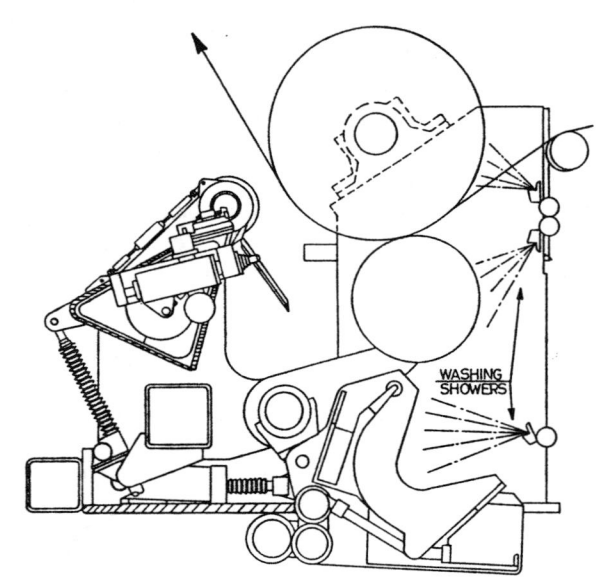

Fig. 3.18. Wash position (courtesy of Valmet Paper Machinery)

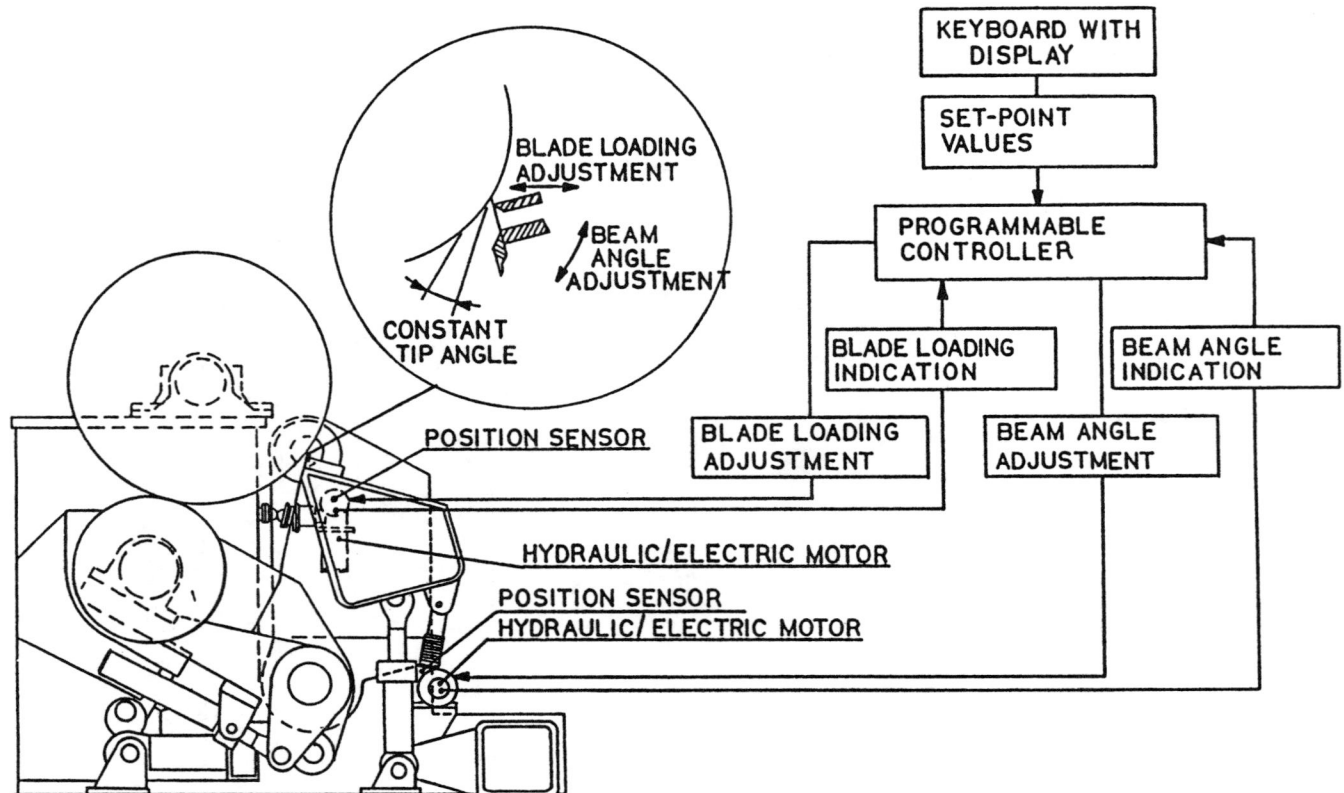

Fig. 3.17. Computerized tip angle control system (courtesy of Valmet Paper Machinery)

Production Uses

The short dwell and roll applicator coaters are used to apply light to heavy coat weights of 3.0-15.0 lb/3300 ft^2 on all types of papers and board. Sheet widths now exceed 340 in., and speeds up to 5500 ft/min are being designed for. Typical products coated with these coaters are printing papers, art grade papers, magazine papers, coated papers for newspaper inserts, and coated board for packaging and promotion.

Puddle-type Applicators

A puddle of coating is formed by the backing roll (covered by the web), the blade and blade holder, and the front wall of the coating head (Fig. 3.19). The edges of the puddle are formed by dams that contact the backup roll. The depth of the puddle is adjustable. Coating, in most installations, is recirculated through screens or filters.

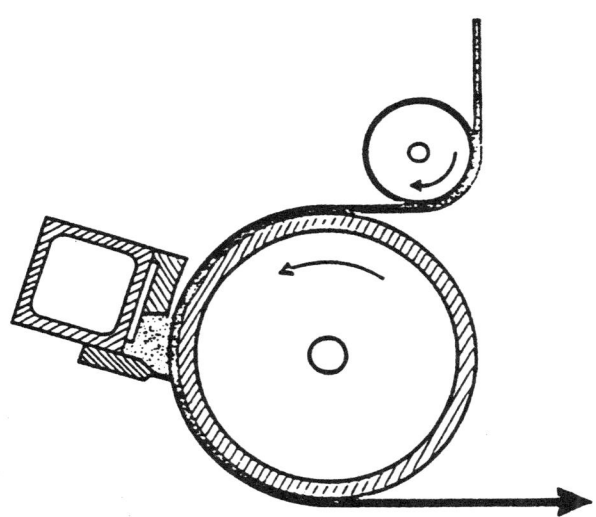

Fig. 3.19. Puddle coater (courtesy of Jagenberg)

Near the blade tip, the recirculation of the coating is very poor. It is believed that a rolling flow near the blade tip is induced by the running web and the entering coating. The excess coating metered off by the blade is at higher solids due to dehydration. This coating flows back to the main volume of the pond, and an equilibrium condition develops. Due to the coater's design, the coating puddle is dumped at each shutdown.

Another example of a puddle coater is the BTG Billblade. The blade side of the BTG Billblade (Fig. 3.20), in the standard form, can also be considered as a puddle coater with a very shallow puddle. The other side of the paper can also be coated using a pond created between the backup roll and the paper. In this configuration, both sides of the paper are coated simultaneously. The circulation is much quicker in this applicator than in a puddle coater. Also the blade angle provides a more open near-tip geometry so this applicator works more efficiently than a puddle coater. All the variations of the BTG Billblade design are aimed at different end products and different coat weight ranges.

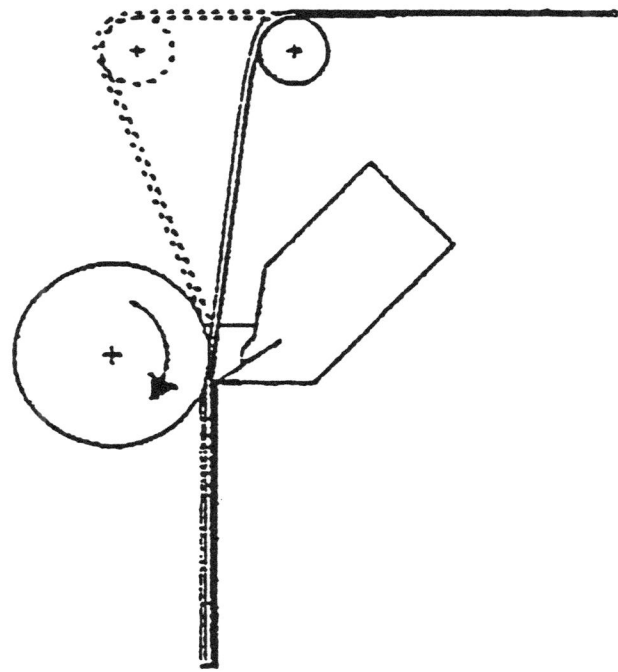

Fig. 3.20. BTG Billblade standard coater

Fountain Applicators

In this applicator (Fig. 3.21), the coating is pumped through a slot, aimed upward toward the web and backup roll creating a fountain of coating. The paper-wrapped backup roll is positioned at an adjustable distance from the lip of the slot. The slot lips can tilt, so the entering opening is larger than the exiting opening. This pre-meters the coating before the blade.

The fountain coater experienced problems in eliminating large air bubbles and did not provide equal application pressure across the slot. This resulted in both skip coating and nonuniform coat weight application. Since the first model, with a 0.5-in. slot, to the recently offered very narrow slot, referred to as a "jet" not a fountain, the problems are reported to be eliminated.

The small space requirement of the fountain allows the designer the flexibility to position the fountain very near the blade or far away from it, thus creating a variable-dwell-time applicator.

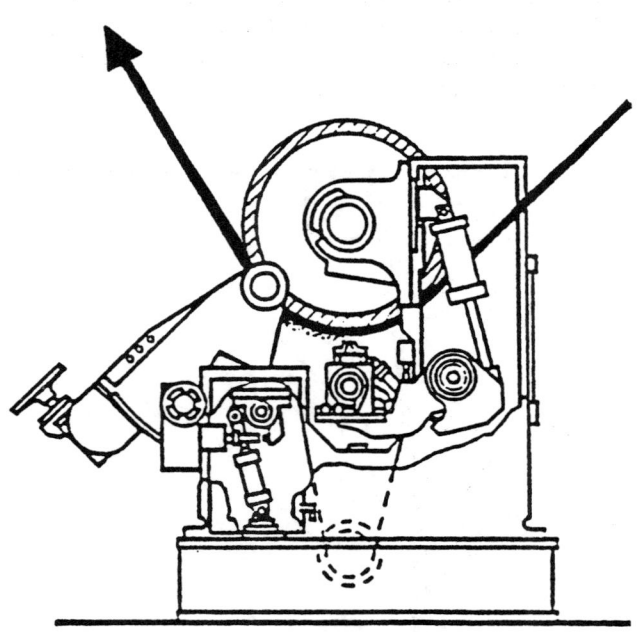

Fig. 3.21. Fountain coater (courtesy of Jagenberg)

TwoStream Coater

Coating Application

On the Valmet TwoStream coater its upward sheet run through the coater, the sheet passes through the nozzle applicators where coating is applied (Fig. 3.22). The amount applied is controlled by the volume of coating supplied by the pump, the angle of the applicator to the sheet, and the distance of the nozzle to the sheet. As is the case with the short dwell applicator, excess color is allowed to overflow the applicator in reverse direction to the sheet to create a seal against the boundary air on the paper surface.

Metering Conditions

As the coating is metered by the roll on one side and the blade on the opposite side, the metering conditions differ from those found on a single-side roll applicator or short dwell blade coater.

As the backing roll rubber covering is more resilient than the blade tip, more coating will pass through on the backing roll side of the sheet than on the blade side. When coating identical formulations on the front and back side of the sheet and equal coat weights are desired, it is necessary to reduce the solid content of the coating color being applied to the backing roll side of the sheet.

As any change in blade pressure will affect coat weight on both sides of the sheet, coat weight changes to

one side only usually must be accomplished by a change in the solid content of the color formulation. An independent coating circulation system is required for each nozzle. By changing the solid content, a corresponding change in the amount of dry coat weight can be achieved.

Production Uses

The TwoStream is usually operated on-machine in place of the size press for application of starch formulations for surface sizing or of pigmented coatings for precoating or for specialty applications such as label paper, carbonless CF paper, or release papers. As the coated sheet must be turned on paper-carrying rolls prior to contact with the steam dryers, additional drying is required to set the coating prior to contact with a roll so that coating pick-off does not occur. This is normally accomplished with the use of electric or gas-fired infrared dryers. Due to difficulties in maintaining nozzle straightness during manufacturing and under operating conditions, the TwoStream is limited to use for sheet widths of 200 in. or less.

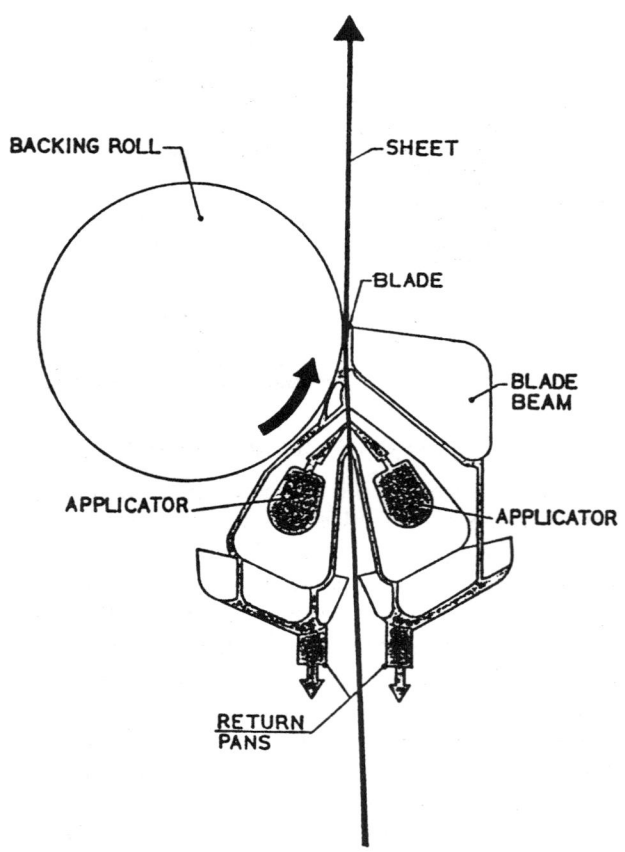

Fig. 3.22. TwoStream principle (courtesy of Valmet Paper Machinery)

Bibliography

Resources

1. Trist, A.R., U.S. pat. 2,368,176.

2. Richardson, C.A., *Pigmented Coating Processes for Paper and Board,* TAPPI Monograph 28, "Trailing Blade Coating of Publication Grades," TAPPI PRESS, Atlanta, 1964.

3. Straenger, K., Gernsbach, 1988.

4. Wight, E.W., Modern Coating Application and blade Metering System.

Chapter 3, Section II

Rod Coaters

George L. Booth, The Black Clawson Co. (retired)

Introduction

History: The Mayer Coating Machinery Co. of Rochester, NY, was founded in 1905 by Charles W. Mayer. The firm began making equipment to manufacture carbon and wax papers, which were new industries. The machines used equalizer bars or doctor rods which were made of carbon steel and wound with different diameters of piano wire. Mayer was issued a series of patents on

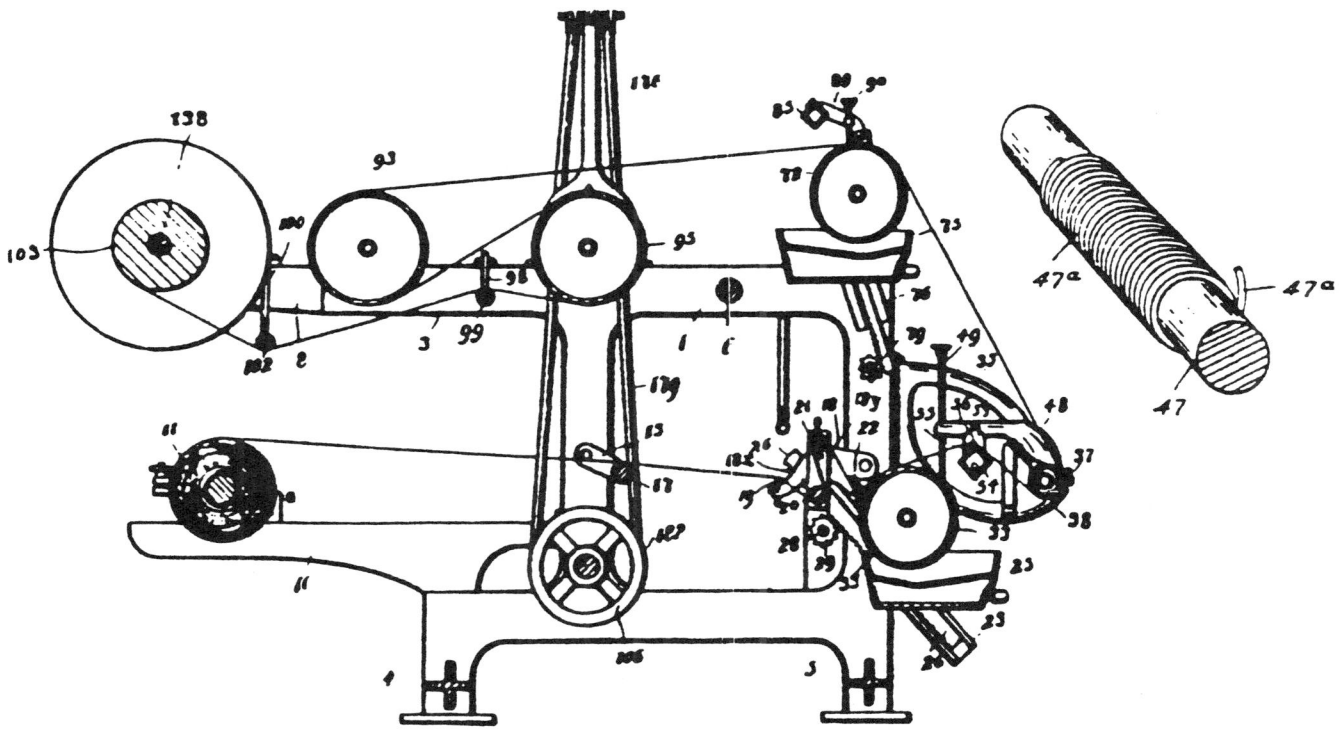

C. W. MAYER.
COATING MACHINE.
APPLICATION FILED NOV. 24, 1911.

1,043,021.

Patented Oct. 29, 1912.

Fig. 3.23. 1912 Mayer patent

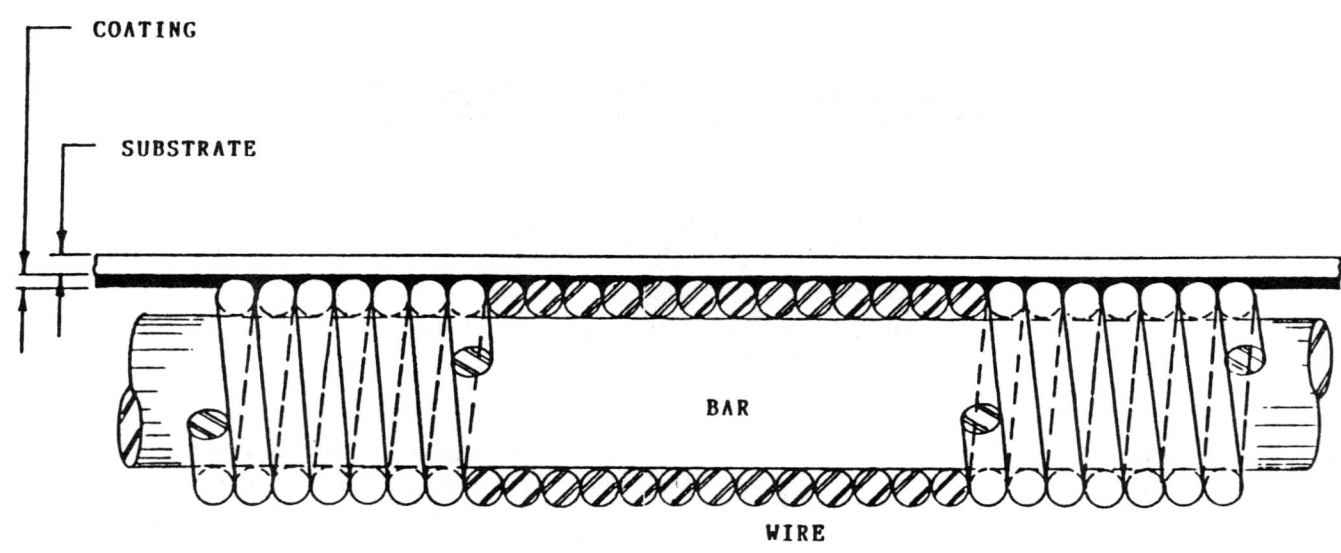

Fig. 3.24. Mayer rod

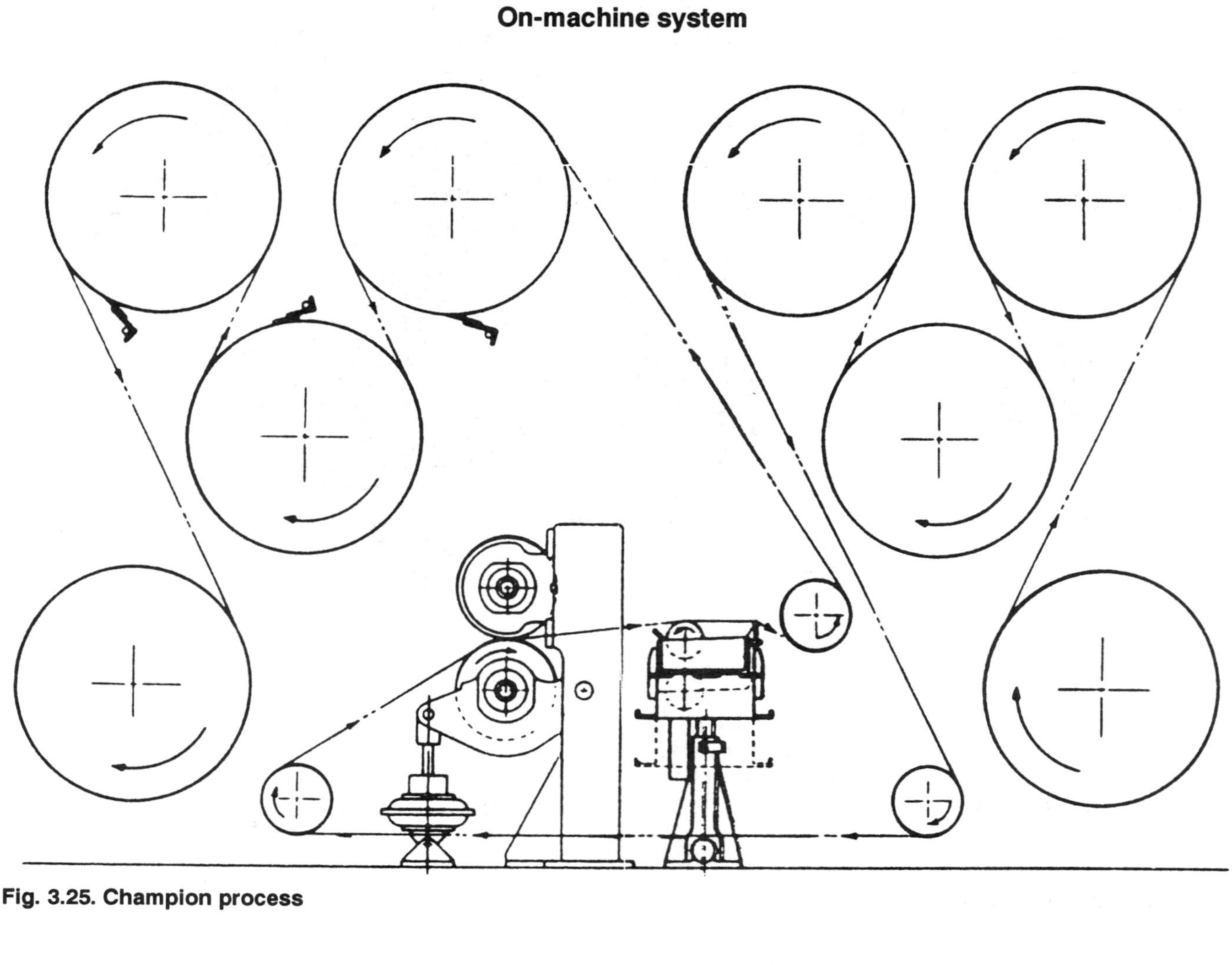

On-machine system

Fig. 3.25. Champion process

his coating machines including one in 1912 which covered his equalizer bars (Fig. 3.23).

Users found they could easily change the thickness of their coatings by switching to rods with different wire diameters, so they began ordering rods with a variety of wire sizes. The early rods rusted and wore out quickly, so Mayer developed a major business in replacement rods to a growing number of coaters that he also built (Fig. 3.24).

Mayer's business success became the target of federal anti-monopoly laws, forcing him to release his designs and patents to others in the 1930s. Under the federal guidelines, Mayer was protected from exposing his proprietary techniques for making his equalizer bars, so his factory continued to produce these rods on specially developed winding machines.

Rod Coaters

Champion Process

In the late 1930s, Champion Paper and Fiber Co. in Hamilton, OH, became very interested in this coater, specifically for use on paperboard machines. They introduced this coating system to the paper industry and protected the process with two patents issued to Bradner in 1941. The Champion process for board coating was the result of this work (Fig. 3.25).

Champion placed coaters in the dryer section on a paperboard machine by removing two or three of the cylinder dryers for each station. Coating was applied to the paperboard when it was approximately 55-60% dry. A second coater was installed, in a like manner, at a point in the dryer section where the sheet was 65-70% dry. Some mills installed a third coater in the dryer section, but the system met with very limited success in further improvement of the finish and the print surface.

Champion made two significant changes in their approach to rod coater development for paper and paperboard. First, they developed and patented a holder which was formed from a stainless steel sheet metal partially encapsulating the rod in a formed curvature (Fig. 3.26). The rod was driven from both ends, since both the width and operating speeds were increased significantly from those used in its original application of wax coating.

The second major deviation was the use of a smooth rod rather than a wire wound rod to meter the coating from the paperboard. Several years later, the paper industry began using a wire wound rod to increase the coat weight. During all of this development, a 1/4-in. (6 mm) diameter rod was used.

The air knife or air doctor coater was introduced to the paperboard machine in the mid-1950s and eventually replaced the Champion process. The air knife produced a superior quality to multiple-rod coating because it ap-

plied heavier coat weights and was able to cover the fibers completely.

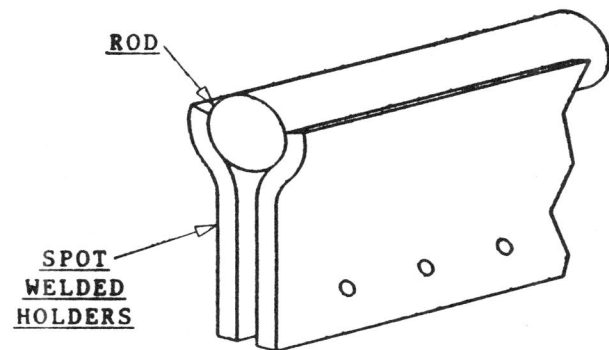

Fig. 3.26. Champion rod holder

With introduction of the air knife coater, paperboard coating gained acceptance quickly. With the variable surface quality of paperboard, it was difficult to achieve the smoothness of a rod coating process. Space was always at a premium on cylinder board machines, leaving little room to install a precoater and a dryer. These problems were solved by making an innovative change that placed a rod coater ahead of an existing air knife coater to precoat the paperboard without drying after the precoat. It worked, so the wet on wet coating systems for paperboard machines was born (Fig. 3.27). It became the standard of the industry since it was an inexpensive system to purchase and it did not require much room. The coater was simple to operate, it improved quality and actually reduced operating costs since the rod coating formulation was less expensive, and the coater could operate at higher solids than air knife coaters.

Champflex Coater

Champion Paper and Fiber Co. also pioneered the use of a backing roll against the rod. This was invented by Warner in Patent 2,729,192 issued Jan. 3, 1956. The captured rod holder system was again used for the metering device operating now against the web supported by a rubber-covered backing roll (Fig. 3.28). This coater was used for paper, bleached board, and reclaimed fiber folding carton board coating operations. Champion freely licensed the coater technology to other companies. It was the original flexible rod coater as we know it today and is now in operation on large, high-speed coaters for paper and paperboard. The same coater is extensively used on reclaimed fiber folding carton board.

Dedicated Rod Coater Systems

Currently dedicated rod coater systems use single-kiss roll applicators, followed by a rod metering device operating against a backing roll. Some of these systems are mounted on the tip of a flexible steel or plastic blade mounted in a clamping device and pneumatically loaded internally as shown in Figs. 3.29 and 3.30. Most blade coaters now in operation are capable of accepting a rod metering device, as shown, in place of the flexible blade.

Figure 3.29 shows the rod holder which is a totally formed plastic piece including the flexible part clamped in the blade holder. The rod is locked in the holder. Loading of the rod against the paper is accomplished in a manner identical to that with which a blade is loaded against the roll. The flexing in the plastic holder itself will level out the pressure of the rod against the roll across the width of the machine. These are direct replacements for the blade in the clamping device on blade coaters. When it is necessary to replace a rod due to wear or

other defect, the entire holder has to be removed and replaced with a second holder. The drive is disconnected on each end of the rod for removal and reconnected once the new holder and rod is put into place. The rod is replaced in the holder and remains in a standby status until it is needed. The plastic bed has a life of several months to a year or more.

Figure 3.30 is a similar design functionally with a slightly different approach. The plastic holder is mounted on the edge of a blade as shown so that the blade is actually the flexing part of the holder. The advantage of this holder is the uniformity of flexing in the blade and a smaller piece that needs to be changed. A concern with this holder is the need to properly trap the plastic ends with the tip of the blade so that it will not come loose under high-speed operation. In order to install the plastic holder on the blade, it does require some effort in seating the blade properly into the plastic holder. Again, from a practical viewpoint, it would be better to remove the

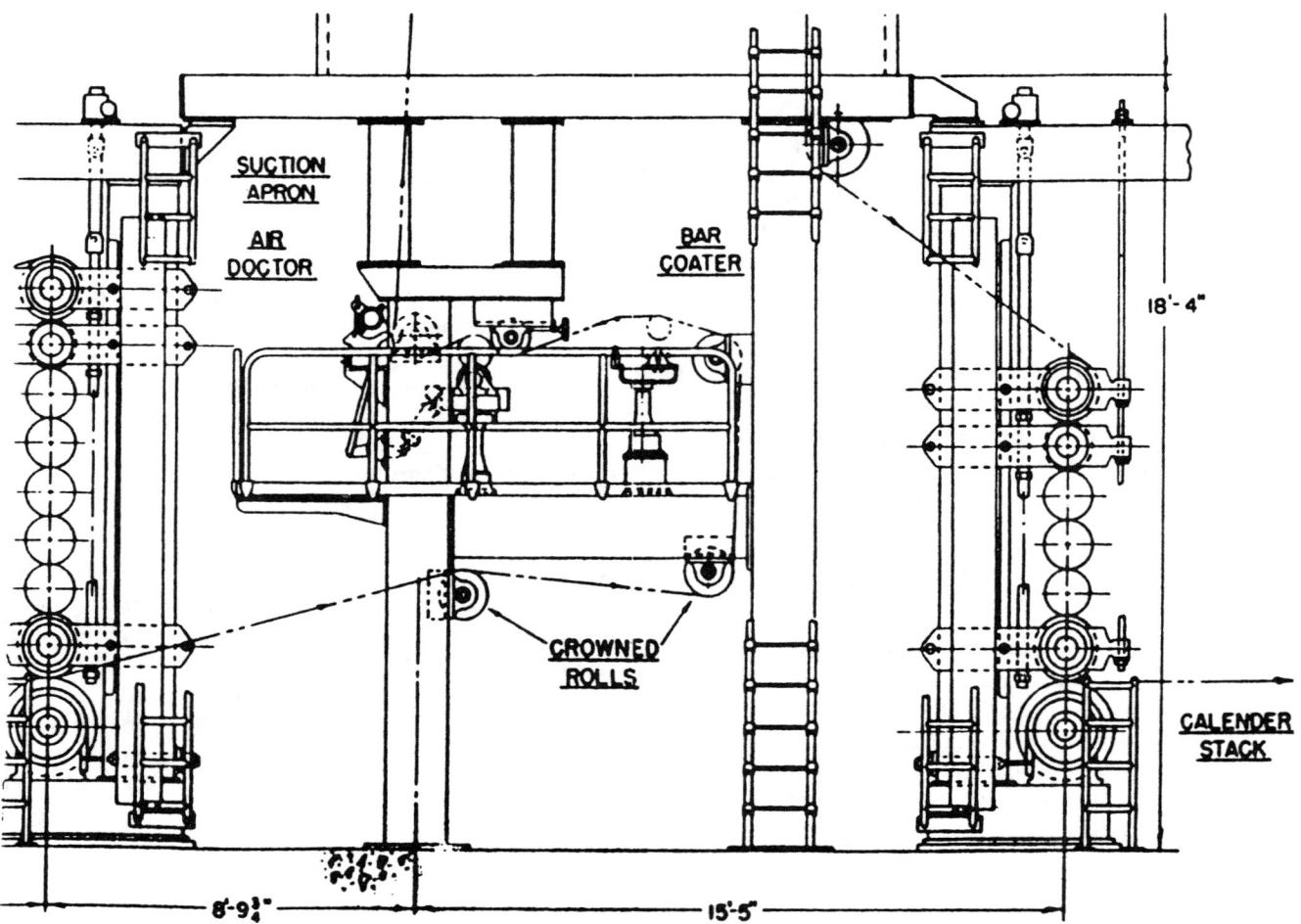

Fig. 3.27. Wet on wet coating

blade and plastic holder as an assembly, replacing it with a new fresh unit when the rod requires changing.

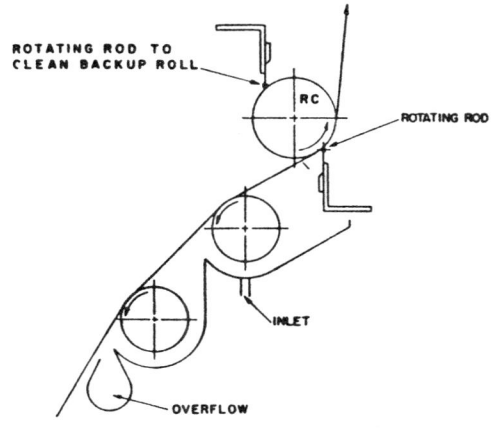

Typical Champflex coater

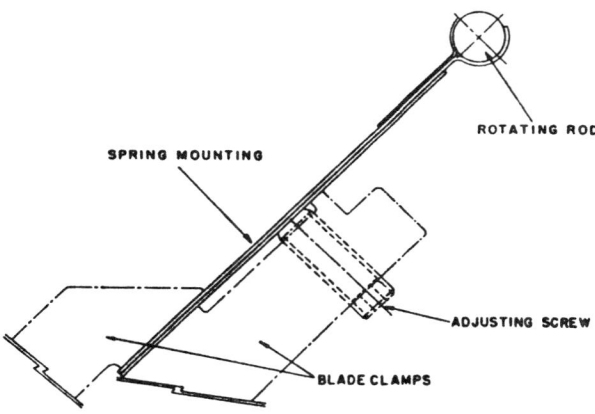

The Champflex coater blade (U. S. pat. 2,729,192)

Fig. 3.28. Champflex coater

In Figure 3.31, the holder is made of a plastic material such as nylon, urethane, or high-density linear polyethylene. The holder is loaded with a single air bladder such that the plastic piece moves up and down within the machined ways of stainless steel bars or loading against the paper. Coat weight in all of these systems is determined by coating rheology, speed, coating solids, and rod diameter. The amount of pressure with which the rod is loaded against the rubber backing drum has very little effect on coat weight. The water channels noted underneath the rod are full of circulating water fed in through a multiplicity of ports and likewise removed at several points across the rod. The number of inlets and outlets are governed by the width of the rod holder itself.

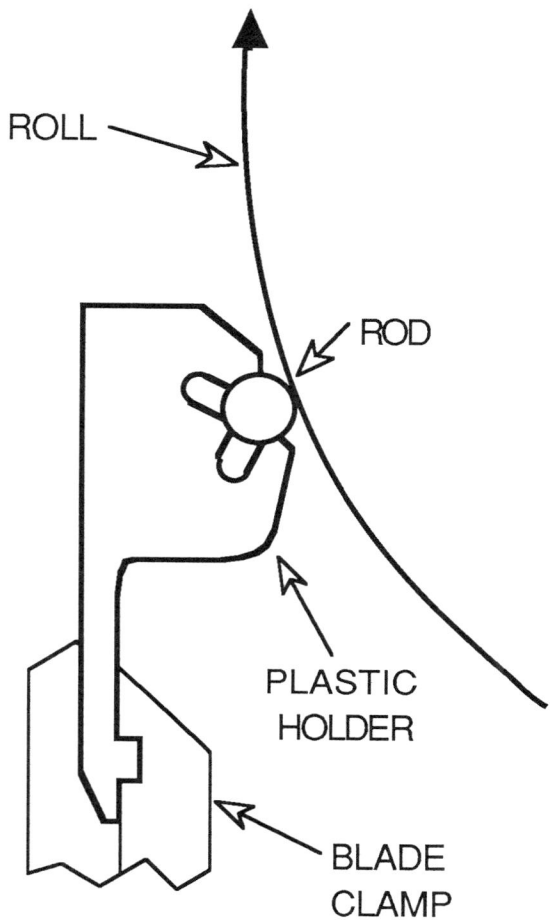

Fig. 3.29. Rod holder blade replacement

Rod

For over 70 years, wire wound rod coaters have been considered a simple, easy-to-operate coating method. The main technological breakthrough has been the development of stainless rods, stainless wire, and precision winding equipment. In the 1980s after patents expired, rods with machine threads have been available in place of wire wound rods. Figure 3.32 shows a wire wound rod, and Figure 3.33 shows a machine threaded rod.

For wire wound or grooved rods, the thickness of the coating is governed by the cross-sectional area of the grooves or space between the wire coils of the rod. The geometry of this system creates a wet film thickness which is roughly proportional to the cross section of the wire or channel used. For example, you can double the coating thickness by doubling the groove or wire size.

The groove between the wires determines the amount of coating material which will pass through. The initial shape of the coating is a series of stripes, spaced apart according to the spacing of the helix. Almost immediately, normal surface tension pulls these stripes together, forming a relatively uniform surface, ready for drying.

The coating that is wiped off is separated from the coating in the pan so it can be processed to remove air and trash before returning to the pan.

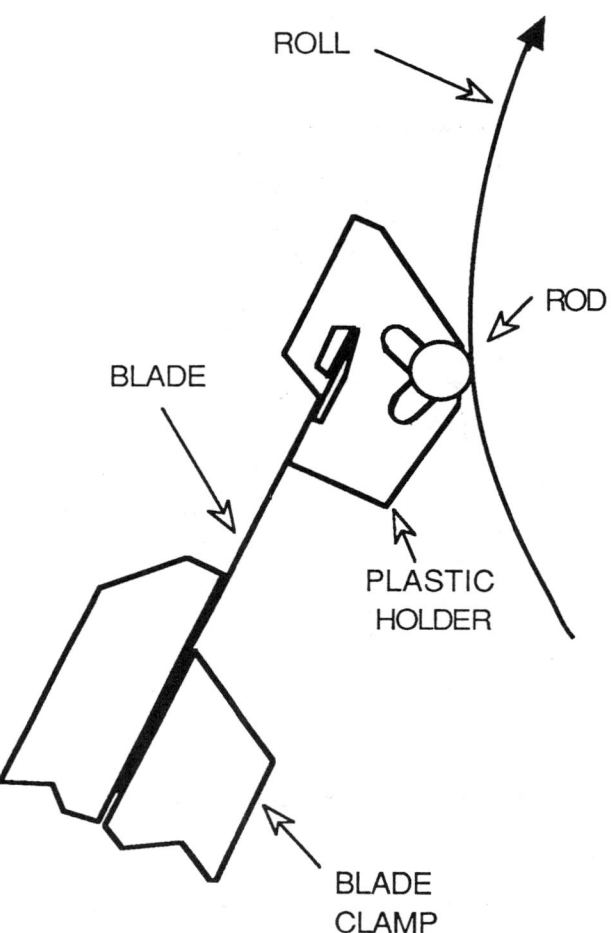

Fig. 3.30. Rod holder blade-mounted

The coating uniformity of rod coaters is dependent on having a completely flat web in intimate, even contact with the surface of the rod. The only force available to ensure contact is provided by web tension. Unfortunately, it is generally impossible to ensure tension uniformity and flatness across the web, so there can be large variations in the coating metered by the rod. This causes coat weight nonuniformity. Additionally, because web tension is constant and the force necessary to resist the hydraulic force that develops under the web increases with line speed or viscosity, the rod coater becomes unstable at speeds about 900 ft/min. The actual point where control is lost depends on the coating

rheology, the web tension, the wrap angle over the rod, and the condition of the web.

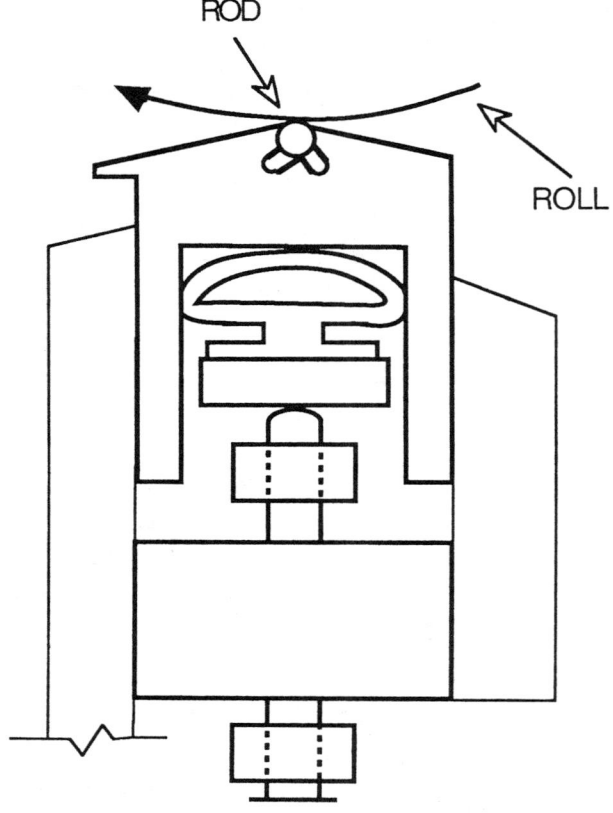

Fig. 3.31. Rod holder with air bladder

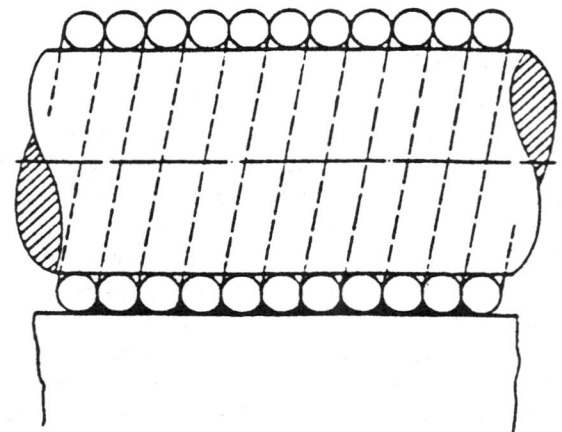

Fig. 3.32. Wire wound rod

Smooth rod coating is accomplished by floating the entire rod on a film of liquid above the surface of the paper. Coat weight depends on the thickness of the hydraulic wedge that develops. This is very similar to a lubricated bearing. To increase coat weight, using the same rod size, the coating can have its viscosity increased by using viscosity modifiers or by increasing solids content, which increases the viscosity as well as the finished coat weight with a fixed wet film deposition.

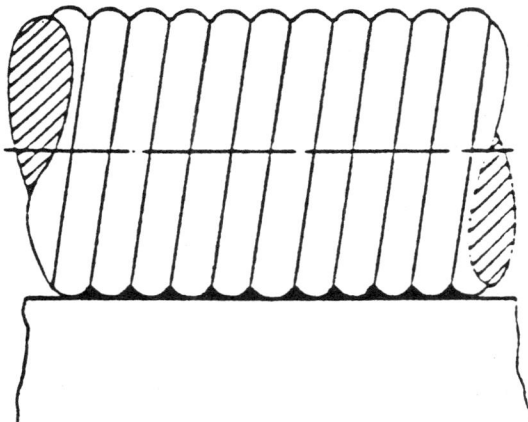

Fig. 3.33. Profile rod

There are several mechanical ways to increase coat weight using a backing roll. The simplest way is to decrease the pressure in the air bladder under the rod. The normal air pressure range is up to 25 psi with the majority of coaters operating between 5 and 15 psi.

In general, as soon as enough pressure is developed to deform the backing roll, increasing the rod pressure decreases coat weight, and decreasing rod pressure increases coat weight. There is a very sharp change in coat weight between 5 and 15 psi; depending on paper absorbency it can be anywhere from 30-70%. Above 15 psi, the effect of increasing rod pressure tapers off quickly until, at approximately 30 psi, no additional reduction is possible.

A very interesting condition can occur at very low rod pressures where the rod is just balanced against the paper without deforming the backing roll. The metering force is equal to the rod load divided by the area of contact. Since there is very little area, we are dividing a small number, the rod load, by a number almost equal to zero. The resultant unit force can be very high. The end result is that, as rod loading is increased slightly, the rubber backing roll deforms which increases the area of contact at a faster rate than the rod loading increases. Thus, for a short increment, coat weight may actually increase with increased rod load. This quickly changes back to the standard conditions where increasing rod loading pressure decreases coat weight. If pressure is maintained above 5 psi, this effect will not be noticed.

Changing smooth diameter size affects coat weight by changing the width of the nip created between the backing roll and the rod. For any air pressure, a larger nip

area causes the metering force—calculated by dividing load by area—to decrease as the rod size is increased. This provides a larger coat weight each time rod size is increased.

Wire or grooved rod coating works by a different principle. Instead of floating on a uniform film, the outside diameter of the rod may in fact contact the paper on the high points. Coat weight is obtained by the amount of coating that can squeeze through the small voids (Table 3.1).

Table 3.1. The relationship between wire diameter and wet film thickness in mils

					Thickness Chart				
Rod No.	Wet Film	Rod No.	Wet Film	Rod No.	Wet Film	Rod No.	Wet Film	Rod No.	Wet Film
3	.27	10	.90	24	2.16	38	3.42	55	4.95
4	.36	12	1.08	26	2.34	40	3.60	60	5.40
5	.45	14	1.26	28	2.52	42	3.78	65	5.85
6	.54	16	1.44	30	2.70	44	3.96	70	6.30
7	.63	18	1.62	32	2.88	46	4.14	75	6.75
8	.72	20	1.80	34	3.06	48	4.32		
9	.81	22	1.98	36	3.24	50	4.50		

$$t = \frac{400B}{WS}$$

where:

	t	= wet film thickness, mils
	B	= weight of dry coating solids, pounds per 3000 sq. ft.
	W	= density of wet coating, pounds per cubic ft.
	S	= percent of dry solids in wet coatings, weight basis

To allow more coating to pass through the opening, you decrease the viscosity. This is illustrated by the standard Zahn cup viscosity test. The lower the viscosity, the sooner the cup empties.

If two wire or grooved rods of different diameters but the same void area are used, a slightly higher coat weight will be provided by the larger rod. However, a one or two thousands of an inch (0.001 in. or 0.002 in.) increase of wire diameter on the smaller rod will usually provide the same coat weight. On fine papers, smooth rods with core diameters of 3/8 in. to 1/2 in. are used, due to the lower cost of the stainless rod core. On highly abrasive kraft paper and paperboard, the increased surface area of the larger rods make 5/8 in. or 7/8 in. a more economical choice. When using wire or grooved rods, the rod pressure is usually set as low as possible to reduce the abrasive wear on the wires by the paper.

For small rod core sizes (3/16 in., 1/4 in., 3/8 in.), the major effect on coat weight is provided by wire size alone. The viscosity required to run a wire rod is low, so no significant hydraulic wedge can develop to float the rod off of the sheet. If there is no wedge to adjust, pressure change will have little effect. When larger rod core sizes are used (5/8-1 in.), the larger nip allows more of a wedge to develop. The force of this wedge is easily over-come. At best, with a 1-in. rod, rod bladder pressure will affect the coat weight by 10-15% with wire sizes from 0.010 in. to 0.020 in. in diameter.

As wire diameters decrease from 0.010 in. to zero, rod bladder pressure has an increasing effect on coat weight. As wire diameter increases above 0.010 in., rod bladder pressure has little or no effect. The general rule is to run a bladder pressure under a wire rod that is high enough to ensure uniformity. It can be increased 3-5 psi to trim coat weight slightly. Any additional pressure will increase rod wear. High pressure may also cause streaking by deforming the paper where it contacts the rod.

Coatings can be specifically formulated to have high solids with low viscosities. If the coating viscosity is too high, the only solution is to dilute the coating until it reaches a low enough viscosity to run correctly with a wire rod.

The typical streaking pattern that a high-viscosity coating makes when used with a wire rod is caused by two things. First, the coating coming through the wire in distinct flow streams has to flow together to form a uniform thickness. If the viscosity is high, the chances are the coating will not level to fill the valleys left by the wire.

A second problem caused by high-solids content is rapid dewatering of the coating so that it reaches the rod with a heavy filter cake that is not mobile and cannot be easily metered. This is a problem when running coatings formulated to dry by losing water into the sheet faster than by normal drying. This problem is solved by adding water-retention agents to the formulation. A viscosity-control unit that monitors the coating being delivered to the coater and adds water to the cycle tank to maintain constant properties is a good solution.

If particle size increases, the wire grooves clog much more quickly, and the wire wears down which decreases coat weight. A coating filter installed on the inlet line to the coating pan will help to reduce streaks. The type and size are an experience decision.

A coater was needed that kept the benefits of rod coating while eliminating its speed and uniformity limitations. The Champflex coater did this by making the rod operate independently of coating parameter listed above. It utilized the concept of an air or mechanically loaded rod pressing against a resilient backing roll.

Two generic types of coating pan designs are used for smooth coating flow to reduce coating buildup.

Figure 3.34 shows the common pan used for coatings that need fast circulation through the pan. The coatings are not susceptible to problems caused by large coating recirculation. The coating enters the main chamber of the pan and is diffused by a baffle flowing to both ends over weirs into a return trough. The returned coating mixes with the coating doctored off by the rod keeping all pan surfaces of continually washed. This circulation eliminates eddies, foam buildup, and dead zones.

Figure 3.35 shows the pan with a reduced recirculation rate. In this configuration, the coating returning from the rod is channeled back to the pan. Lower density caused by air entrainments is the first to overflow the end weirs. Since the returned coating is reintroduced to the pan outside the coated area of the web, a majority of fresh coating is applied to the web. By increasing residence of the coating in the pan, the recirculation rate is drastically reduced.

There are now several design variations of this equipment in paper and film converting plants, paper mills, and paperboard mills. Some differences exist in coater design of the rod holder, loading system and general arrangement. The web with excess coating wraps an elastomer-covered backing roll, which supports the web as it passes through a nip created by the rod and the backing roll.

In one loading system, the rod is held in a precisely machined plastic holder element which is free to float in a retainer slot. It is loaded against the web supported on the backing roll using a flexible air bladder or plastic tube that can be adjusted from 0 to 50 psi air pressure (Fig. 3.31).

The rod, plastic holder, and air bladder are limber enough so that at pressures above 3-5 psi, the rod is uniformly loaded against the backing roll regardless of backing roll deflection. Since the web is on the roll prior to the application of metering force, web tension has no effect on coat weight which is a major consideration.

A second holder was described earlier and is shown in Figs. 3.29 and 3.30.

The backing roll is driven at line speed by its own drive and becomes a positive nip to isolate web tension from equipment upstream or downstream from the coater. For a given coating formulation, coat weight can be varied using the applicator roll speed, rod size, or rod loading pressure. Of these three, rod diameter, wire size, or both are the most reactive.

The applicator roll is generally run in the direction of web travel at 15-70% of web speed applying an excess amount of coating in the ratio of 3:1 to 10:1 measured against final coat weight. Running in the direction of the web at slow speed reduces film split pattern and tends to provide a more uniform excess by shearing the coating on the web. In general, the applicator roll should run just fast enough to allow some laminar run back on the incoming side of the roll.

The rod size determines the coat weight range over which the coater can operate. For converting machines, smooth chrome plated rods between 1/8 in. and 1 in. provide coat weights between 0.2 and 25 lb/3000 ft^2 with a

Recirculation rate 5:1 and greater   (plan view)

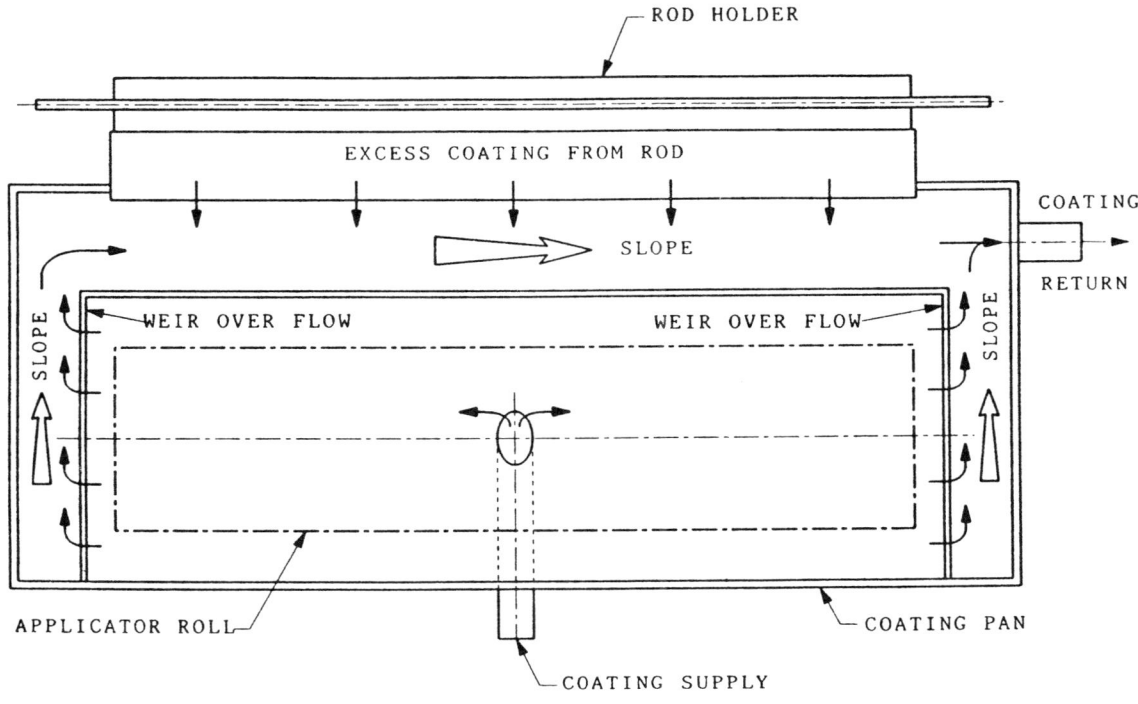

Fig. 3.34. Normal recirculation rate

Recirculation rate up to 4:1   (plan view)

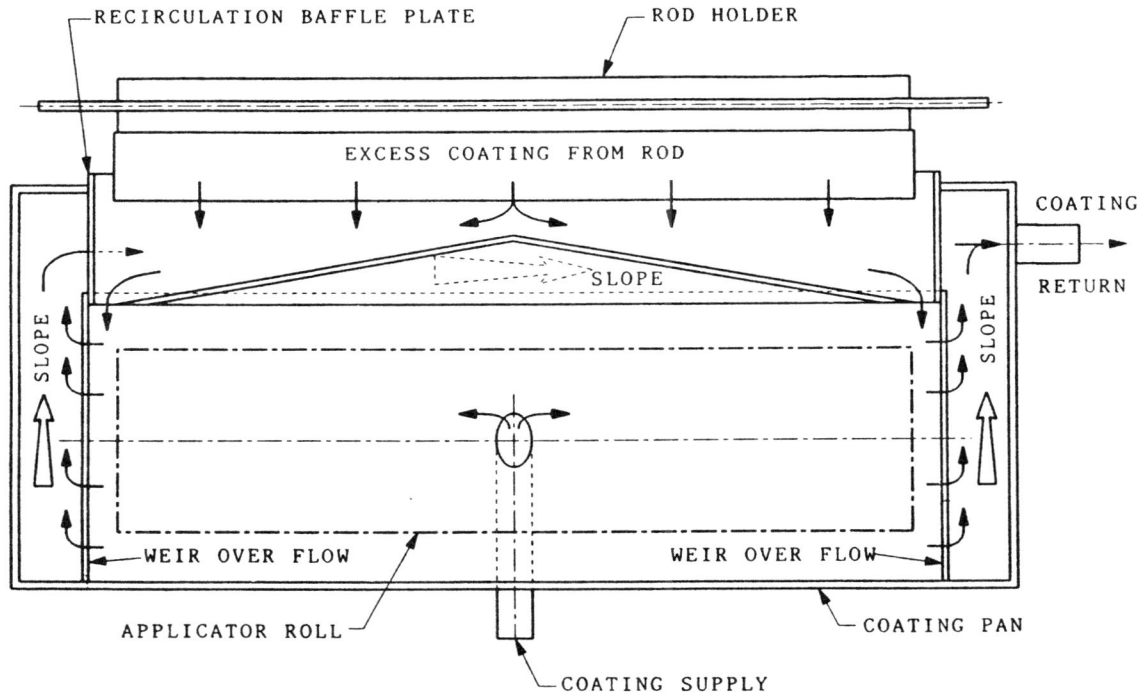

Fig. 3.35. Reduced recirculation rate

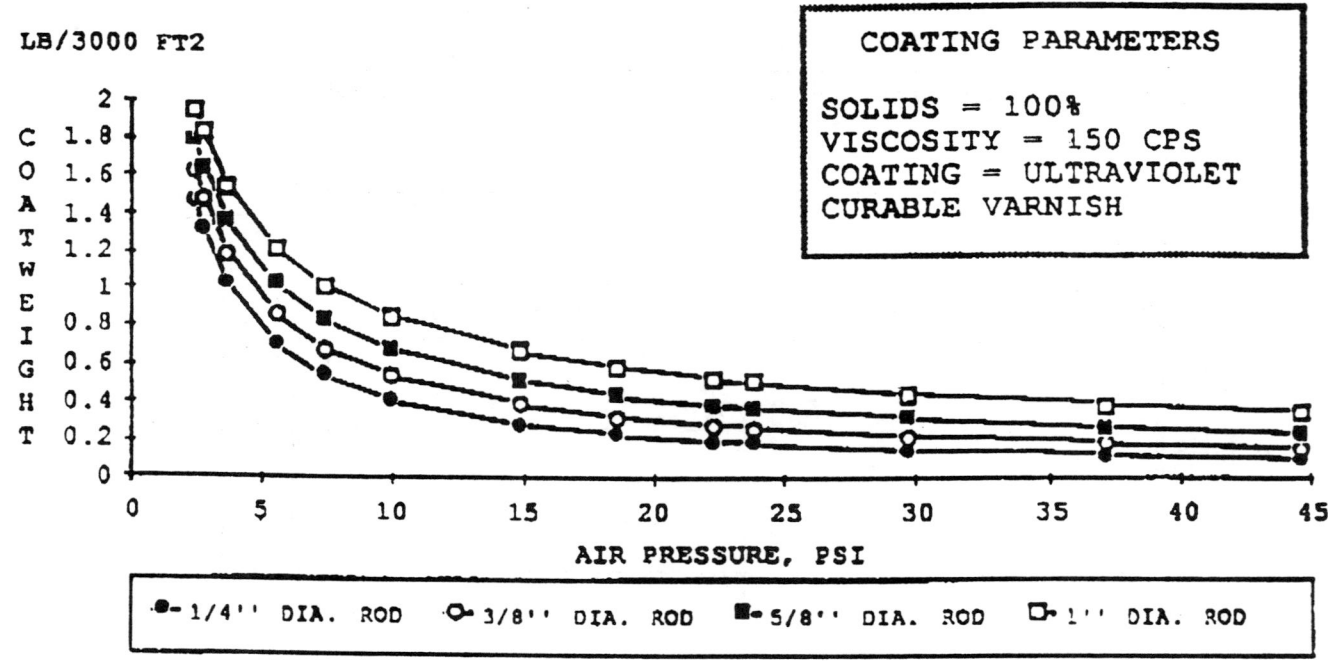

Fig. 3.36. Air pressure vs. coat weight: 100% solids, 150 cps

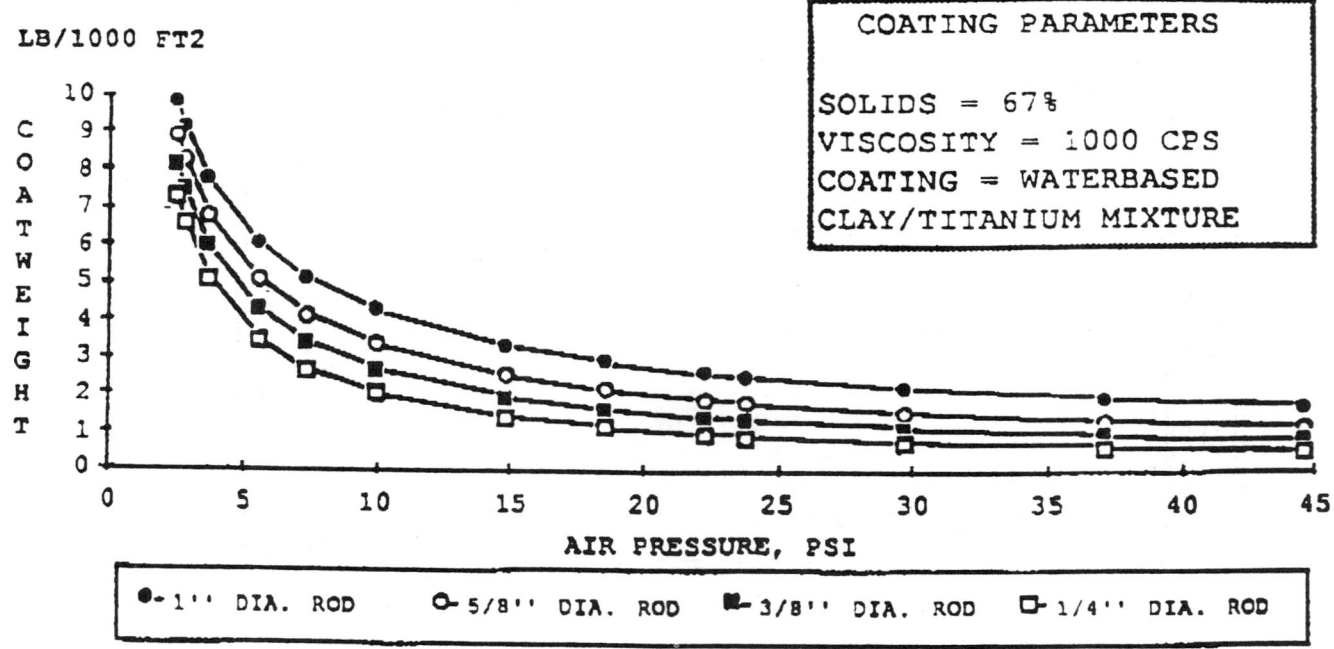

Fig. 3.37. Air pressure vs. coat weight: 67% solids, 1000 cps

wide variety of coatings. Much heavier coatings can be achieved by using wire wound rods rather than smooth rods.

Figures 3.36 and 3.37 illustrate pressure or coat weight curves for two dissimilar coatings. To date, machines are in operation coating a variety of water-based, solvent-based, and solventless coatings. Viscosities from 20 cps to 20,000 cps are run successfully at speeds of 50 ft/min to 3000+ ft/min. Coatings have included silicones, adhesives, fax, carbonless, varnishes, clays, titanium, waxes, and other proprietary coatings.

Both Fig. 3.36 and Fig. 3.37 illustrate the wide adjustment range available for any one rod size. This allows the operator to dial in almost any coat weight and compensate for process changes simply and easily. Closed loop coat weight control is easily adapted by varying bladder pressure under the control of a process computer.

Coat weight uniformity is generally between 1% and 4%. The rougher the substrate, the larger the percentage of error.

By changing from one rod to another, a completely different coater is available to react to the needs of a changing marketplace.

There are several other benefits to smooth rod coating. The rod turns slowly against the sheet direction. This allows it to continually clean itself to run without streaks. Additionally, the smooth rod has a polishing effect that tends to orient coating particles in the same direction, which improves coated surface smoothness. With the exception of rigid blade coaters, no other coating method delivers as smooth a coating. Because the backing roll supports the sheet under the rod, coating can be accomplished at higher solids and viscosities than ever before.

Patent Review

The following pages list 17 patents that trace the development of rod coating as it relates to the paper coating industry. A synopsis of each patent is included to explain briefly the subject of the patent.

U.S. Patent No. 2,229,620 (Fig. 3.38)
Issued to D.B. Bradner
Date issued: Jan. 21, 1941

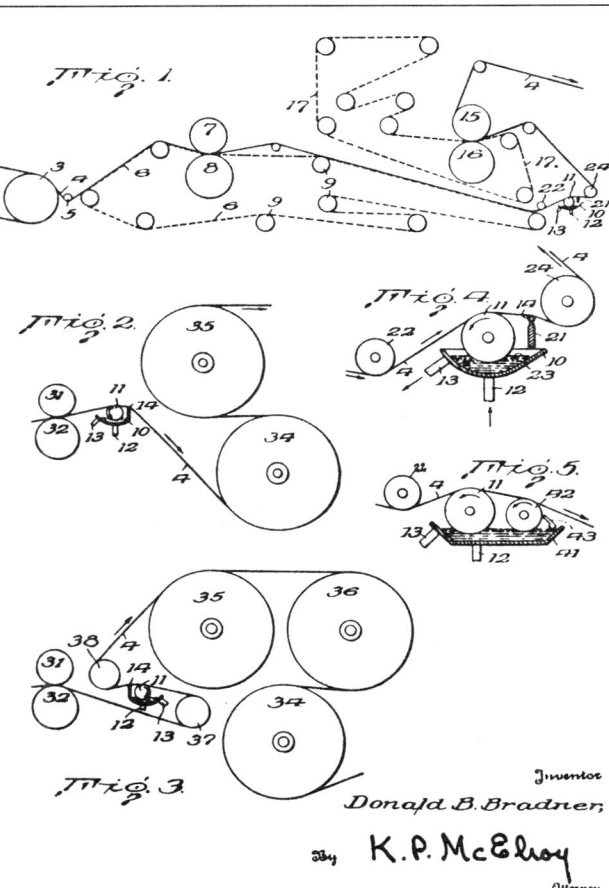

Fig. 3.38. Method of applying coating materials to paper

This is one of the first patents issued using the rod coater in the Champion Board Coating Process. It specifically relates to applying coating to the paper after wet pressing primarily to fill the surface of the paper. The method of application is an applicator roll followed by a rigid knife having a rounded surface as the metering device. Such rounded surfaces of small radius could be construed to be a rod as well as a stationary knife.

U.S. Patent No. 2,229621 (Fig. 3.39)
Issued to D.B. Bradner
Date issued: Jan. 21, 1941

U.S. Patent No. 2,534,320 (Fig. 3.40)
Issued to W.P. Taylor
Date issued: Dec. 19, 1950

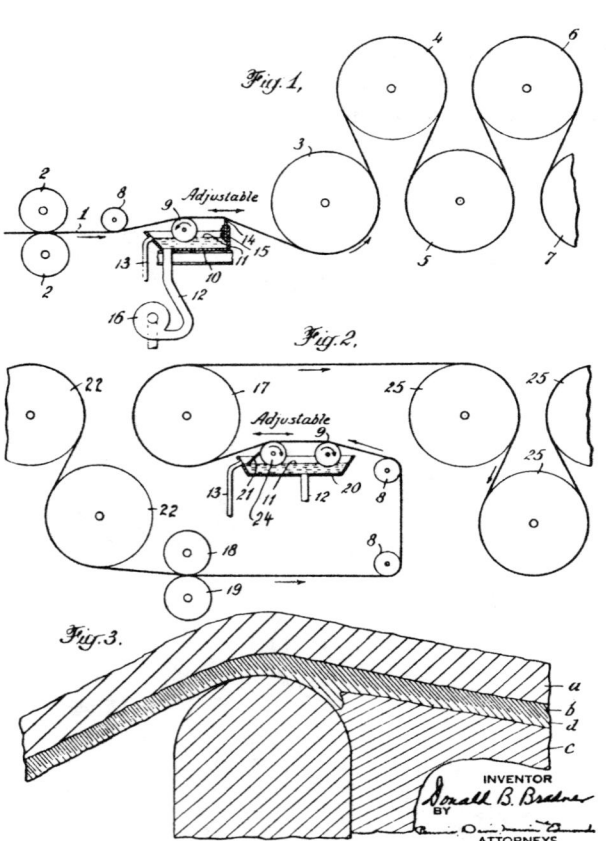

Fig. 3.39. Method of coating paper

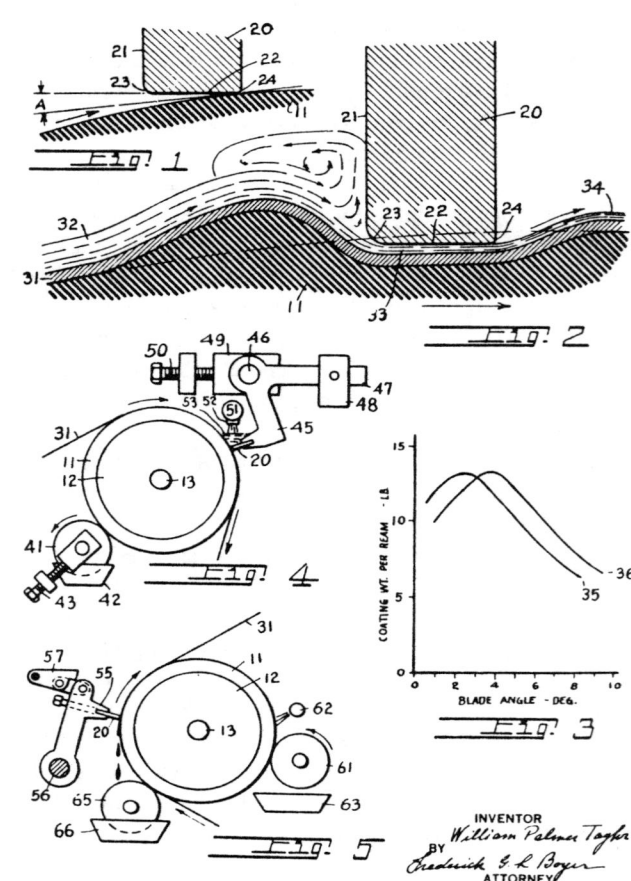

Fig. 3.40. Apparatus for coating paper

This patent is the complement to the Patent No. 2,229,620 except that it gets into more detailed process claims. It is a coating process rather than a filling process described in Patent No. 2,229,620 where the surface of the paper is covered with the coating from the coater. It discusses the formation of a filter cake and the metering device "wiping off substantially all said fluent coating composition without removing substantially any of the filter cake . . ."

This patent is one in which a non-rotating machined blade is used for removing coating from a flooded surface, leaving the finished coating on the web after passage underneath the bar. Said bar is limited to 0-10° angle in its operation, but would appear to be only a sophisticated knife-over-roll coater since the knife does operate against a relatively soft backing roll. The patent is in the family of sharp metering devices such as a rod.

U.S. Patent No. 2,598,733 (Fig. 3.41)
Issued to E. Warner
Date issued: June 3, 1952

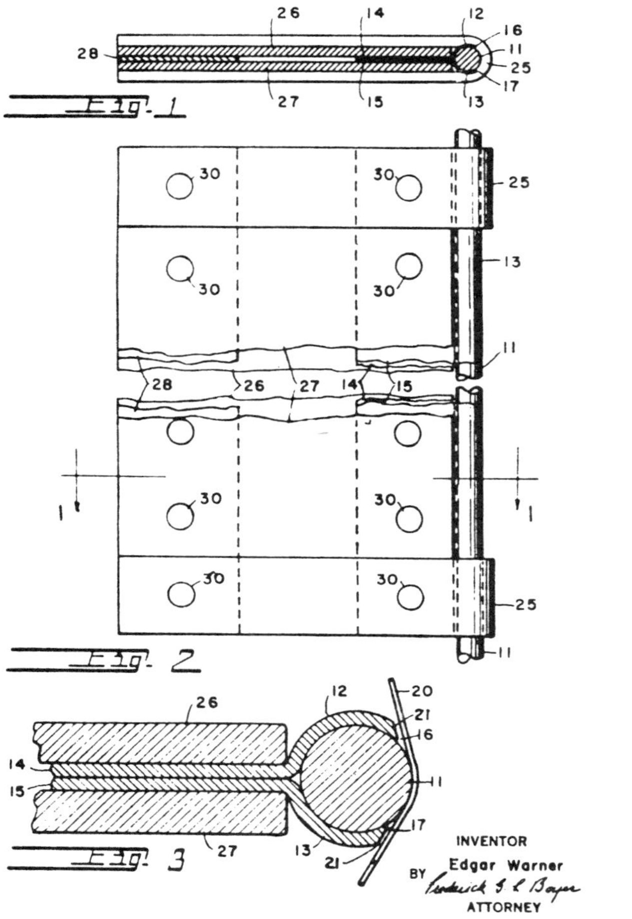

U.S. Patent No. 2,774,329 (Fig. 3.42)
Issued to R.V. Smith
Date issued: Dec. 18, 1956

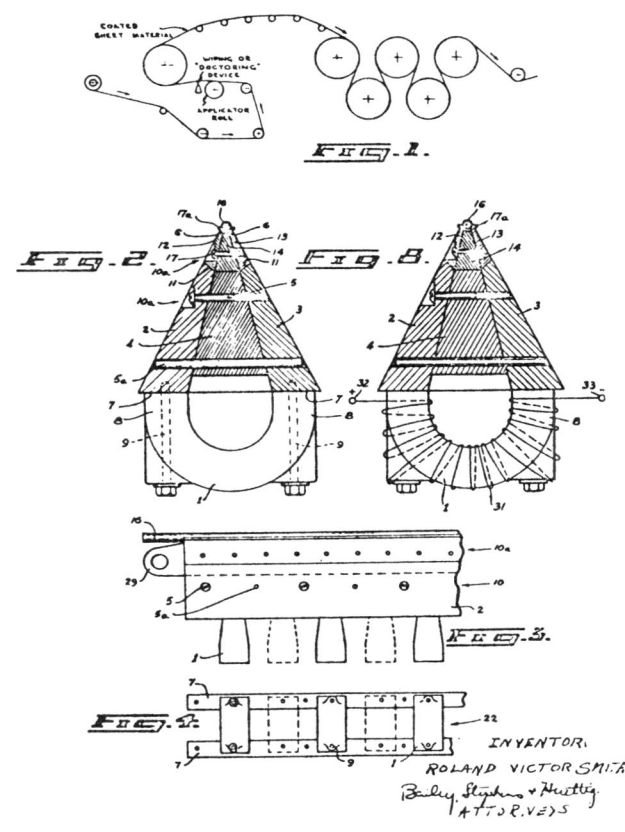

Fig. 3.42. Wiping or doctoring devices for removing excess coating from sheet material

Fig. 3.41. Wiping blade for coating devices

This patent covers the design of the rod holder which has been referred to historically as the Champion rod holder. It is two pieces of thin gauge sheet metal stock formed in a way such that the rod is trapped at one end exposing approximately 140 degrees of the rod for the metering function. There is no lubrication or flexibility in this particular rod holder.

The patent was issued to the E.B. Eddy Co. and is the base patent for the magnetic rod holding system that was used so successfully for many years. In addition to covering the use of magnetic force for holding the rod in place, it also was the first holder to have water lubrication and cleaning of the rod in a pocket underneath the rod.

U.S. Patent No. 2,729,192 (Fig. 3.43)
Issued to E. Warner
Date issued: Jan. 3, 1956

U.S. Patent No. 3,084,663 (Fig. 3.44)
Issued to E. Warner
Date issued: April 9, 1963

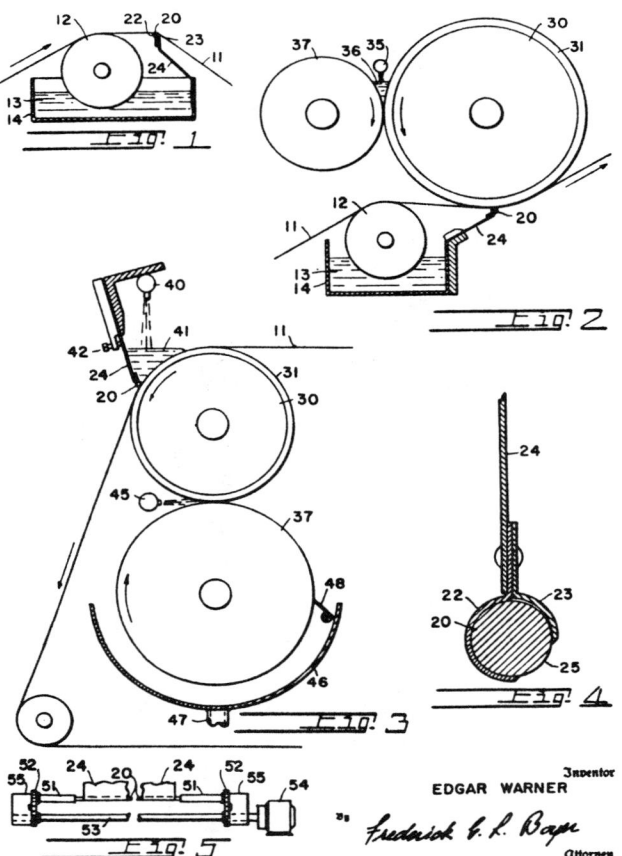

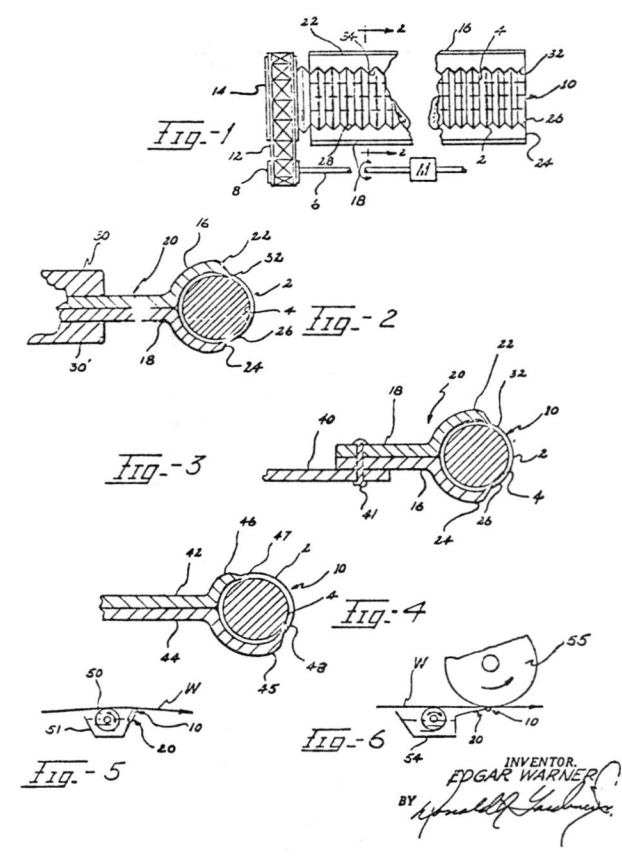

Fig. 3.44. Doctor blade for coating apparatus

Fig. 3.43. Doctor blade for paper coating apparatus

The patent covers the development of the Champion rod holder on a flexible steel blade. The holder was the foundation for the Champflex coater in that the holder, as well as the means of loading against a resilient backing roll, are covered in this patent.

This patent covers the use of a concentrically grooved rod in a spot welded rigid or nonrigid holder which could be used alone with rigid construction or in conjunction with a backing roll using a flexible holder. The thrust of the patent is a spiral grooved rod. It is inner related to the Canadian Patent 724,578.

Canadian Patent No. 724,578 (Fig. 3.45)
Issued to E. Warner
Date issued: Dec. 28, 1965

U.S. Patent No. 3,304,910 (Fig. 3.46)
Issued to E. Warner
Date issued: Feb. 21, 1967

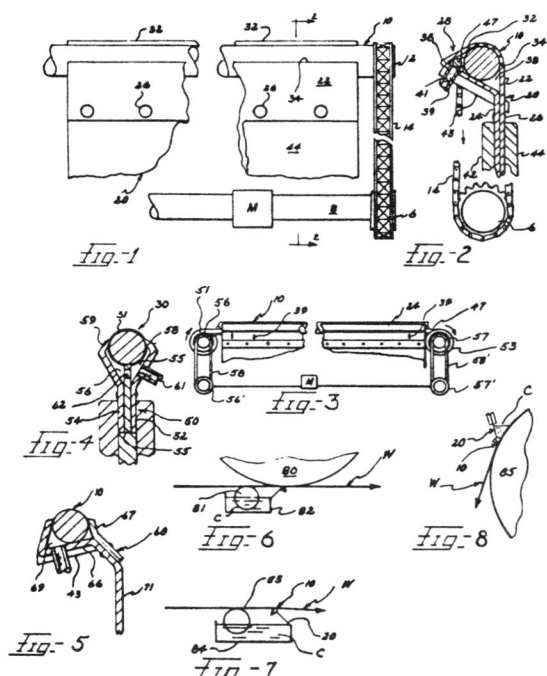

Fig. 3.45. Doctoring device

The main thrust of this patent is the water chamber underneath the rod for both flushing and keeping the rod clean. It also mentions the use of a grooved rod as well as a smooth rod used in a flexible or rigid holder. It is interrelated with the U.S. Patent 3,084,663.

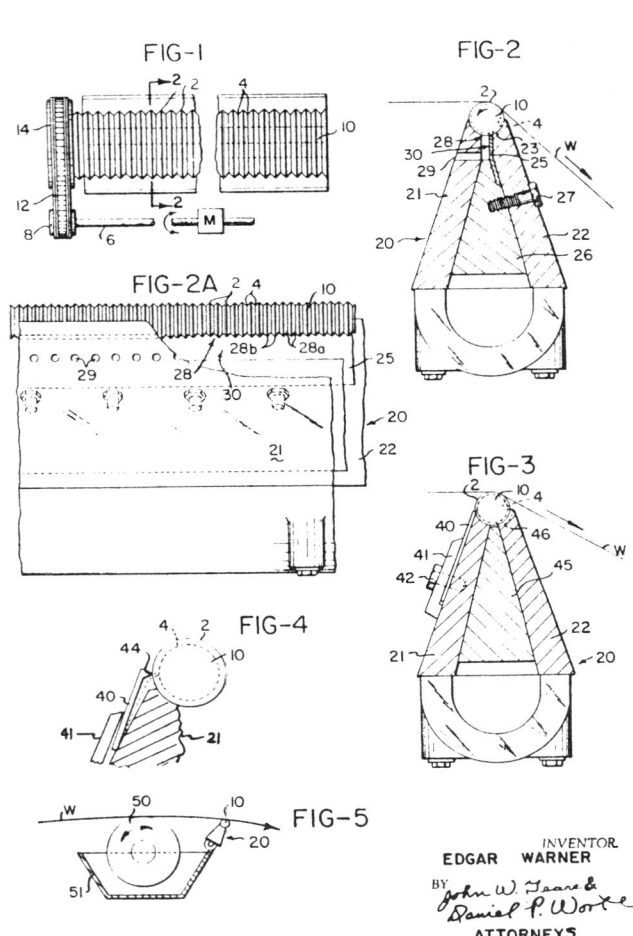

Fig. 3.46. Doctor blade for coating apparatus

This patent is very similar to the Smith patent, 2774329, in its use of magnetic forces to retain the rod in the holder. The patent covers the use of a grooved rod with another member that is complementarily shaped to engage the ribs and grooves on the doctor rod surface. The Smith patent is cited several times in the description of this unit.

U.S. Patent No. 3,453,137 (Fig. 3.47)
Issued to J.E. Penkala, et al.
Date issued: July 1, 1969

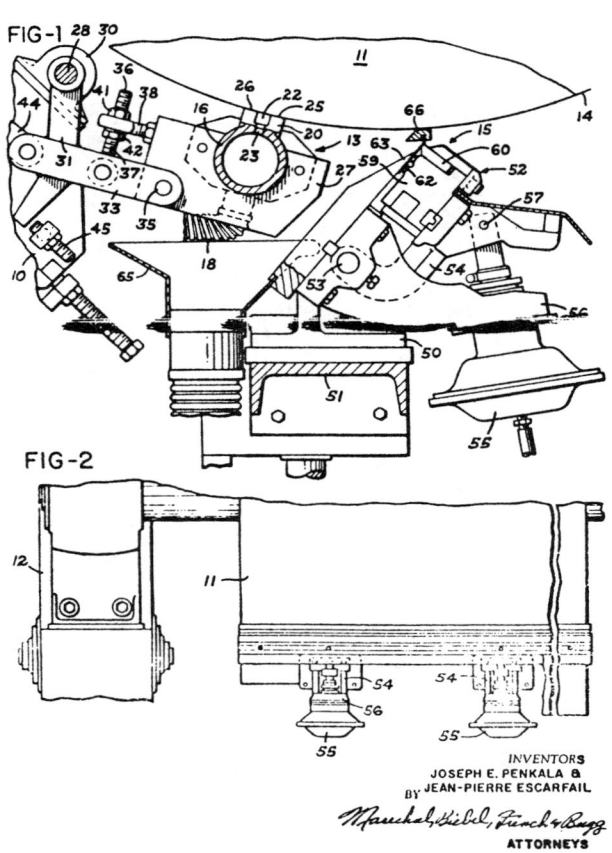

Fig. 3.47. Paper machinery

The patent and claims covered herein are the basis for using a nonrotating member as a metering device against a resilient backing roll. A specially shaped shoe was mounted on a flexible steel blade in a blade coater for removing excess coating and leaving a smooth surface on the freshly coated paper. Several patents have followed this particular concept in design for metering systems.

U.S. Patent No. 4,263,870 (Fig. 3.48)
Issued to H. Saito, et al.
Date issued: April 28, 1981

Only three claims are listed for this patent. In fact, it is a short-dwell rod coater where the coating is supplied immediately behind the rod so that the excess is run off down the surface of the holder on the incoming side and the exiting paper smoothed by the relative speed and shear force imparted by the rod itself.

---

[54] **COATING PROCESS**

[75] Inventors: **Hiroki Saito; Toshiro Tahara; Akihiko Nagumo; Toshio Shibata; Kimio Yukawa; Minoru Minoda,** all of Minami-ashigara, Japan

[73] Assignee: **Fuji Photo Film Co., Ltd.,** Ashigara, Japan

[21] Appl. No.: **824,133**

[22] Filed: **Aug. 12, 1977**

[30] **Foreign Application Priority Data**

Aug. 12, 1976 [JP] Japan 51-96676

[51] Int. Cl.³ B05C 1/08
[52] U.S. Cl. 118/259; 118/258; 118/414; 427/428
[58] Field of Search .... 427/428, 434 A; 118/414, 259, 258

[56] **References Cited**
U.S. PATENT DOCUMENTS

1,649,960	11/1927	Kratz	118/258 X
2,066,780	1/1937	Holt	118/414 X
2,172,326	9/1939	Wittich	427/428 X
2,432,074	12/1947	Jennings	427/428 X
2,464,040	3/1949	Huebner	118/258 X
2,946,307	7/1960	Warner	118/119
3,231,418	1/1966	Muggleton	427/428
3,473,955	10/1969	Moriarty	427/428
3,526,536	9/1970	Spengos et al.	427/434 A
3,535,157	10/1970	Steinhoff et al.	427/428 X
3,941,897	3/1976	Vecchia	427/428 X
4,046,924	9/1977	Tanguy	427/428 X

FOREIGN PATENT DOCUMENTS

980510	5/1951	France	427/428
4590	of 1907	United Kingdom	427/428

Primary Examiner—Shrive P. Beck
Attorney, Agent, or Firm—Sughrue, Rothwell, Mion, Zinn and Macpeak

[57] **ABSTRACT**

A bar coating process for coating a coating liquid on a continuously travelling web which comprises the steps of supplying a coating liquid so as to form a liquid reservoir immediately before a position of contact between a bar and the web, and coating the coating liquids on the web using the bar, wherein the bar is axially positioned perpendicularly to the travelling direction of the web, is supported on a supporting member and is rotated in the same direction as that of the web while coming into contact with the web. The diameter of the bar being 6 to 25 mm.

3 Claims, 6 Drawing Figures

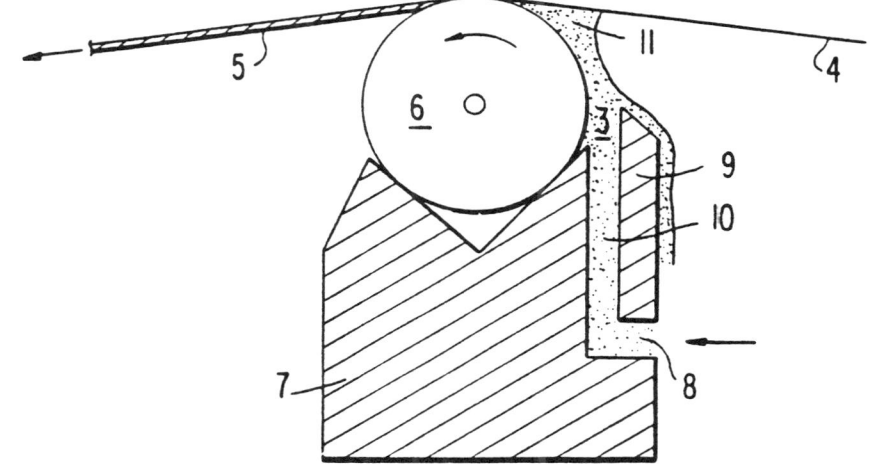

Fig. 3.48. Coating process

U.S. Patent No. 4,282,826 (Fig. 3.49)
Issued to G. Wohlfeil
Date issued: Aug. 11, 1981

In this patent the use of a plastic or flexible roll bed is described mounted on a member which in itself is flex-ible for loading. There is pneumatic loading of this movable plastic in which the roll doctor is mounted. Profilers are also described in the claims.

---

[54] **APPARATUS FOR REGULATION OF THE COATING THICKNESS IN THE COATING OF CONTINUOUS WEBS OF MATERIAL**

[75] Inventor: **Gerhard Wohlfeil**, Monheim, Fed. Rep. of Germany

[73] Assignee: **Jagenberg-Werke AG**, Dusseldorf, Fed. Rep. of Germany

[21] Appl. No.: **96,357**

[22] Filed: **Nov. 21, 1979**

[30] **Foreign Application Priority Data**

Nov. 24, 1978 [DE] Fed. Rep. of Germany 2851015

[51] **Int. Cl.³** B05C 11/02
[52] **U.S. Cl.** 118/118; 15/256.5
[58] **Field of Search** 118/118, 119, 110, 262, 118/123, 126; 15/256.52, 256.5; 427/361

[56] **References Cited**

U.S. PATENT DOCUMENTS

3,029,779   4/1962   Hornbostel .... 118/119 X

3,450,098   6/1969   Williams, Jr. 118/126
3,683,851   8/1972   Nolden 118/126
3,701,335   10/1972   Barnscheidt 118/119
3,817,208   6/1974   Barnscheidt et al. 118/119

FOREIGN PATENT DOCUMENTS

2307404   8/1974   Fed. Rep. of Germany 118/126

Primary Examiner—John P. McIntosh
Attorney, Agent, or Firm—Sprung, Felfe, Horn, Lynch & Kramer

[57] **ABSTRACT**

An apparatus for the regulation of the coating thickness in the coating of continuous webs of material has a doctor-roll which bears on the coated side of the web and which is supported in a shape-retaining, wear-resistant doctor-roll bed. The doctor-roll bed is elastically joined to a stationary frame of the apparatus in such a manner that it is removably mounted for replacement independent of the remainder of the elastic joint.

12 Claims, 5 Drawing Figures

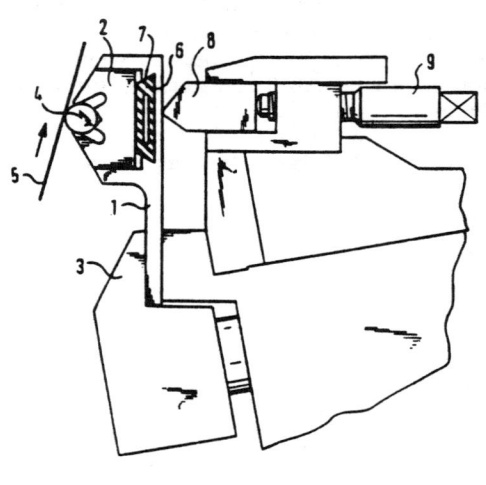

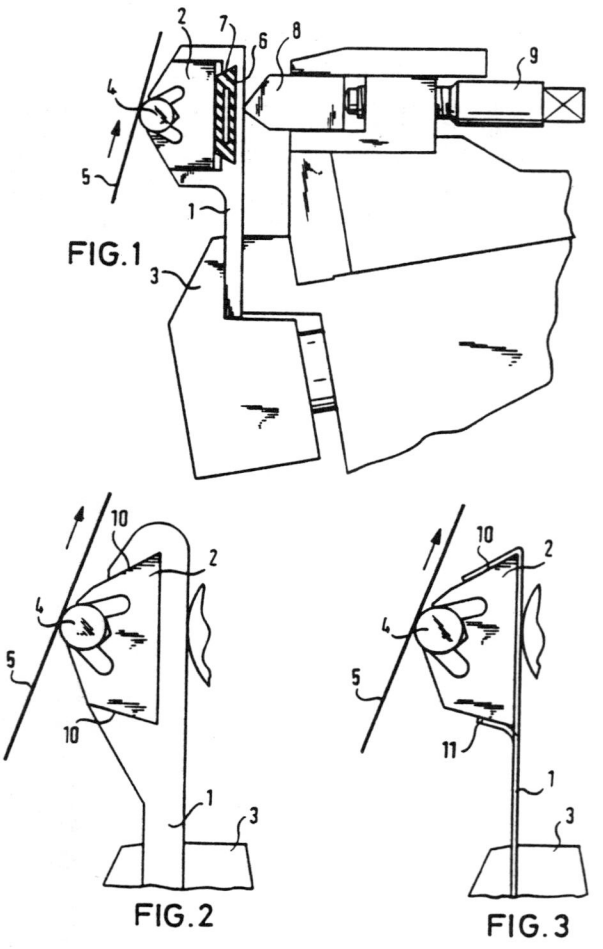

FIG.1

FIG.2   FIG.3

Fig. 3.49. Apparatus for regulation of the coating thickness in the coating of continuous webs of material

U.S. Patent No. 4,465,015 (Fig. 3.50)
Issued to F. Osta, et al.
Date issued: Aug. 14, 1984

This is an interesting patent in that it applies to extrusion coating, but shows a means of deckling either at the ends or to stripe coat having uncoated areas across the web. The patent is based on means of stripe coating using a rod smoothing system after which it can be coordinated in undercutting of the rod to conform with the coated and uncoated areas across the web.

---

[54] **DEVICE FOR SPREADING A SUBSTANCE ONTO A MOVING WEB OF MATERIAL**

[76] Inventors: Francesco Osta, Tenuta Avallano, 1-15039, Ozzano Monferrato; Giuseppe Cattana, Vorso Manacorda 45, 1-15033, Casale Monferrato, both of Italy

[21] Appl. No.: 477,734

[22] Filed: Mar. 21, 1983

[30] **Foreign Application Priority Data**

Apr. 5, 1982 [IT] Italy 67442 A/82

[51] Int. Cl.³ B05C 1/08; B05C 1/16
[52] U.S. Cl. 118/222; 118/249; 118/259; 118/414
[58] Field of Search .... 118/249, 262, 259, 248, 118/258, 244, DIG. 15, 221, 222, 224, 212, 410, 411, 412, 414, 216; 427/286; 425/376 R, 376 A, 376 B, 379 R, 466, 467

[56] **References Cited**

U.S. PATENT DOCUMENTS

2,875,728	3/1959	Kasak	118/410
3,818,861	6/1974	Turner	118/221
3,980,043	9/1976	Pomper	118/414
4,051,807	10/1977	Graf et al.	118/410
4,055,389	10/1977	Hayward	425/466
4,263,870	4/1981	Saito et al.	118/259
4,293,517	10/1981	Knox	425/467

Primary Examiner—Michael R. Lusignan
Assistant Examiner—Robert J. Steinberger, Jr.
Attorney, Agent, or Firm—Young & Thompson

[57] **ABSTRACT**

A device for spreading a substance onto a moving web of material, which comprises an extrusion head and a rotating bar which is carried by a support member applied to the extrusion head near the outlet of a delivery slit. This latter has an inner profiled recess in which engage limiting blade segments which extend up to the outlet of the delivery slit in order to laterally limit the regions in which spreading is carried out. The rotating bar has its outer surface operatively contacting the web only in the regions in which the substance is spread, and it is recessed or interrupted where no spreading is wanted. This device allows to spread a substance onto a width of web reduced with respect to the width of the extrusion head, or even onto separate strips of the web.

14 Claims, 8 Drawing Figures

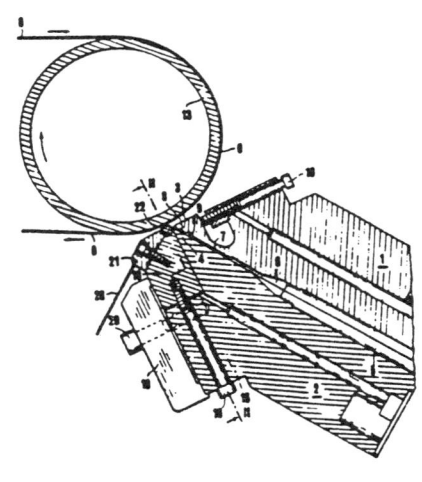

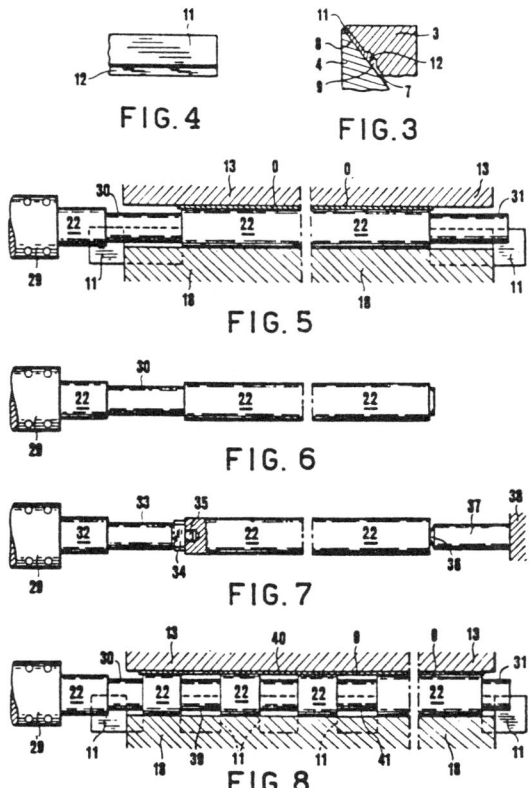

Fig. 3.50. Device for spreading a substance onto a moving web of material

U.S. Patent No. 4,521,459 (Fig. 3.51)
Issued to H. Takeda
Date issued: June 4, 1985

This patent is in effect an extension of Patent No. 4,263,870, both of which were assigned to the Fuji Photo Film Co., Ltd. The coating is fed in immediately ahead of the rotating rod which is trapped in a bed and is driven rotatably in either direction. Following the smoothing using the wire wound rod, the web passes over a stationary knife device with a convex surface designed to smooth the coating as it exits the active application and smoothing section. Recirculation of the coating is required as indicated on the patent.

[54] **COATING METHOD AND APPARATUS**

[75] Inventor: **Hideo Takeda**, Kanagawa, Japan

[73] Assignee: **Fuji Photo Film Co., Ltd.,** Kanagawa, Japan

[21] Appl. No.: **495,929**

[22] Filed: **May 19, 1983**

[30] **Foreign Application Priority Data**

May 19, 1982 [JP] Japan 57-83236

[51] Int. Cl.³ B05D 3/12
[52] U.S. Cl. 427/359; 118/118; 118/119, 118/206; 118/410; 118/414; 427/371; 427/434.3; 427/434.4
[58] Field of Search 427/359, 434.3, 371, 427/434.4, 118/206, 414, 118, 119, 410

[56] **References Cited**
U.S. PATENT DOCUMENTS

2,285,041 6/1942 Mayer et al. 118/118

FOREIGN PATENT DOCUMENTS

1964908 8/1970 Fed. Rep. of Germany 118/414
2046138 11/1980 United Kingdom 118/414

Primary Examiner—Norman Morgenstern
Assistant Examiner—Janyce A. Bell
Attorney, Agent, or Firm—Sughrue, Mion, Zinn, Macpeak & Seas

[57] **ABSTRACT**

In a coating method, a coating section is arranged immediately before a coil bar while a smoother is disposed immediately after the coil bar. Immediately after a coating solution is applied to a web which is run continuously a surplus of coating solution is scraped off and the web is subjected to smoothing by the smoother directly without undergoing a gaseous phase.

10 Claims, 3 Drawing Figures

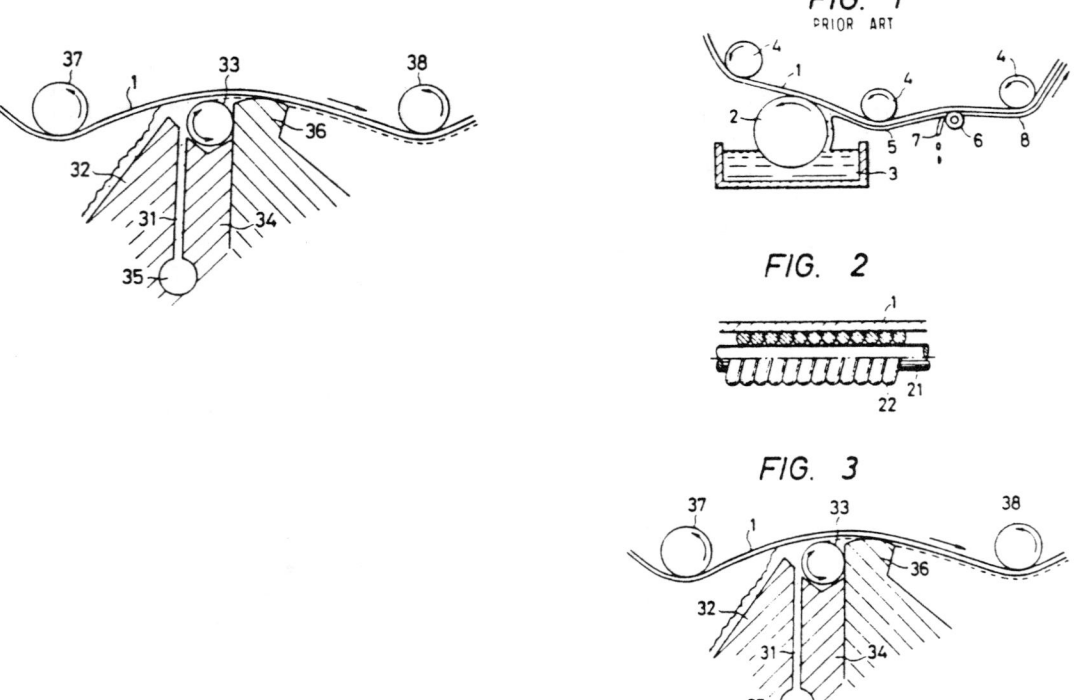

Fig. 3.51. Coating method and apparatus

U.S. Patent No. 4,651,672 (Fig. 3.52)
Issued to H. Sommer
Date issued: March 24, 1987

This device is the Hydrabar System developed in Germany. It is a very similar system to that which was developed and patented by Penkala, Patent No. 3,453,137, described earlier. In this case, the shoe is pneumatically loaded directly against the web rather than flexibly against the web on the tip of a flexible blade or other device.

---

[54] **DEVICE FOR COATING CONTINUOUS WEBS**

[75] Inventor: Herbert Sommer, Dusseldorf, Fed. Rep. of Germany

[73] Assignee: Jagenberg AG, Dusseldorf, Fed. Rep. of Germany

[21] Appl. No.: 554,490

[22] Filed: Nov. 23, 1983

[30] **Foreign Application Priority Data**

Nov. 23, 1982 [DE] Fed. Rep. of Germany 3243317
Oct. 21, 1983 [DE] Fed. Rep. of Germany 3338323

[51] Int. Cl.⁴ B05C 11/04
[52] U.S. Cl. 118/126; 118/261
[58] Field of Search 118/126, 119, 122, 123, 118/261

[56] **References Cited**

U.S. PATENT DOCUMENTS

343,345	6/1986	Sparks	118/122 X
1,312,034	8/1919	Jones	118/126
3,029,779	4/1962	Hornbostel	118/119 X
3,450,098	6/1969	Williams, Jr.	118/126
3,882,817	5/1975	Zink	118/126
4,259,379	3/1981	Britton	118/126 X
4,458,376	7/1984	Sitko	118/126 X

Primary Examiner—John P. McIntosh
Attorney, Agent, or Firm—Sprung, Horn, Kramer & Woods

[57] **ABSTRACT**

A device for coating continuous webs with regulable coating strength by means of a doctor rod that rests against the coated side of the web, that is mounted in a doctor bed, and that can be elastically forced with regulable pressure against the web being coated. The doctor rod is a doctor batten that moves with little friction along at least one overflow-side slide face of a rigid doctor bed. The face of the doctor batten toward the web that is being coated is rectangular and demarcates in conjunction with the web a narrowing coating-material application gap with a geometry that depends on the degree of contact pressure. The upper edge of the face constitutes a straight and sharp stripping edge. The face and stripping edge of the doctor batten are flexible. Its face is highly resistant to wear. The contact pressure is exerted by an inflatable means of exerting pressure positioned between the batten and the bed.

22 Claims, 7 Drawing Figures

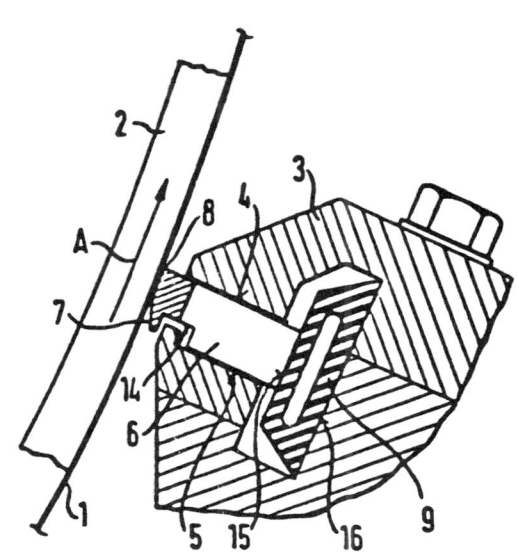

Fig. 3.52. Device for coating continuous webs

U.S. Patent No. 4,658,753 (Fig. 3.53)
Issued to D. Eklund
Date issued: April 21, 1987

This patent is a mechanical design patent showing a way of mounting a plastic bed containing a rotatable rod on a flexible device for pneumatically loading against the web supported by a resilient backing roll.

---

[54] **ROLL BLADE ASSEMBLY FOR A PAPER COATING MACHINE**

[75] Inventor: Dan Eklund, Grankulla, Finland

[73] Assignee: Oy Wärtsilä AB, Värtsilä, Finland

[21] Appl. No.: 749,099

[22] Filed: Jun. 26, 1985

[30] **Foreign Application Priority Data**

Jun. 29, 1984 [FI] Finland 842625

[51] Int. Cl.⁴ B05C 11/04
[52] U.S. Cl. 118/119; 118/118; 15/256.52
[58] Field of Search 118/123, 126, 118, 119, 118/110; 15/256.51, 256.52

[56] **References Cited**

U.S. PATENT DOCUMENTS

3,817,208	6/1974	Barnscheidt et al.	....	118/119
4,367,120	1/1983	Hendrikz	....	15/256.51 X
4,375,202	3/1983	Miller	....	118/119 X

Primary Examiner—John P. McIntosh

Attorney, Agent, or Firm—Birch, Stewart, Kolasch & Birch

[57] **ABSTRACT**

A roll blade assembly for a paper coating machine, comprising a frame (5), a support mounted on the frame (5), a blade member (8) mounted on the support (3), a set of mounting accessories (1, 6, 7) for attaching the inner longitudinal end of the blade member (8) to the support (3), a bar unit (9) provided with a rotating metering bar (10) and, attached to the blade member (8), and a pneumatically inflatable loading bag (2) arranged to push the bar unit (9) outwards from the support (3) when the inflation pressure is raised. The blade member (8) is flexible, and the support (3), the bar unit (9), and the blade member (8) form a recess (4) compliant with the loading bag profile and extending in the lateral direction of the blade member (8) over at least 50 percent of the unobstructed blade width. The large inflated bag contact surface makes a more accurate control of the metering bar pressing force possible by directly contacting the blade member.

10 Claims, 2 Drawing Figures

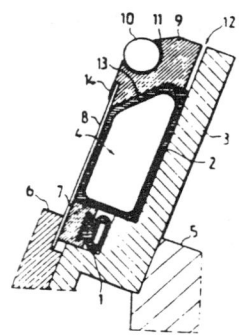

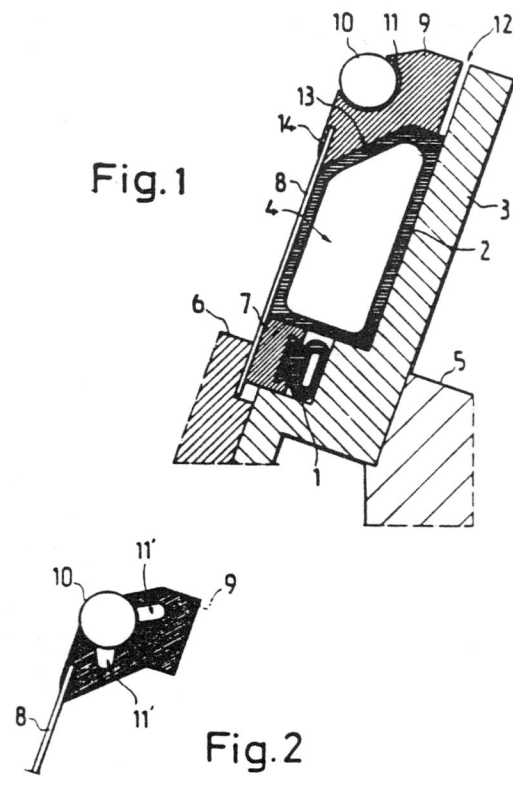

Fig.1

Fig.2

Fig. 3.53. Roll blade assembly for a paper coating machine

U.S. Patent No. 4,757,782 (Fig. 3.54)
Issued to M. Pullinen
Date issued: July 19, 1988

This is a machine or device patent showing a means of pneumatically loading a shoe-like device, a rod, or a blade against the web in a short-dwell coater configuration. It also shows a means of coating feed using a serrated or fragmented doctor blade.

---

[54] APPARATUS FOR COATING A WEB

[75] Inventor: Martti Pullinen, Jyväskylä, Finland

[73] Assignee: Valmet Oy, Finland

[21] Appl. No.: 892,838

[22] Filed: Aug. 4, 1986

[30] Foreign Application Priority Data

Aug. 7, 1985 [FI] Finland 853041

[51] Int. Cl.⁴ B05C 5/02
[52] U.S. Cl. 118/411; 118/410; 118/414
[58] Field of Search 118/414, 410, 413, 411, 118/419; 427/356

[56] References Cited

U.S. PATENT DOCUMENTS

2,729,192	1/1956	Warner	118/414 X
3,855,927	12/1974	Simeth	118/261 X
4,063,531	12/1977	Zitzow	118/411 X
4,245,582	1/1981	Alheid et al.	118/119
4,405,661	9/1983	Alheid	118/413 X
4,465,015	8/1984	Osta et al.	118/414 X
4,558,658	12/1985	Sommer et al.	118/413 X

FOREIGN PATENT DOCUMENTS

1964908 8/1970 Fed. Rep. of Germany 118/414

Primary Examiner—John P. McIntosh
Attorney, Agent, or Firm—Steinberg & Raskin

[57] ABSTRACT

Apparatus for coating a web, such as paper or board web, comprises a coating nip formed between a rotating counter roll and a smoothing member, or between two smoothing members, through which the web to be coated is passed with a coating substance being applied to the web before it passes through the nip. The apparatus comprises an arrangement by which the linear load and/or the distribution of the linear load in the coating nip is adjustable to control the quantity and/or the distribution of the quantity of the coating substance applied to the web in the transverse direction thereof. A smoothing member comprises an elongate member extending transversely to the direction of web movement through the apparatus and is flexible with respect to its longitudinal dimension so that the profile of the load distribution in the nip can be adjusted to be relatively steep, i.e. the load in the nip may differ relatively widely at points which are relatively close to each other. An arrangement is provided for selectively variably adjusting the load applied to the smoothing member along its longitudinal dimension.

8 Claims, 12 Drawing Sheets

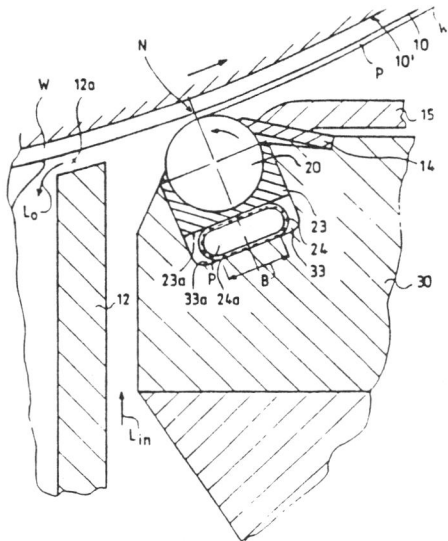

FIG.1

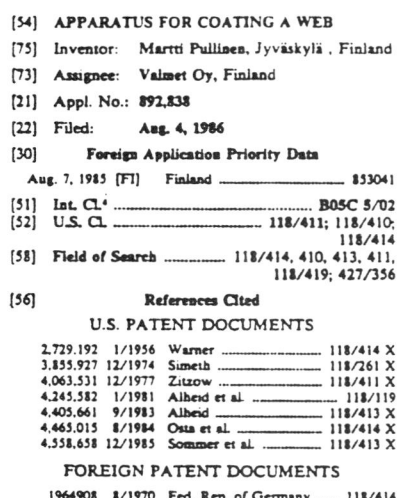

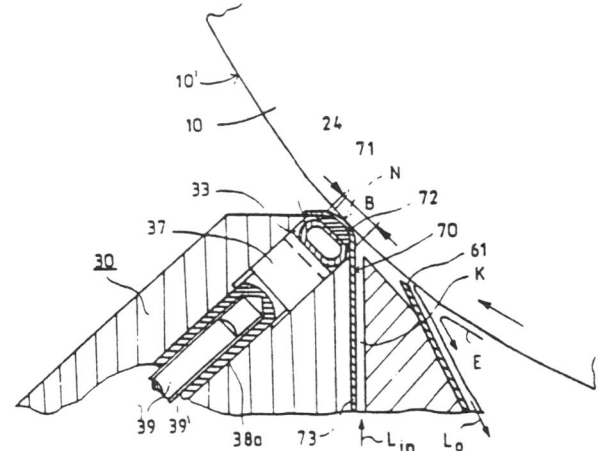

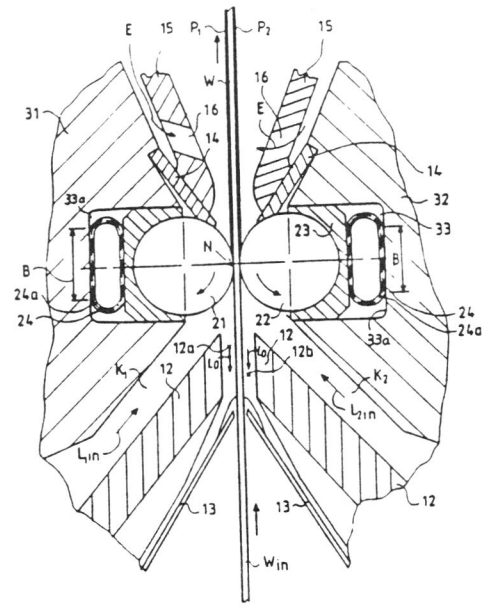

FIG 2

Fig. 3.54. Apparatus for coating a web

Bibliography

Literature Cited

1. Booth, G.L., *Pigmented Coating Process for Paper and Board,* TAPPI Monograph 28, Mack Printing Co., Easton, Pennsylvania, 1964.

2. Booth, G.L., "Coating Systems On Board Machines," *Paper Mag. Can.,* May 1960.

3. Bushchman, T., "Metering Bar Tests Improve Quality and Reduce Scrap," *Paper, Film and Foil Converter,* p. 54, March 1986.

4. Kohler Coating Machinery Corp., Technical Bulletin For Iso-Bar Coaters, 1990.

5. Kohler, H.B. and Lathrop, P.H., The Kohler Coating Machinery Corp., The Iso-Bar Rod Coater, unpublished 1990.

6. MacLeod, D. and Fahrendorf, P.M., "Rod Coating Comes of Age," 1988 Polymers, Laminations and Coatings Proceedings, TAPPI PRESS, Atlanta.

Patents

Bradner, D.B.,U.S. Pat. No. 2,229,620 (Jan. 21, 1941).

Bradner, D.B.,U.S. Pat. No. 2,229,621 (Jan. 21, 1941).

Taylor, W.P., U.S. Pat. No. 2,534,320 (Dec. 19, 1950).

Warner, E., U.S. Pat. No. 2,598,733 (June 3, 1952).

Smith, R.V., U.S. Pat. No. 2,774,329 (Jan. 25, 1954).

Warner, E., U.S. Pat. No. 2,729,192 (Jan. 3, 1956).

Warner, E., U.S. Pat. No. 3,094,663 (April 9, 1963).

Warner, E., Canadian Pat. No. 724,578 (Dec. 28, 1965).

Warner, E., U.S. Pat. No. 3,304,910 (Feb. 21, 1967).

Penkala, J.E., et al., U.S. Pat. No. 3,453,137 (July 1, 1969).

Saito, H., et al., U.S. Pat. No. 4,263,870 (April 28, 1981).

Wohlfeil, G., U.S. Pat. No. 4,282,826 (Aug. 11, 1981).

Osta, F., U.S. Pat. No. 4,465,015 (Aug. 14, 1984).

Takeda, H., U.S. Pat. No. 4,521,459 (June 4, 1985).

Sommer, H., U.S. Pat. No. 4,651,672 (March 24, 1987).

Eklund, D., U.S. Pat. No. 4,658,753 (April 21, 1987).

Pullinen, M., U.S. Pat. No. 4,757,782 (July 19, 1988).

Chapter 3, Section III

Air Knife Coaters

Herbert B. Kohler, Kohler Coating Machinery Corp.

Introduction

Air knives are metering devices used to reduce an excess of previously applied coating material on a moving web to a uniform thickness.

There are two types of air knife operating methods. What we call "air brushing" is a gentle jet of low velocity, impinging on a vertical running coated substrate at approximately a right angle, providing a pressure dam which limits the amount of coating liquid passing by. Excess liquid runs back down the strip, under the influence of gravity, to a pond or a drip off point. The jet does not remove any liquid from the substrate. For coating paper, this method has largely fallen into disuse and will not be discussed further.

We call "air knifing" the method where a jet of moderate- to high-velocity air impinges on a coated substrate supported by a roll, at an angle on the order of 45° opposed to substrate movement. The jet shears the liquid film, and removes the excess as liquid, spray, or mist. This excess is then collected in a blowoff containment system.

The theory in removal of excess coating by the air doctor is called the filter cake theory. A coating is applied in excess to the sheet at the applicator section. Water in the coating immediately begins to migrate at the interface of the wet coating and the paper web so that the coating at this point immediately becomes semi-dry or plastic. As the sheet of paper passes under the air-knife jet, the fluid coating is removed from the sheet by the air doctor and is sheared at the point where the filter cake begins. There is a zone in the coating cross section where the coating makes the transition from the fluid to a semi-plastic coating, and it is in this area that the air doctor shearing takes place.

The exact point at which the shear takes place varies with the amount of energy that the air blast has. At a very high pressure or velocity of air from the air doctor, the air penetrates more into the filter cake or plastic area, leaving less coating on the sheet. It should be noted that the normal air pressure used in the air doctor is from two to nine pounds; however, when a plastic non-absorbent web is coated by the air doctor system, pressure in the range of two to ten ounces is used in the air doctor. This would substantiate the filter cake theory in that there is no filter cake formed in this case, and the air doctor is only shearing to a depth in the coating for which there is energy to penetrate it.

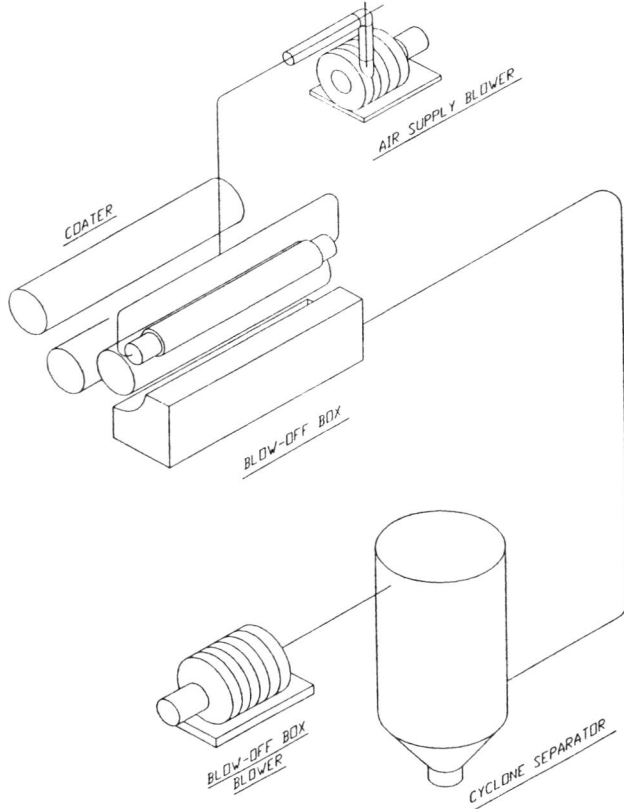

Fig. 3.55. Air-knife coating system

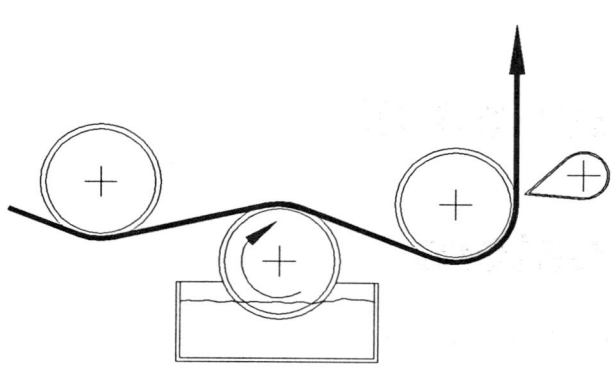

Single roll

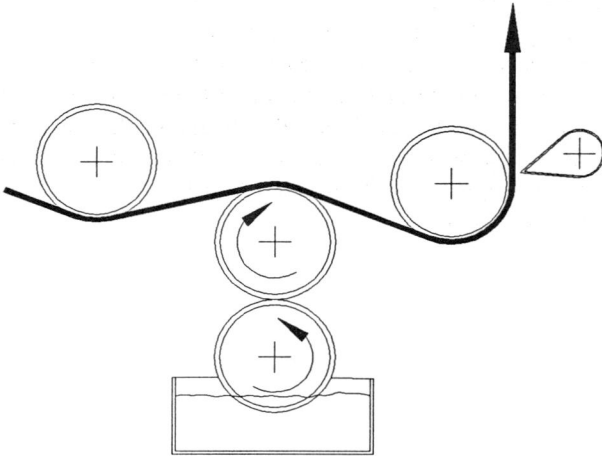

Two roll

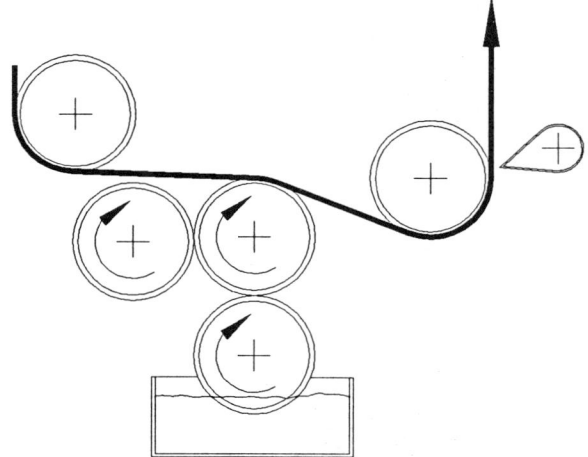

Three roll

Fig. 3.56. Air-knife applicators

Referring to Figure 3.55, there are four main components in an air-knife coating system: (1) the coating application system, (2) the air knife metering element, (3) the compressed air delivery system, and (4) the coating exhaust and recovery system.

Coating Application System

Since air knifing was invented in the 1930s, many different applicator systems have come and gone. Most methods used rolls, although a number of die or fountain applicators have been tried.

Today three systems predominate. These are shown in Fig. 3.56. Roughly 70% of the applicator systems used today are single-roll systems, 25% are two-roll systems, and the remaining systems are three-roll or other systems.

Single Roll

The single-roll system is exactly that. It is a single roll normally rotating in web direction at a speed of 10-40% of the web speed, which picks up a crudely controlled amount of coating and applies it in excess to the web that wraps the roll. Usually this wrap will be from 3 in. to 4 in. and, if well designed, it will maintain a puddle of coating at the nip point formed by the web as it touches the roll. The runback of coating should be laminar in nature and not a cascading effect in order to get a good smooth application of coating to the web and to prevent foam generation caused by turbulent flow and cascading of the coating off the roll. This system is the simplest available and also the crudest one in operation.

It is used by the vast majority of air-knife coaters in operation and is used almost exclusively at speeds below 250 meters/min.

Two Roll

The two-roll system provides an additional metering nip to reduce and level the amount of coating applied to the web. It is the application unit of choice for moderately high-speed, wide machines. These would be machines over 200 meters/min in speed or widths greater than 3.8 meters.

Three Roll

The three-roll system was developed for high-speed air knifing (over 600 meters/min) to further reduce and even the coating applied to the web. The air knife is generally limited to pressures below 9.0 psi (62.1 kPa), which limits the amount of coating which can be metered off at high speeds. The three-roll system extends the operating range of the air knife, but its tight tolerances and complexity have limited it to narrow width coaters (less than 3.8 meters wide).

Smoothing Roll

On many high-speed systems, it is a common practice to use a smoothing roll on the coated side of the web.

This roll is run reverse to web direction and is usually designed with as small a diameter as is structurally possible. Wrap angle is about 5°.

The primary purpose of this roll is to assist in the removal of the film split pattern on the web; however, the breaking of the foam generated from this process is also of significant importance. The roll also serves to reduce the effect of dwell time from the point of application to the point of action of the air-knife metering device. Up to 90% of the coating is redistributed, preventing rapid dewatering and preventing the formation of a filter cake. The danger of long dwell distances is water loss into the web to the degree that the air knife becomes a poor metering device, due to its viscosity sensitivity and a need to run at higher air pressures. The redistribution effect of the smoothing roll before the air jet keeps the coating in suspension longer, allowing better coat weight control and reducing the effect of the long dwell distance.

Air-knife Metering Element

There have been numerous designs of knives over the past 60 years. The basis for most designs have their roots in either the Terry (S.D. Warren) or Pomper (John Waldron Corp.) patents. The Terry patent was predicated on asymmetrical lips, whereas the Pomper patent used symmetrical lips or a classical nozzle design.

The elements of adjustment and control of the air knife include the slot opening of the lips and the relationship of the top and bottom blade tips, i.e., the amount of overhang of the top lip relative to the bottom lip.

Modern air knifes have very good adjustments to maintain the slot with one lip always fixed in position. Most knives use a pushpull adjustment to "wrap" the adjustable lip and maintain a uniform slot regardless of air-knife length. Operation slot openings range from 0.020 in. to 0.035 in.

The overhang, or relative position of the top lip to the bottom lip is important in asymmetrical lip designs. The top lip usually will extend 0.005 in. to 0.008 in. beyond the bottom lip. This gives some stability to the fluid stream and directs the stream downward from the bisector of the angle formed by the lips.

All air knives fall into one of the three categories shown in Fig. 3.57.

Non-opening air knives were developed first. Opening style and dual air knives were developed later to reduce cleaning time on paper machines that could not stop the paper when making scrap.

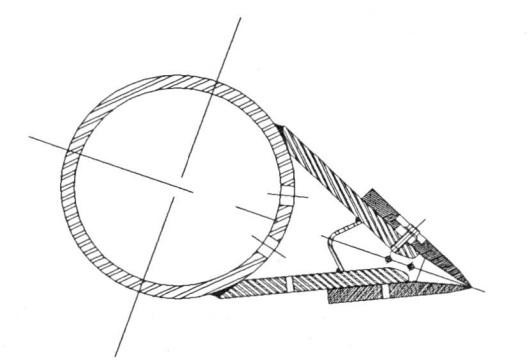

Non-opening airknife

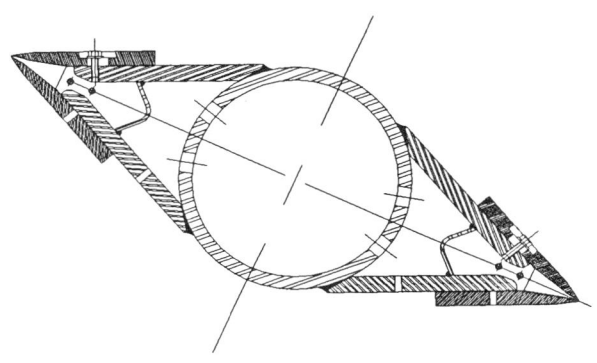

Dual rotating airknife

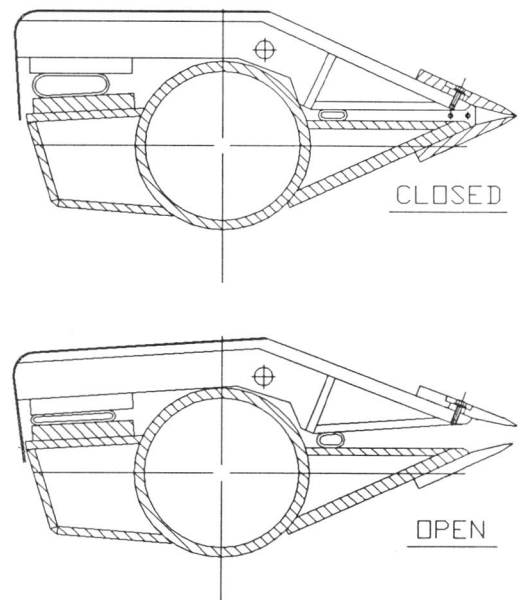

Opening airknife

Fig. 3.57. Air-knife metering elements

Air Delivery System

The air delivery system of a coater consists of four basic elements: (1) the blower, (2) the aftercooler (if used), (3) the piping, and (4) a means of adjusting pressure and flow.

Blowers

The blower for an air knife should always be a centrifugal type. These units deliver clean air, free of water and oil and completely free of pulsation.

A blower should be sized according to the cubic feet per minute (CFM) required at the largest gap setting and highest pressure at which you expect to operate. Allowance should be made for pressure drop through the piping system and the aftercooler (if used).

Air volume requirements vary with air-knife length, air-knife gap, and pressure requirements. The following table shows the requirements per inch of opening at various air-knife gaps for a range of pressures, as well as the formulas from which they were calculated.

Blower Requirements

Cubic feet per minute (CFM) of free air at standard pressure of 14.7 psia and 70°F per inch at 0.040-in., 0.035-in., and 0.030-in. nozzle openings.

Table 3.2. Blower requirements

PSIG	0.040" OPENING	CFM/Inch 0.035" OPENING	0.030" OPENING
1	5.9	5.1	4.4
2	8.3	7.3	6.2
3	10.1	8.9	7.6
4	11.6	0.2	8.8
5	12.9	1.5	9.8
6	14.2	2.6	10.8
7	15.3	3.6	11.6
8	16.6	4.5	12.4
9	17.3	5.4	13.2
10	18.6	6.2	13.9

Note:

CFM/in. (0.040-in. opening) $F = 5.86 \times \text{SQRT} (P)$

CFM/in. (0.035-in. opening) $F = 5.133 \times \text{SQRT} (P)$

CFM/in. (0.030-in. opening) $F = 4.40 \times \text{SQRT} (P)$

Example: You have a 163-in. wide coater and you wish to increase the blower size to allow you to operate in a range from 3 psi to 5 psi at a gap of 0.035 in. The system has an aftercooler with a maximum pressure drop of 0.5 psi, and the loss through the piping system is 0.5 psi or less. From Table 3.2, we read the value corresponding to 0.035-in. gap and 5 psi to be 11.5 CFM/in. of length.

Therefore, at 0.035-in. gap and 5 psi, the air knife requires 163 (11.5) = 1875 SCFM. Note: 1 SCFM has the mass of 1 ft^3 of air compressed from inlet conditions of 68°F and 14.7 psia pressure. Any variation from these conditions requires that modifications be made to the blower calculations.

Since the system has an inherent pressure drop of 1 psi, we require a machine capable of delivering 1875 SCFM at 6 psi. This requires a 75-hp blower.

The Aftercooler

The requirement for an aftercooler depends largely on the amount of pressure boost provided by the blower. Air temperature increases roughly 16°F for each pound of pressure increase. Generally, any air-knife system operating over 3.5 psi (24 kPa) could benefit from an aftercooler.

Two methods of cooling the blower air are generally used, water-cooled heat exchangers and water-injection systems. Water-cooled heat exchanger systems are used in most installations today. The air is cooled without contacting the cooling water. A moisture separator is usually used at the exit of the cooler to remove any condensed moisture.

Water-injection systems cool by injecting a small amount of water mist into the airstream. The air is cooled by the massive heat absorbed as the water evaporates.

The drawback to this method is that any impurities in the water remain in the airstream. If the injection method is to be used, the preferred water source is chilled steam condensate. A moisture separator removes any unevaporated water.

The main benefit to this system is that it dramatically reduces water evaporation by the air jet and reduces drying in the blowoff pan. Since the air is at almost 100% relative humidity, it cannot absorb any more water. The air coming out of the heat exchanger has low relative humidity and as a result can absorb more water.

A properly designed injection system is less expensive initially, cheaper to operate, and provides more benefits than the heat exchanger system.

Piping

The general rule with piping systems is to keep them as short as possible, with no sharp bends. To keep pressure loss to a minimum, piping from the blower should be at least 20% larger than the inside diameter of the blower outlet, until it splits off to each end of the air knife. The tee should be the same size as the main supply line, and made of hard pipe. Flexible hose at each end of the air knife should be sized for flow equivalency.

Pressure and Flow Adjustments

Blowers can be throttled to adjust the discharge pressure or inlet volume, or both, to any selected point within the operating capability of the blower. This is usually accomplished by installing and adjusting a blast gate or butterfly valve on the inlet or discharge opening of the blower.

When the blower is throttled at the discharge to a selected volume under standard performance curve conditions, the blower will deliver the full discharge pressure for the inlet volume as shown on the standard performance curve. However, the throttling device will supply sufficient resistance to air flow to provide the desired effective overall discharge pressure beyond the throttling device as required by the process. When the discharge of a blower is throttled, any effective discharge pressure and inlet volume above the surge range can be attained, within the capability of the blower. However, the full input horsepower for the corresponding inlet volume shown on the standard performance curve is required.

When the blower is throttled at the intake to a selected SCFM and discharge pressure, the throttling device serves to create sufficient resistance to air flow to drop the absolute pressure at the blower inlet, so that the desired discharge pressure above ambient is obtained.

The air volume (SCFM) entering and passing through the throttling device, however, will increase in volume because of the drop in pressure across the throttling device, so that the inlet volume to the blower (ICFM) exceeds the SCFM. This condition lowers the input horsepower requirements. Therefore, it is recommended that the throttling device be placed on the inlet to the blower.

The third and most energy-efficient method of pressure and flow variation is by varying the speed of the blower motor.

The volume entering a blower varies directly with the speed in revolutions per minute (rpm), but the input horsepower varies with the cube of the speed in rpm. Fairly large energy savings can be made by this method.

Another advantage of variable-speed control is that it lowers the output temperature of the blower in comparison to the other two methods.

The chart shown in Fig. 3.58 shows the fan volume output plotted versus input power for inlet throttling, outlet throttling, and two types of variable-speed controls.

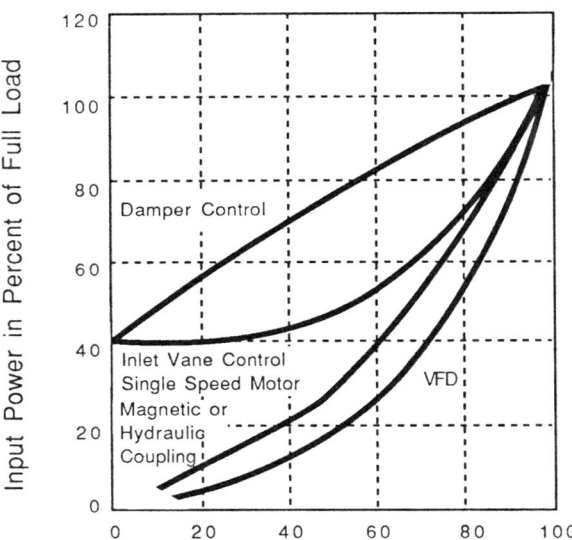

Fan Volume Output in Percent of Full Load

Fig. 3.58. Comparison of fan controls

Coating Exhaust and Recovery Systems

As with air knives, there have been many different designs of exhaust and recovery systems. Modern systems all share the following characteristics:

1. Double-wall stainless steel construction for water circulation. Chilled water circulates between the walls to chill any surfaces that contact air or coating. The moisture that condenses on these walls prevents coating buildup.

2. A vacuum exhauster system that removes 3-5 times the amount of air produced by the air knife, from the pan

3. Internal surfaces are smooth. Coating is separated out of the air stream gradually, without sharp velocity transitions that cause coating buildup. Older designs used many internal baffles that removed coating by changing the air direction enough times to expend its velocity.

The major differences between modern systems concern the basic separation philosophies. Some systems use very large pans that expand the air within the pan until its velocity will only allow it to carry very minute coating particles.

Other designs use smaller pans that remove most of the liquid in the pan and separate the remaining coating in a water-cooled cyclone separator. Another system integrates the cyclone principle within the pan itself.

Almost all modern systems pass the exhausted air through a light water mist to remove submicron particles.

In many cases, this water is recycled as coating makeup water.

Air-knife Setup

One of the least understood areas of coater operation is setting the air-knife geometry.

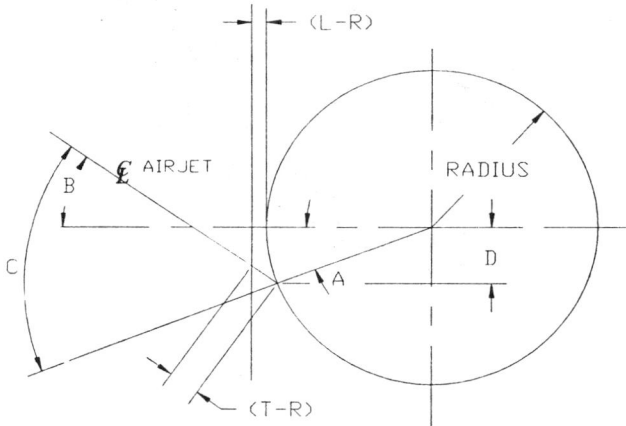

Fig. 3.59. Air-knife geometry

Referring to Fig. 3.59, the following information is required:

R = the radius of the backing roll

D = the distance of the impingement point below backing roll center line

A = the angle below the horizontal center line formed by a radial line and a horizontal line passing through the impingement point

B = the angle above horizontal of the air-jet center line (This is not the metering angle!)

C = the angle between the center line of the air jet and a radial line passing through the impingement point (This is the metering angle.)

LR = the horizontal distance between the backing roll and the top lip

TR = the distance along the air-jet center line from the tip of the lips to the roll. On most air knives this is almost impossible to measure. TR, the metering angle C, and air pressure, control coat weight, and appearance.

Step 1. Measure backing roll diameter, divide by 2 for roll radius R.

Step 2. Check distance D on both sides of machine. Make sure both sides are at the blowoff pan manufacturer's recommended height and are equal. To establish impingement point on roll, check with air turned on.

Step 3. Calculate A, the angle of the impingement point below the center line, by taking the inverse sin of distance D divided by the roll radius R.

Step 4. Find the center line of the air jet by taping a piece of string to the top lip of the air knife. Measure the angle from horizontal up to the string with the air on. Note: Air knives with asymmetric lips have an air jet that issues 5-7 below their mechanical center line. Measure and record angle B so that you know its relationship to the angle measurement device on the air knife. This angle measurement device usually reads out the angle between the mechanical center line of the air knife and horizontal. Apply your correction factor to get angle B for any setting.

Step 5. Most air knives adjust lip to roll distance along the horizontal axis. The problem with this design is that if you move the air knife in or out, it decreases or increases, respectively, the air knife metering angle C by allowing the impingement point to move up and down the roll. Whenever you make a change of one parameter (angle or distance) with this design, you must correct the other to avoid conflicting results.

Some air knives move in and out along the air-jet center line which allows either angle or distance to be changed without affecting the other.

Most air knives operate with Angle C in a range of 52° ±3°. Calculate angle B from angle C minus angle A. Set even on both sides.

Step 6. Set lip to roll distance LR between 3 mm and 4.5 mm (0.120-0.180 in.) as measured by dropping feeler gauges vertically between the lip and the roll. Remember to add the amount of your thickest paper to this distance if you coat board. Set equally on both sides.

Fine Tuning

Step 7. Of the two adjustments, angle and distance, you have the most latitude with distance. In general, operate as far back as possible, using about 90% of your available pressure at top machine speed. This will keep the lips as clean as possible for the longest time and still leave you some adjustment for process changes.

Step 8. The angle adjustment has the most effect on surface quality. In general, operating at too high an angle can cause vertical lines in machine direction. Operating at too low an angle can cause cross-machine lines that are ½-1 in. long. Both conditions gradually fade away as you approach the optimum angle.

Between these two extremes are a number of conditions that can appear or disappear with a frustrating lack of consistency.

Mottle, orange peel, over-spray, and other conditions are all affected to some extent by angle, distance, coating solids, coating viscosity, sheet porosity, operating speed, sheet temperature, applicator roll speed, binder system, and sheet tension.

When making a change in angle or distance, be sure other adjustments are compensated correctly to avoid masking the effect of the change.

The best troubleshooting procedure is to have a regular monitoring program in place to record the above variables for all grades during normal operation. Before adjusting the air knife, check to see if any other parameter is out of specification.

Contour Coating

The air knife is known as a contour coater. By that, we mean that it applies a relatively uniform coating thickness regardless of surface roughness. This has made it especially useful for board grades that require even coverage rather than improved smoothness. It has also proven quite useful in applications of coatings that cannot accept mechanical contact or are themselves abrasive or corrosive.

The air knife has had a longer operating history than most coating methods. Its major market is now under attack. New rod coaters are coming on-stream that are being used on the same grades that air knives have traditionally coated.

These new coaters use an air-loaded rod against a backing roll. The contour coating is provided by the use of threaded rods that meter low-viscosity, high-solids coatings volumetrically. These coatings flow through the grooves of the rod, and then over the surface of the sheet. They form a slightly smoother contour coating than the air knife.

The major drawback with these systems has been the wear of the threaded rods. Titanium dioxide coatings (the major market for air knives) are extremely abrasive. Recent advances in abrasion resistance coating application systems have provided the solution to this problem. Today, threaded rod systems can replace almost any air knife with benefits ranging from higher solids to better uniformity and less off-quality paper.

Summary

The air knife is and will remain a valuable coating tool for many years to come. Recent inroads to its core markets have dramatically reduced the number of new installations that are made with air knives.

Bibliography

Resources

1. Booth, G., *Coating Equipment and Processes,* Lockwood Publishing Co., New York, 1970.

2. Booth, G., "Applicators for Air Knife Coating," *1985 Air Coating Seminar Report,* TAPPI PRESS, Atlanta.

3. Barton, A.K., "Air Knife Coating of Natural Kraft," *1985 Air Knife Coating Proceedings,* TAPPI PRESS, Atlanta.

Chapter 3, Section IV

Transfer Roll Coaters, A History

George L. Booth, The Black Clawson Co. (retired)

Introduction

Transfer roll coaters were first used in the late 1930s to coat paper as a reaction to market pressure to develop a low-cost, two-side clay coated paper for use in magazines with multi-color pictures. The natural instinct for printers, who required this paper, was to apply the principles they knew best, namely the distribution of ink on rollers for inking offset and letter press plates.

It was with this background that Peter Massey worked in conjunction with Consolidated Water Power and Paper Co. to develop the Consolidated Process or transfer roll coater. Shortly thereafter, Harry Faeber, another printer, worked with the St. Regis Paper Co. to develop the Faeber coater and later the St. Regis-Faeber process. Both of these systems revolutionized paper coating for low-cost magazine papers. We know them today as No. 4 and No. 5 grade coated papers.

Transfer Roll Coaters

The Consolidated Process

The original Consolidated process had a series of rollers in the top deck, as shown in Fig. 3.60, are copied from the distributor rolls used in inking systems on printing presses. The rolls are labeled with their type and function. As originally conceived, the rolls came in a variety of coverings such as rubber with differing hardness, chrome plated rolls, some rolls oscillating back and forth, and some rolls running at a preset speed relative to the other rolls. Initially this coater was designed to coat only one side with a second coater following intermediate drying to coat the second side. Very soon this coater took on the configuration as shown in Fig. 3.61 where the two trains were stacked one on top of the other in order to coat both sides simultaneously.

Drying was done with air tubes just after the coating application and then followed by steam cylinders. These steam cylinders are under very close temperature control, for both top and bottom decks, in order to dry the coated paper.

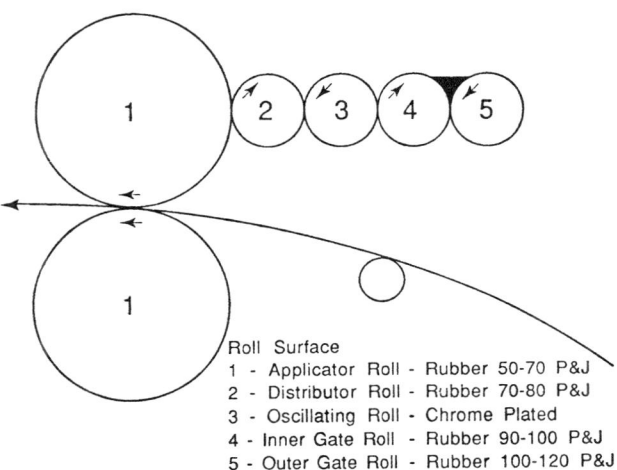

Roll Surface
1 - Applicator Roll - Rubber 50-70 P&J
2 - Distributor Roll - Rubber 70-80 P&J
3 - Oscillating Roll - Chrome Plated
4 - Inner Gate Roll - Rubber 90-100 P&J
5 - Outer Gate Roll - Rubber 100-120 P&J

Fig. 3.60. Massey coater

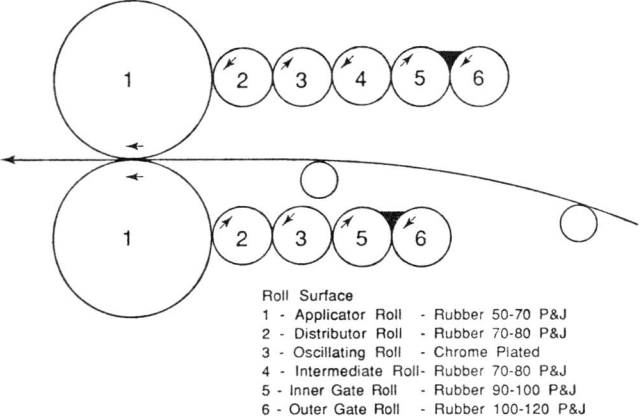

Roll Surface
1 - Applicator Roll - Rubber 50-70 P&J
2 - Distributor Roll - Rubber 70-80 P&J
3 - Oscillating Roll - Chrome Plated
4 - Intermediate Roll - Rubber 70-80 P&J
5 - Inner Gate Roll - Rubber 90-100 P&J
6 - Outer Gate Roll - Rubber 100-120 P&J

Fig. 3.61. Consolidated (Massey) coater, two sides coated

Modifications were made of this coater as shown in Figs. 3.62 and 3.63. In these later modifications, care was given for better web handling in threading and broke handling. A vertical web lead helped the drying configuration, although web handling and broke handling suffered.

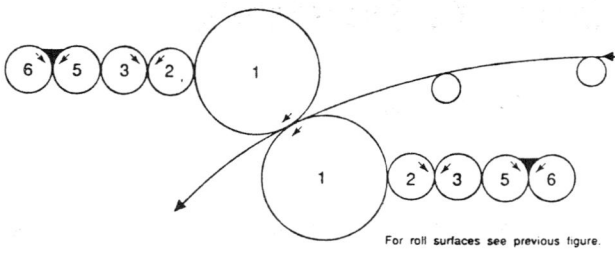

For roll surfaces see previous figure.

Fig. 3.62. Modification of Massey coater

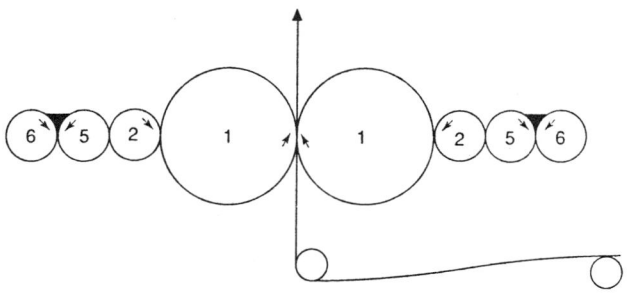

Fig. 3.63. Modification of Massey coater

The Faeber Coater

The Faeber coater, conceived and built for the St. Regis Paper Co. process, was an exaggeration of the printing press inking systems where 26 rolls were originally used in each inking system per side. Only two of these coaters were ever built for an off-machine operation. Neither one proved to be very successful because of the nightmarish operation and drive control for all the rolls (Fig. 3.64).

However, two very important concepts were displayed in the Faeber process. First, the coating was applied in a wet on wet fashion with the coating coming from a common gate roll system. The concept was to apply two thin films, one on top of the other, in order to get rid of the patterning that was inherent in the transfer roll coater. In theory, this approach was sound, but in actual practice, it was not practical because of coating pick and the need for more precise coating formulation. In the era of this coater, the needed adhesives and pigments were not available to make this process viable.

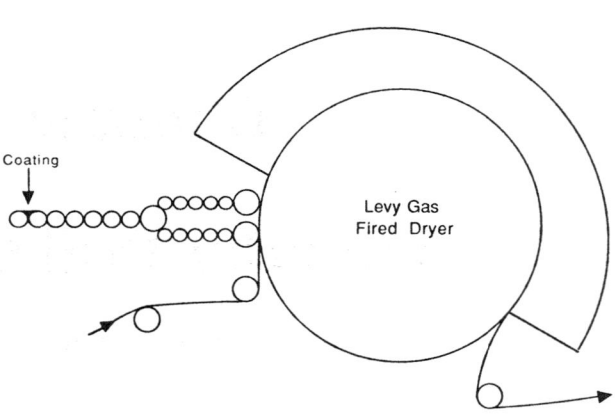

Coating

Levy Gas Fired Dryer

Fig. 3.64. Faeber coater

The second important lesson learned from the Faeber coater was the Levy high-intensity dryer. The drum was gas-heated internally and capable of getting to high temperatures with direct gas flame impingement on the inside of the drum. The coating would dry much more rapidly due to the high surface temperature when compared to a drum heated with steam. An air exhaust system shrouded the drum to remove the moisture. This might be considered the first high-intensity dryer ever used. As we know today, contact drying creates problems in the coated paper such as case hardening, adhesive migration, and railroad tracking. Nevertheless, the concept was a correct attempt to dry the coating more rapidly.

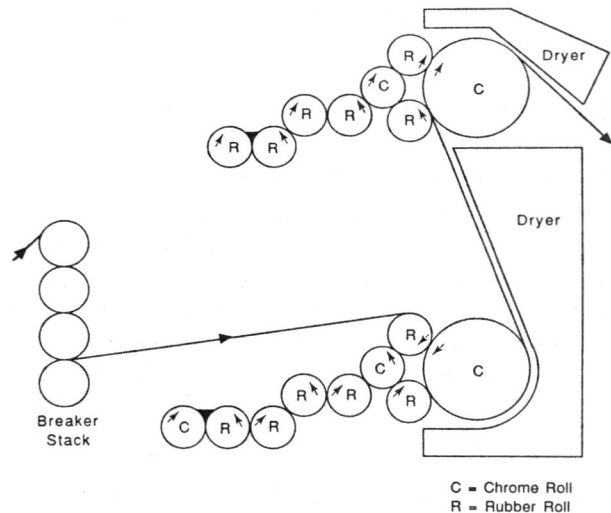

Dryer

Dryer

Breaker Stack

C = Chrome Roll
R = Rubber Roll

Fig. 3.65. St. Regis Roll Coater

St. Regis Roll Coater

Following the Faeber coater, St. Regis Paper Co. modified the machine in the form of the St. Regis roll coater as invented by William Zonner. This machine is shown in Fig. 3.65.

The machine had considerably fewer distributor rolls than what was originally designed into the Faeber coater. However, the concept of two-roll applicators against the drum was maintained, one immediately on top of the other coating two thin films and prior to drying.

In order to circumvent some of the patents of other designs, the machine was split into two separate coater and drying systems as shown in Fig. 3.65. The concept of stacking these coaters in the paper machine saved room and reduced the web lead for threading the paper machine.

Combined Locks Process

During the early to mid-1940s, another process was developed by Gerald Muggleton which came to be known as the Combined Locks coating process. The concept was an interesting approach for a series of coating applications for double and triple coating.

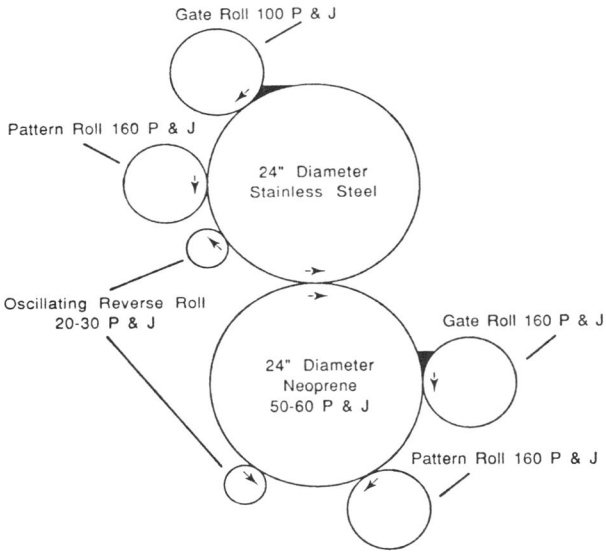

Fig. 3.66. Combined Locks Coater

Initially the coater took the form shown in Fig. 3.66. The machine was developed around a size press with two large vertical rolls used as the main applicator rolls and the web passed through them horizontally. A gate roll was put on both sides with the applicator roll for puddling coating in the nip formed by the gate roll and the applicator roll. The gate roll ran at a very slow speed. By using both pressure and speed, the coating was metered onto the applicator roll. The second roll, called a pattern roll, was used to reduce the film split pattern coming from the gate roll. This roll turned at a different speed and different pressure in order to reduce the patterning.

Following the pattern roll, a rubber-covered and oscillating reverse running roll actually smoothed the coating on the applicator roll before transferring the coating to the paper. This was the first machine to use a reverse running smoothing roll to reduce the pattern on the applicator roll. It was subsequently followed by the Champion Hamilton coater and West Virginia Pulp and Paper Co. Haywood coater to be described later. All of the smoothing was good, but the final transfer of the coating from the roll to the paper did result in redevelopment of a film split pattern occurring at the point of transfer.

Muggleton improved his initial coating process by adding two other points of clay application on the paper machine. The initial attempt was applying some coating or pigment at the dandy roll on the fourdrinier former. This took the name of the dandy roll coater and was only marginally successful (Fig. 3.67). In another attempt, coating was applied at the press. Coating was metered and applied as a film onto the press roll and then transferred to the paper. Again, this was marginally successful as a means of application (Fig. 3.68). The complete Combined Locks process did teach one thing about coating applications, namely to apply two or three coats of pigment to the surface of the sheet develops a better product.

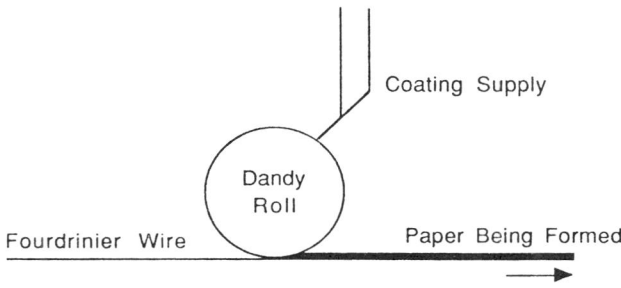

Fig. 3.67. Dandy roll coater

Kimberly Clark-Mead (KCM) Coater

The Kimberly Clark-Mead (KCM) roll arrangement for a single side is shown in Fig. 3.69, and a simultaneous two-side coating operation is shown in Fig. 3.70. Both Mead Corp. and Kimberly Clark developed the coater independently without knowledge of the other company's activity. When it was realized they were developing this coater through patent disclosure, they agreed to allow each other to use the coater without penalty. This is how the coater became known as the KCM Process.

Normally, all of the rolls in the KCM process are operated at essentially the same surface speed, although there is a slight speed differential at the application nip as shown on the diagram. Again, the speed differential is needed to stabilize the web lead from the applicator rolls.

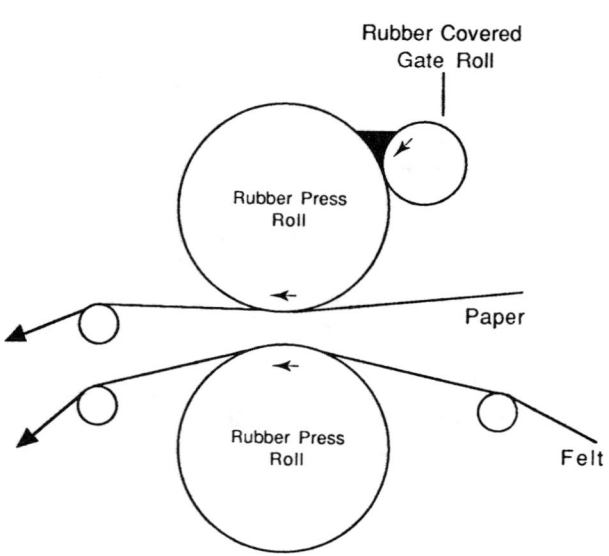

Fig. 3.68. a. Press roll coater (straight)

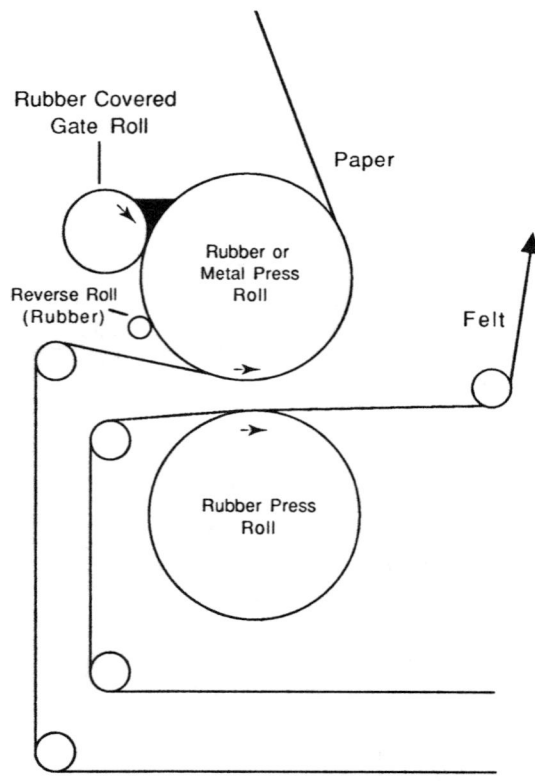

Fig. 3.68. b. Press roll coater (reverse)

It is necessary in many cases to crown the metering roll in the KCM process although slight differences in nip loading pressure can be compensated by vertical

movement of the 10-in. diameter distributing roll. Nip pressure approaches 250 pli in this coater to maintain a minimum pattern in the coated surface and to meter the coating properly. The applicator roll nip is usually maintained at 50-100 pli to prevent undue impregnation of the coating into the web.

Speed differentials at the nip are not used for coat weight adjustment, since speeds are critical for pattern control. Manipulation of the pressure in the nip, coating solids, and the coating rheological properties are the methods used for coat weight control.

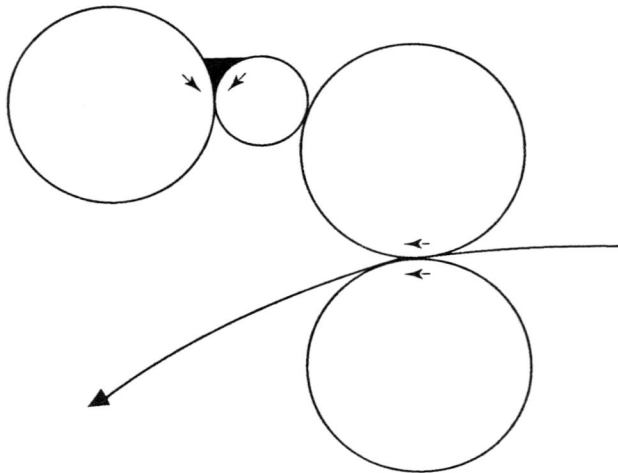

Fig. 3.69. Single-side KCM coater

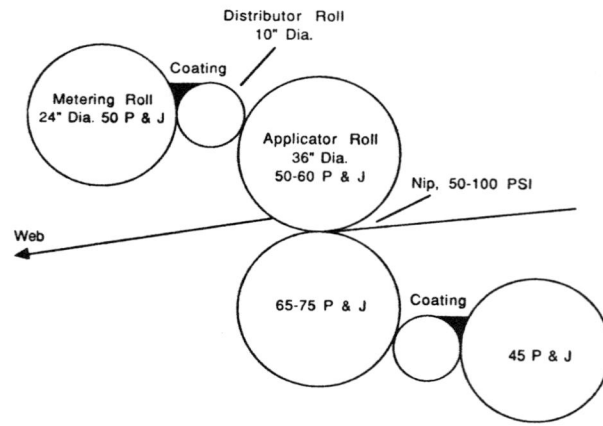

Fig. 3.70. Two-side KCM coater

Champion Hamilton Coater

The Champion Hamilton roll coater is shown in Fig. 3.71. The coater was developed using nonmetering gate rolls and is dependent on the reverse rotating small-diameter metal rod for coating weight control. The coating is applied in excess to the applicator roll, and the metering roll mounted against the applicator roll effectively meters and smooths the coating prior to the sheet.

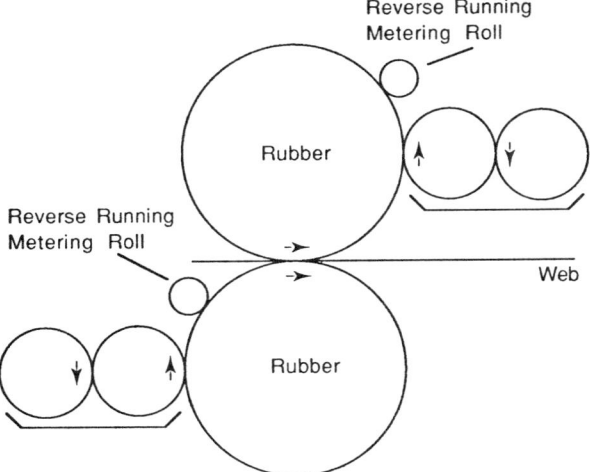

Fig. 3.71. Champion roll coater

West Virginia Coater

The West Virginia coater is shown in Fig. 3.72. This coater is basically the same as the Consolidated coater except for the addition of a reverse smoothing roll, operating on the distributor roll. The latter reduces the pattern in the coating on the distributor roll. However, as the film is again split between the distributor roll and the applicator roll, a film pattern tends to develop. Nevertheless, advantages have been found with the use of a smoothing roll.

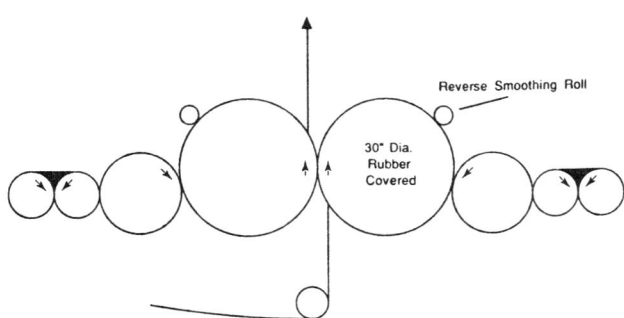

Fig. 3.72. West Virginia coater

Offset Gravure Coater

The name offset gravure coater leaves no doubt that it is an adaptation of a printing method. The coating color is picked up by a furnish roll and applied to the knurled or etched roll (Table 3.3, Fig. 3.73). A doctor removes the excess material, leaving coating in the cells. This is transferred to a rubber-covered applicator roll and from there to the paper. The method is good for one-roll two-sided coating and deposits a very uniform layer.

Table 3.3. Coat weight and gravure cells

Coat Weight lb/3300 sq. ft.	Cells per in.	Cells per sq. ft.	Cell depth, in.
9	48	2304	0.0040
8	56	3136	.....
7	64	4096	.....
6	74	5476	0.0028

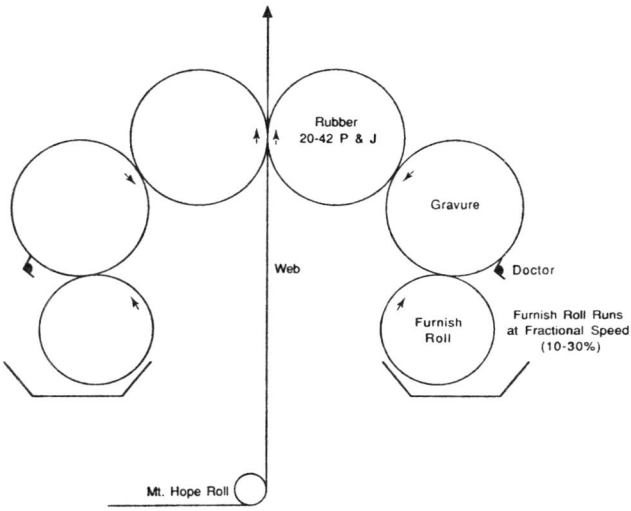

Fig. 3.73. Offset gravure machine coater

It has run at speeds up to 1200 ft/min with uniform coating weight control. Provided the gravure pattern is properly selected and the coating formulation is properly designed for release from the gravure cells, there is little coating weight variation.

The offset gravure coater has the disadvantage of its lack of versatility to apply a wide range of coating weights at approximately the same coating solids. To do this, the machine is stopped and a roll with a different gravure pattern is installed. The gravure coater is a very good machine for high-speed production on the same item day-in and day-out.

Table 3.4 gives a comparison of some of the important features of the various transfer roll coaters just discussed. As the table reveals, there are some differences in design, but all the coaters are similar, for they were developed by different paper companies to serve the same purpose, the coating of paper for magazines.

Table 3.4. Transfer roll coater comparisons

Type	Max Speed fpm	Coat Weight Range Lt. 3300 sq ft	% Coating Solids	Metering (Gate)	Distributor	Smoothing	Applicator	Base Stock % Moisture
Consolidated Massey	2000	2-15	40-65	2 90-100	2-3 160-180		1 55-60	4-8
Kimberly Clark-Mead	3000	2-15	50-65	1 40-50	1 Metal		1 50-75	4-6
West Virginia	1500	2-15	50-65	2 40-50	1 80	1 Metal	1 60	4-7
Combined Locks	1500	3-12	50-65	1 60	1 170	1 80	1 60	4-9
St. Regis Faeber	1400	5-11	50-65	2 Metal & 60	2-3 Metal & 225		1 60	4-10
Champion Hamilton	1700	2-14	50-65	2 40-50		1 Metal	1 60	4-8
Offset Gravure	1500	3-10	50-65				1	

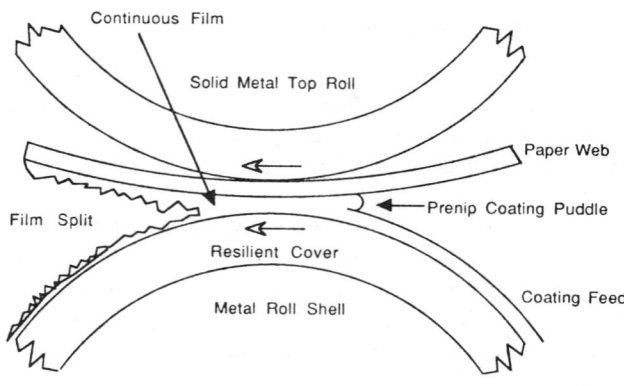

Fig. 3.74. Film split pattern

The Transfer Roll Coater Challenge

Film Split

Film split was mentioned more than once in the previous section. Film split as it normally occurs in the transfer roll coater is shown in Fig. 3.74. This is an ever present phenomenon in the transfer roll coater. It is for this reason that the transfer roll coater was eventually replaced by the trailing blade coater system. Much work was done to get rid of the problems associated with film splitting. One of the early studies of this was done by Smith, Trelfa, and Ware and published in the May 1950 issue of Tappi Journal. A thorough study was made of the rheological characteristics of the coatings and how they operate on a transfer roll coater. It was from this study a formula was developed for the leveling index, I, of coatings.

Following are excerpts from the paper as presented by Smith, Trelfa, and Ware:

Rheology in Roll Coating

In general, the properties of coating colors govern the weight and surface pattern relationships which result when the color is applied on the coating machine. With a satisfactory mechanical adjustment of the machine, a rheologically operable coating formulation gives definite, characteristic weight pattern relationships. At very light coating weights there may be insufficient coating applied to cover the paper, and the surface of the paper will be striped with parallel bands of coating. Above the point where the paper is covered lies a range of coating weights with relative freedom from pattern; the more versatile the color, the wider is the pattern-free range. At the upper limit of this range the paper surface becomes roughened or grained, and further increases result in a moire or stipple pattern sometimes referred to as orange peel.

With rheologically inoperable colors, there is no pattern-free range. In some cases the tops of the ridges on a striped pattern may have heavy moire. The flow properties of a color seem to control the existence and extent of the range of coating weights free from pattern. To be significant, though, the flow properties must be measured at rates of shear above the point at which radical changes in fluid behavior occur.

Newtonian, pseudoplastic, and plastic materials have not been made to operate with satisfactory freedom from surface pattern at any coating weight on the machine coater unless considerable thixotropy was also exhibited. In addition, any detectable dilatancy apparently precludes freedom from pattern, no matter how great the thixotropic effect. Pseudoplastics, however, require much less thixotropy for acceptable performance than Newtonians and plastics.

Leveling Index

During the early phases of correlating the flow properties of the color with its patterning tendency on the machine, it was readily apparent from a visual inspection of the rheograms that the hysteresis (or thixotropic effect) was the dominant factor in eliminating surface pattern on the coater.

The first attempt to determine the influence of thixotropy quantitatively was a direct comparison of the Green and Weltmann (4) coefficient of shear thixotropy with patterning. This coefficient is a number which is a measure of the tendency of a material to break down under increasing rates of shear. More specifically it is the loss of viscosity resulting from a definite increase in rate of shear. It has the advantage of being a constant for a given material and, within limits, independent of the conditions of measurement. The attempted correlation of this coefficient with patterning tendency was unsatisfactory, chiefly because a given operable coating mix would generally remain operable at all solids contents below the dilatant level, while the values of the coefficient would increase from very small to very large as the solids increased.

It was observed that an increase of solids, however, also increased the plastic viscosity of the material and apparently at about the same rate as the coefficient increased. The ratio of the coefficient to the plastic viscosity was then found to be substantially constant. This dimensionless ratio we have called the leveling index, and it is the most practical measure of the patterning ten-

dency of a color of which we are aware. The index I can be easily determined for thixotropic Newtonians and plastics from the following equations: $I = M/U_2$ where M is the coefficient of shear thixotropy and U is the plastic viscosity.

Theoretical Treatment of Roll Shear Conditions

Based on the concept of a uniform, symmetrical velocity gradient across the roll nip, the approximate rate of shear developed can be considered to be of the order of 10^4 to 10^5 reciprocal seconds. Color circulating in the nip of these rolls (Fig. 3.75) obtains a downward vector from the roll motion, and the portion nearer the roll faces begins to move toward the nip. The volume contained between the roll faces decreases as the nip is approached and, since the liquid is substantially incompressible, pressure is developed. The portion of the color farthest from the rolls reverses line halfway between the roll faces, there must be a point at which the coating is substantially stationary. Vertically above this point, color is moving upward; below it everything is moving downward through the nip and out of the other side.

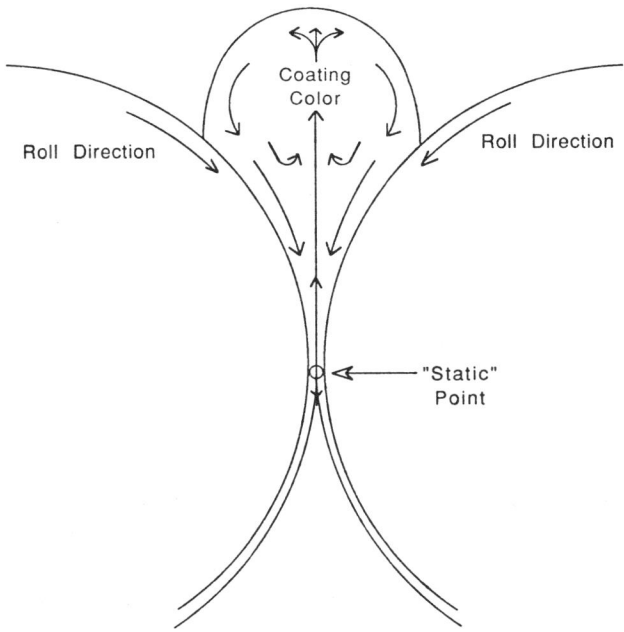

Fig. 3.75. Coating circulating in a nip

At and immediately below this point, the color must flow in a manner analogous to flow through a capillary.

Hypothetical Mechanism of Patterning

The original hypothesis developed to explain the superior non-patterning performance of thixotropic materials as analogous to the well-known after-leveling of thixotropic paints. This hypothesis is not now considered a major factor in elimination of residual pattern because of the limited time available for leveling before the absorption of water sets the coating and prevents further flow. Furthermore, after-leveling has not been verified experimentally. A more complex theory is required and may be described as follows:

When paper is introduced into a nip containing coating color, the paper immediately begins to absorb water from the color. The flow pattern is consequently distorted, but the time of contact between the paper and the color in the nip is short and the loss of water is probably greatest in the segment of color adjacent to the paper.

Another factor must now be introduced. Henry Green (2, 7) has shown that the relative tack of a material (compared to a standard) is proportional to its plastic viscosity if the application is made at high speeds. By tack, he implies pull resistance or resistance to rupture. Also, the tendency to form "necks," and the length and diameter of the necks are obviously involved and will increase with the plastic viscosity. With thixotropic Newtonian, thixotropic plastic, thixotropic pseudoplastic, or pseudoplastic materials, viscosity is least where the rate of shear is greatest, and this point occurs close to the face of the roll; therefore, resistance to rupture is also least at that point.

It has been observed that as the color leaves metering and distributing nips, where no absorbent surface is present, it is drawn out into threads before rupturing. Superficially, it would seem that this effect is in contradiction to that anticipated from the theoretical analysis just described. Actually, it is entirely in accord with the theory. Necking down involves lateral flow of the color along the roll face prior to the creation of the two new air color interfaces because the work of creating the new surfaces is additive to the work involved in overcoming flow resistance. The film must, therefore, flow as long as possible in preference to rupturing. The lateral flow supplies an additional quantity of the material so that necking down can proceed, and as it begins, stress concentration develops. The highest rate of shear is then transferred from the face of the roll to the center of the film. This stress concentration causes films of any materials that are not dilatant to rupture in the center rather than at the face of the roll.

When the paper is present in the nip, its absorption of water in effect solidifies a layer of coating to the paper so that lateral flow cannot proceed and stress concentration cannot develop so readily. Cleavage of films of thixotropic or pseudoplastic materials then proceeds at or close to the roll face. The patterning tendency of these materials on paper, therefore, is expected to be slight for thin films.

In the case of Newtonian and non-thixotropic plastics the situation is different. With the exception of the region in contact with the paper, the film possesses the same viscosity throughout its thickness, regardless of sharing conditions. There is no point of minimum viscosity or relative tack, hence no point of preferential rupture. The film may break at any point, the average height and diameter of the necks are larger, and the resulting surface pattern is heavier. With dilatant materials the sequence is reversed. With paper in the nip, the preferred cleavage point might be in the center of the film where the rate of shear is lowest, and the pattern might be even heavier. In addition a deposit might slowly form on the applicator roll, although this formation is not inevitable because dilatant effects are instantly reversible. Other behaviors, resembling dilatancy except for instantaneous reversion, may inevitably cause accumulation.

Regardless of where rupture occurs, it is almost certain that some necking down results. Pseudoplastics require less thixotropy than plastics to yield pattern-free paper according to this theory because a portion of the viscosity decline required to produce the point of preferential cleavage at the roll face would result from the pseudoplasticity itself, despite the change in direction of flow at break. Practice has confirmed this expectation.

Following this work, International Printing Ink did studies with high-speed photography on the exact mechanism of film splitting. This was presented at the TAPPI Coating Conference in 1957. The actual mechanism was slowed down photographically to see exactly how this process worked mechanically. It did substantiate the work that had been done earlier by Smith, Trelfa, and Ware.

Transfer Roll Coater Uses

Transfer roll coaters have been used primarily to apply mineral pigmented coatings to magazine stock. In these cases, they have also been developed to apply a proteinaceous adhesive or a combination of protein and latex adhesives to the number two merchant grade papers and enamel coated papers.

Transfer roll coaters are also used as prime coaters for subsequent coating operations, such as blade coating operations or air-knife coating operations. The inherent ability of the roll coater to print a uniformly continuous film of coating makes them ideal as prime coaters resulting with a uniform thickness, all over coating.

The introduction of blade coaters in the 1950s retarded the expansion of roll coating installations. However, those that were in use at that time continue to be used either as primary coating units or as the complete coating system.

Summary

In conclusion, it can be stated that transfer roll coaters are now in the history book. The modified KCM press is still used quite extensively for pre-coat, specialty coatings such as CF or carbonless coating systems, and to some degree in the Pacific Rim countries for an ultracoat system. The "ultracoat" system is only three to five grams coating per side for use in low-cost printing papers for advertising and catalogs. It is unique to the Pacific Rim countries and is not generally used in any other place in the world.

Bibliography

Literature Cited

1. Massey, P.J., U.S. Pat. No. 1,921,369 (Aug. 8, 1953).

2. Massey, P.J., Thiele, W.F., and Raprager, B.F., U.S. Pat. No. 2,105,488 (Jan. 18, 1938).

3. Muggleton, G.D. and Piepenberg, A.F., U.S. Pat. No. 2,398,843 (April 23, 1946).

4. Muggleton, G.D. and Piepenberg, A.F., U.S. Pat. No. 2,398,844 (April 23, 1946).

5. Muggleton, G.D., U.S. Pat. No. 2,426,043 (Aug. 19, 1947).

6. Muggleton, G.D., U.S. Pat. No. 2,772,604 (Dec. 4, 1956).

7. Muggleton, G.D., U.S. Pat. No. 2,798,414 (July 9, 1957).

8. O'Connor, J.J., Savage, R.H., and Schwalbe, H.C., U.S. Pat. No. 2,565,260 (Aug. 21, 1951).

9. Hoel, L.W., U.S. Pat. No. 2,606,520 (Aug. 12, 1952).

10. Heywood, G., U.S. Pat. No. 2,560,572.

11. Faeber, H., U.S. Pat. No. 2,456,495.

12. Zonner, W., U.S. Pat. No. 2,645,199.

13. Montgomery, W.J., et al., U.S. Pat. No. 2,676,563.

14. Booth, G.L., *Pigmented Coating Processes for Paper and Board,* TAPPI Monograph No.28, TAPPI PRESS, Atlanta, 1964.

15. Booth, G.L., "A Survey of Machine Coating Methods," Tappi 39(12): (1955).

16. Hirakawa, M. and Iwase, H., "Mill Experience of Gate Roll Coaters," 1988 Coating Conference Proceedings, TAPPI PRESS, Atlanta.

17. Smith, J.W., Trelfa, B.T., and Ware, H.D., TAPPI 33:212 (1950).

Chapter 3, Section V

Cast Coated Papers

George L. Booth, The Black Clawson Co. (retired)

The process of manufacturing coated papers by casting on a flawless, chrome plated drum was conceived by Donald Bradner of Champion Coated Paper Co. in Hamilton, OH, and who received a patent which was issued in July 1929, U.S. Patent No. 1,719,166.

Cast Coating Processes

There are three generic methods of cast coating practiced today for the manufacture of coated paper and coated paperboard. These methods are the wet process, the gel process, and precast coating.

Wet Process

The original Champion patents describe the wet coating process as a process wherein coating is applied to the basestock after which it is laid on a heated chromium plated drum. The drum has a highly polished finish. The coating is dried slowly on the surface of the drum and, when dried, the coated paper is stripped from the surface. The coated paper takes the mirror image of the drum surface, thus the term "cast coating." Donald Bradner coined the term "cast coating" to define this coating process. Casting is considered to be the process of hardening the pigmented coated surface against the polished metal drying drum to produce a surface corresponding to that of the drying drum. The drying drum surface is called the casting surface.

The process is widely practiced today although it is relatively slow at 30-60 meters/min. The process requires a fairly large diameter chromium plated drum ranging in size from 2 meters to 3.5 meters in diameter. Speed of the coating machine is directly related to drum size, coating weight, and coating solids.

Gel Process

The coating used in the process is gelled so that it is rendered relatively immobile prior to contact with the drum surface. A high-pressure nip squeezes the coating against the drum surface to ensure a glossy surface on the coating.

S.D. Warren initiated commercial practice of this process calling it Lustercote III. An acid bath was used in conjunction with a metal chelate to gel or coagulate the protein adhesive in the coating formulation. Later, a gelling mechanism was developed using heat to trigger the coagulation or gelling. These mechanisms have improved and become much more sophisticated as coating and chemicals have improved.

Precast Coating

The original method of precast coating for paperboard was developed by Interstate Folding Box Corp. In this method, the coating was air-knife coated on a heated chromium plated drum and then dried under a high-velocity dryer. In order to develop film forming characteristics, polyvinylidine chloride adhesives were used.

The dried film of coating was laminated to the adhesive coated paperboard and then stripped from the drum. The precast coating process was the only one performed in-line on a paperboard machine for many years. Recently, other finishing processes that approach cast coating quality have been developed to be performed on paper machines.

Other forms of precasting are practiced in the converting industry using a stainless steel belt as the carrier for the coating. This system is well suited to short specialty runs and is particularly well adapted to coatings using volatile solvent vehicles by running the web through an enclosed dryer system.

Wet Process

Discovery of cast coating by Bradner was an outgrowth of work he was doing to try to gelatinize starch and pigments together against a hot drying surface. The essential feature of cast coating, as described in Bradner's patent, involves first bringing the coating surface, while the coating is in a plastic or flow condition, into contact with a nonadhering surface of high-finish drum substantially equal in surface quality to that required on the finished coating, then rendering the coating nonplastic, and thereafter removing the coated paper from the nonadhering surface. The claims cover any coating composition that is convertible from a plastic to a solid state and thus would include non-aqueous coatings which are plastic while hot and solidified by cooling, as well as aqueous coatings that can be transferred from the plastic state by drying. The process today is used for pigmented and coated papers as was Bradner's intention.

For cast coated paper, the coating is not supercalendered but, as explained above, is dried against a highly polished metal surface which imparts the ultrahigh finish. Thus the coating has a much higher bulk and may have greater ink absorbency than paper made by a conventional process involving supercalendering, which tends to compress and densify the coating.

A sketch of the cast coating process is shown in Fig. 3.76. It requires a paper precoated with a porous coating to allow the water vapor to pass through during the drying process. In effect, part of the paper surface quality is a direct result of the quality of the base coating. It is almost as if the top cast coating were translucent showing through to the base coat.

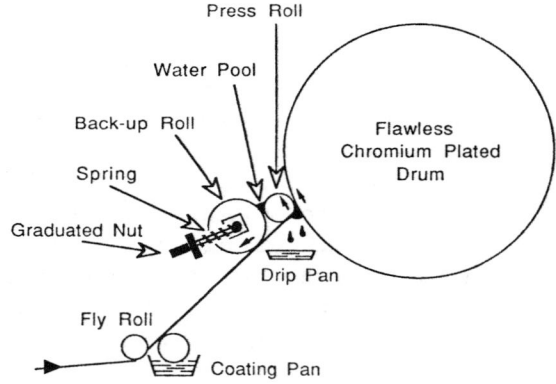

Fig. 3.76. Cast coating system

As an alternate process, Montgomery and Bradner obtained a variation on the original patent in which the web coating is applied to the drying surface rather than to the base paper. Two methods of transferring the coating to the paper were claimed: (a) by contacting the paper with the wet coating before the drying was completed or (b) by drying the coating on the casting surface and then cementing the paper to the dry coating. In the latter case, wet coating applied to the base paper can be used as the bonding agent, in effect producing a double coating.

Prevention of Air Bubbles

One of the problems in cast coating is the occurrence of imperfections in the coated surface caused by entrapment of air between the coating and the casting surface which produces nonglossy or noncast areas. If the surface of the coating, as it reaches the casting drum, does not wet the entire casting surface, nonglossy or noncast areas result. If the casting surface is wetted completely, no air is entrapped. The original Bradner process works satisfactorily when the water content of the coating mixture is just right, but it requires careful adjustment. Several methods have been proposed and used to some extent to overcome the difficulty of incomplete wetting of the coating surface. Among these are:

1. Maintaining a small pool of liquid coating at the point of contact with the casting surface so the coating is rolled into contact with the casting surface and entrapped air is forced out

2. Flooding the nip formed by the coating surface and the casting surface with water. The water may be applied to the coating surface just before where the paper comes in contact with the casting surface

3. Steaming the coating surface just prior to its entry into the nip formed by the coating surface and the casting surface.

Treatment for Release Properties

In order to produce satisfactory cast coated paper, the coating must adhere to the casting surface firmly enough to remain in intimate and undisturbed contact during the critical period of drying and setting, but it is also essential that the coating release from the casting surface, readily and cleanly after drying. Simple coating mixtures containing only pigment and adhesive applied to a clean metal surface do not release satisfactorily and, therefore, special techniques are required to obtain release properties.

Gel Process

Acid Insolubilization of the Coating

The series of patents issued to Hart, Frost, and co-workers, Freeman and Thurlow, during the 1950s, described the insolubilization or gelling system of cast coating developed at S.D. Warren. The name Lustercote was given to the S.D. Warren process. Each new development was sequential with the series extending to number XII at last count.

In the earlier practice of these inventions, the coating was applied, usually with an air-knife coater, and then run through an acid bath to cause a reaction in the adhesive to gel or immobilize the coating. The coated paper was then applied to a drum surface under high pressure in order to dry it and develop a mirror image of the drum on the surface of the coating. The paper did not require precoating and was capable of receiving the total cast coating directly onto its surface.

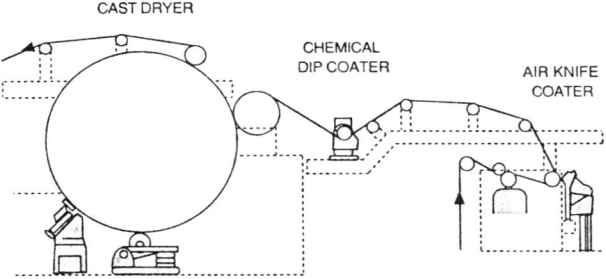

Fig. 3.77. Chemical insolubilization cast coating

Erickson also developed a similar process using hydroxyethyl cellulose as a coating to replace parchment paper and to enhance the use of parchment paper as a barrier material.

Thermal Insolubilization

As a logical sequence in the Lustercote processes, it was found that elevated temperature could also insolubilize the adhesives. This was the result of work completed in Japan and is covered in the U.S. patent under the name of Imura.

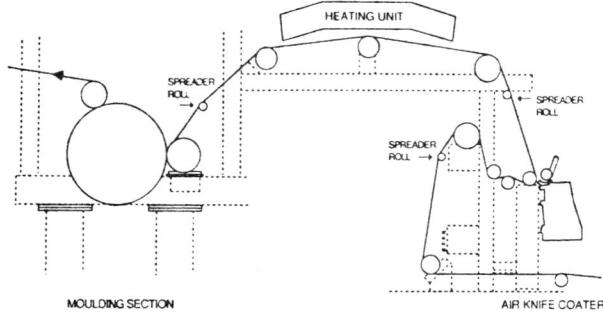

Fig. 3.78. Thermal insolubilization

The coating was subjected to high temperature, normally infrared, in order to trigger the reaction insolubilizing the coating on the paper. The degree of heat required varied depending on the chemistry of the system. It is important not to develop too "hot" a system since this could create premature coagulation of the coating when in the process of application and if subjected to shearing would generate heat from the hysteresis effect. This method improves the speed potential of the cast coating process and greatly reduces the required diameter of the

casting drum since some of the drying is initiated using the infrared.

Precast Coating

Precast coating is a variation of cast coating. It differs from conventional cast coating in that the coating is cast on a polished metal surface, dried, and then transferred to the paper, where it is cemented in place by means of an adhesive. So far, the process has been used only for the coating of paperboard. A sketch of the precast coating process is shown in Fig. 3.79.

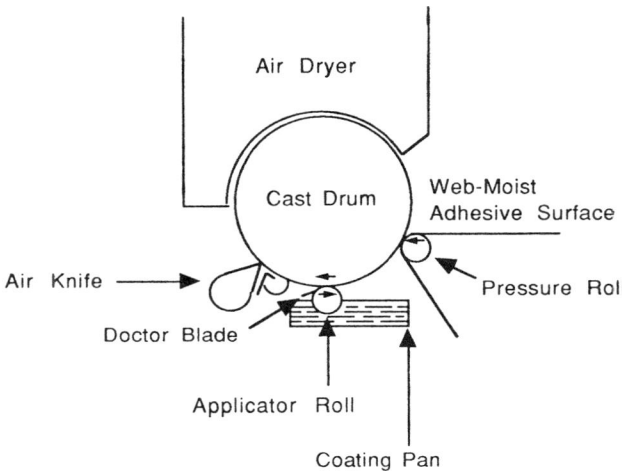

Fig. 3.79. Precast coating

The advantages of precast coating are: (a) improved efficiency of drying of the coating layer, (b) more efficient use of the coating material in that none soaks into the paper, (c) uniformity in coating layer of a controlled thickness, (d) reduction in associated problems with spots and dirt specks in base paper, (e) improved flexibility and bending qualities of the coating layer, (f) uniformity of color and printability, and (g) ability to use any base body stock, even those that are impervious to water, such as foil-laminated and asphalt-laminated papers.

The coating mixture is applied to the surface of a highly polished metal drum. The coating mixture may be of conventional composition, containing approximately 80% pigment and 20% binder with the binder system generally consisting of either protein or synthetic resin emulsions or latex. Drying of the coating takes place in seconds since the moisture is driven off directly through the porous coating layer. Drum temperatures of 180°F to 200°F are most suitable.

In a separate operation, adhesive is applied to the base paper. This adhesive may be a dextrin, polyvinyl acetate emulsion, casein, or similar adhesive. The adhesive layer is then partially dried and brought into contact with the dried coating on the surface of the casting drum. It is important that the moisture content of the adhesive film be controlled exactly at the point of contact to prevent its disturbing the coating layer. Under the correct conditions of moisture and pressure, the coating layer is stripped from the casting surface and firmly adhered to the base paper.

Although precast coating can be applied to any base body stock, the smoothness of the base sheet has a direct effect on the smoothness of the final coating.

Discussion of Important Patents

Use of Coagulants

In this process, the coating surface is treated prior to drying with a coagulant which converts the adhesive in the coating mixture into a deformable gel that releases readily from the casting surface after drying.

Erickson first described the use of coagulants for modifying the coating. The process described the use of a coating mixture containing hydroxyethyl cellulose dissolved in alkali as the adhesive. After the coating is applied to the paper, it can be dampened with a salt solution which precipitates the hydroxy-ethyl cellulose from solution. Squeeze rolls can be used to remove excess salt solution. The final coating surface is quite level and needs no calendering. To date, this process has been little used.

The predominant process involving the use of coagulants is the one described by Hart, although other patents have been issued. In Hart's process, the coatings, while still wet, are treated with a coagulant and then pressed against a highly polished drying surface heated above 212°F. The temperature can be far above 212°F if the paper is pressed against the drying surface hard enough to prevent steam bubbles from forming between the drum and the coated surface. This makes possible the use of smaller casting drums.

The gelling agent may be acid alone, acid plus metal chelate in the coating mixture, or heat in conjunction with a metal compound in the coating.

Among the coagulants claimed for use with protein adhesives are water-soluble compounds containing calcium, zinc, barium, lead, magnesium, cadmium, and aluminum salts of various acids, or the acids themselves. Water-soluble nonionic surface active materials may be incorporated in the coagulant bath to aid in the release of the coating from the casting surface. Frost and Smith claim advantages from the use of water soluble resins dissolved in a pool of aqueous moistening liquid.

The essential detail is the high drying temperature. As the coating is heated above 212°F, a large portion of the water in the coating almost instantaneously evaporates.

Hart also disclosed that, contrary to prior belief, it is possible to produce paper of cast coated quality without the necessity of having the wet coating immovably adhere to the drum until dry. If pressed firmly against the hot casting drum and held there until substantially dry, any gelled but still plastic or moldable coating that wets the drum surface produces a "cast" finish, whether or not it actually sticks to the drum. If the coating is removed from the casting surface before it is dry and further dried in some way, it will not look the same. If the base sheet has a rough surface, the final coating will have a rough, "stippled" appearance, even though the gloss is high.

Pretreatment with Oleaginous Materials

In this patent, Montgomery claims the use of oleaginous materials for bringing about release properties. This patent calls for the application to the casting surface of an invisible continuous film of oleaginous materials of a thickness substantially less than 2 microns. The casting surface can be preconditioned or seasoned by rubbing with oleaginous material. The drum is thereafter kept in suitable condition by the incorporation of additional oleaginous material to 0.1-1.0% by weight of the pigment in the coating mixture. This oleaginous material is absorbed from the coating onto the casting surface and maintains a film of suitable thickness to bring about release properties. Among the suitable oleaginous materials are oils, fats, fatty acids, partially saponified oils or fats, fatty acid amines, and dimers of aliphatic ketenes. Sulfonated oils and phosphate are not satisfactory.

Treatment of Chromium Casting Surface to Impart Passivity

Release of cast coatings from a mirror finish chromium surface can be accomplished by chemical treatment of the chromium surface. By applying dilute aqueous oxidizing acids, such as 2-10% acetic acid, or nitric acid solution at pH 1-2, to the chromium surface which is hot, but below the boiling point of water, the chromium can be induced to assume the chemically nonreactive condition known as passive. In this condition, coatings cast and dried against the chromium surface do not adhere, regardless of whether a release agent is present.

The difficulty with this process is that certain chemical combinations, especially at elevated temperatures, destroy the passive-releasing condition of the chromium surface and convert it to an active or nonreleasing surface. Discharge of static electricity from an over-dried sheet against the passive chromium surface likewise converts it to the active condition. To *maintain* or restore the releasing condition of the chromium, a thorough treatment with the dilute, aqueous oxidizing acid is needed. Application by a cloth impregnated with the solution is most effective, and mechanical aids, such as a powered buffing wheel, are timesaving.

The durability of a chromium surface in the passive condition varies with the operating conditions. Agents incorporated in the coating color may extend the running life of the passive chromium surface by protecting it from chemical conversion to an active, nonreleasing surface. Tributyl phosphate is one such agent.

It has been reported that when the chromium is in the active condition no release agents known are effective. It may be argued that some of the effectiveness of drum treatments ascribed to the creation of oily films, especially when applied with abrasives, is actually due to the conversion of chromium surface to a passive condition.

Effect of Coating Composition

Composition of coating for cast coated paper is similar to other high-grade coated paper. The adhesive ratio is higher than the usual grade of supercalendered coating paper to resist the strong pull when the paper parts from the drum, and to obtain reasonable ink hold because the coating has very high bulk and is not densified or rendered nonabsorbent, as in conventional supercalendered coated paper. Montgomery and Bradner obtained a patent claiming the use of cast coatings containing adhesive in excess of 30% by weight as compared to adhesive content less than 20% for ordinary coatings. The effect of this exceptionally high percentage of adhesive used in cast coating is opposite to its effect in conventional coated paper. For example, excessive adhesive in ordinary coated papers tends to impair the gloss and smoothness after calendering, but when cast coated the gloss is enhanced and the finish improved. The adhesive commonly used is casein, either alone or in combination with latex. The coating mixture is usually applied at high solids and the pH is generally under 9.

Use of Special Pigment, Adhesive or Additives

In a Japanese patent application, improved release properties are claimed by pretreatment of the pigment with a hydrophobic material such as soap, sodium resinate, or stearin. The hydrophobic material may be added as a sodium salt and precipitated on the clay by means of alum to coat the clay particles. Or the hydrophobic material may be formed in the manufacture of calcium carbonate pigment.

In another special cast coating process, a copolymer of a vinyl ester of a high fatty acid, such as stearic acid, is used as the adhesive. No release agent is required.

Effect of Coat Weight

Generally high coat weights are applied in cast coating, ranging from 15 to 30 g/m^2 per side. Coating thickness is sufficient to give complete coverage of the fiber in the base paper. Heavy coating, in general, is a disadvantage because it is one of the factors contributing to the high cost of producing cast coated papers.

Other Factors in Cast Coating

To prevent defects in the finished cast coated paper, it is advantageous to limit water penetrating from the coating into the base paper to reduce swelling of the fiber. This is usually accomplished by using high solids coating mixtures and a well sized base paper.

In cast coating, water from the coating must pass through the paper web and escape out the back of the sheet. Consequently, a fairly open base sheet is preferred. Another helpful technique is to blow air on the back of the coated paper during drying to increase the rate of water evaporation from the back of the paper. Wetting the back of the paper is also claimed to improve the rate of evaporation.

Gloss Coating Supercalendered Paper

A special case of cast coating, sometimes referred to as cast calendering, is a process by which an ultrahigh gloss is produced on a supercalendered coated paper. The surface of an already coated surface is rewetted sufficiently to cause the hydrophobic adhesive in the coating to swell. The wetted surface is brought into contact with a highly polished, metal heated roll under pressure.

This process differs from cast coating since the coated paper is generally densified by supercalendering before it receives a high finish against a highly polished roll. Thus, the paper has a more dense coating layer and higher ink holdout than conventional cast coated papers. For this reason, the paper can be printed by regular printing inks used on supercalendered papers, although the paper looks and is marketed as regular cast coated paper.

The gloss coating of supercalender paper involves the following steps:

1. Pigment coating is applied to paper and dried in the conventional manner.

2. The paper is supercalendered with enough pressure to densify the coating. It is not essential that high gloss be obtained at this stage.

3. After supercalendering, an aqueous liquid is applied to the surface of the coated paper. A pool of aqueous liquid may be maintained in the nip where the coating surface contacts the gloss calendar.

4. Immediately after this application, the wet surface is pressed against the surface of a highly polished heated metal roll until dry.

The adhesive used in gloss coating of super-calendered paper must be hydrophobic so the moistening step will cause the coating layer to swell and become moldable to the surface of the polished roll. Conventional adhesives are suitable and include casein, starch, soy protein, glue, and polyvinyl alcohol alone or mixed with synthetic resin emulsions or latexes.

Only a very small quantity of moistening fluid is used in this process making the drying less difficult, thus the process runs at higher speeds than in regular cast coating. Water alone can be used as a moistening fluid, or it is advantageous in some instances to include a small quantity of an anti-stick or release agent such as ammonia, soaps, polyethylene glycols, glycol monolaurate, or ammonium stearate.

Variables affecting the finish obtained in gloss coating of supercalendered paper include: (a) the degree to which the hydrophobic adhesive swells, (b) the external pressure applied in forcing the coating surface against the polished drying surface, and (c) the condition of the metal drying surface. The most suitable drying surface is a highly polished chromium plated roll. The paper is pressed against this roll by a rubber-covered pressure roll. The temperature of the surface of the drying drum generally preferred is 160-195°F.

Polymer Coatings

Polymer coating systems that are cast coated are used for functional rather than decorative purposes. The polymer coating is applied to a casting surface such as a stainless steel belt or coated casting paper and then transfer and laminated to the web. The process is shown in Fig. 3.80, using a stainless steel belt. In the diagram, a three-roll reverse roll coater is used to apply the coating to the belt. The coating is dried, then stripped from the belt. The unsupported coating film is laminated to the web by passing through a set of nip rolls, and then fused to the web by passing through a fusing or curing oven.

Transfer laminating or coating the web by the process described above is commonly used in the textile industry. A large amount of clear plastic film is also made this way. This type of cast coating requires the coating have film-forming properties in order to produce an impervious coating that can be transferred to a web. A good example of such a coating is a vinyl dispersion, either organosol or plastosol.

Another common method of cast coating is the use of a specially coated release paper onto which the coating is applied and then removed through wet lamination to the web. In many cases, the casting paper remains with the coated web to act as protection for the coating as in pres-

sure-sensitive adhesive coatings on labels, stickers, and signs.

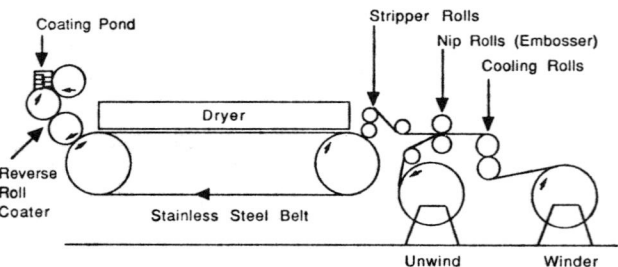

Fig. 3.80. Stainless steel belt cast coater

The paper is coated with an alkyd resin compound or a silicone coating to give a smooth release surface. Most alkyd coatings are applied with a reverse roll coater. Silicones, in solvent form, are best applied by the direct gravure process. In water-based silicone coating systems, the air knife is the best method. All three coating methods are contour coating methods; that is, they apply a uniformly thick coating to the web. This is important in order to obtain uniform release properties.

It is possible to obtain as many as eight re-uses of the carrier web. Once the paper is damaged either by heat on the uncoated edges or poor release of the coating, it is necessary to discard the entire roll. This makes the process slightly more expensive than the stainless steel belt method, but the initial investment and maintenance are less.

Application of Cast Coating

The mineral pigmented cast coated products are used as high-grade printing papers and paperboard for boxes. Because the cost is high, its use is limited to items such as candy boxes, greeting cards, prestige labels, covers for financial reports, cosmetic boxes, and other luxury items. It is possible to cast coat both sides, but because of the difficulty in drying, is very expensive.

The polymer coatings, on the other hand, are used for pressure-sensitive labels, leatherette, jersey knit plastic-coated fabrics for garments, automobile interior trim, and packaging materials.

The products in both systems are of a premium quality and as such are expensive. Because of the unusually smooth and glossy surface, however, their uses are many and growing.

Bibliography

Literature Cited

1. Alex Cowan & Sons, Ltd. Scotland, Belgian Pat. No. 609,600.

2. Bergstein, F.D., U.S. Pat. No. 2,934,467 (April 26, 1960).

3. Bradner, D., U.S. Pat. No. 1,733.525 (Oct. 29, 1929).

4. Erickson, D.R., Tappi 32 (7): 289 (1949).

5. Freeman, E.W., U.S. Pat. No. 2,950,989 (Aug. 30, 1960).

6. Frost, F.H. and Smith, R. L., U.S. Pat. No. 2,819,184 (Jan. 7, 1958).

7. Frost, F.H., et al., U.S. Pat. No. 2,759,849 (Aug. 21, 1956).

8. Frost, F.H., et al., U.S. Pat. No. 2,759,847 (Aug. 21, 1956).

9. Frost, F.H. and Lane, R. E., U.S. Pat. No. 2,780,563 (Feb. 5, 1957).

10. Hart, R.T., U.S. Pat. No. 2,776,226 (Jan. 1, 1959).

11. Hart, R.T., U.S. Pat. No. 2,919,205 (Dec. 29, 1950).

12. Hart, R.T., U.S. Pat. No. 2,769,725.

13. Hart, R.T., U.S. Pat. No. 2,919,205.

14. Imura, M., Patent Application 10,004 (filed July 3, 1956).

15. Leighton, P.S., U.S. Pat. No. 2,678,800 (May 18, 1954).

16. Montgomery, W. and Bradner, D., U.S. Pat. No. 2,029,273 (January 1936).

17. Montgomery, W., U.S. Pat. No. 2,508,288 (Sept. 18, 1961), Re. 23,637 (March 24, 1953).

18. Montgomery, W., Re. 23,637 (March 24, 1953).

19. Montgomery, W. and Bradner, D., U.S. Pat. No. 2,214,566.

20. Montgomery, W. and Bradner, D., U.S. Pat. No. 2,214,564.

21. Montgomery, W. and Bradner, D., U.S. Pat. No. 2,337,013 (Dec. 14, 1943).

22. Robinson, J.V. and Brookbank, E.B. Jr., U.S. Pat. No. 3,020,176 (Feb. 6, 1962).

23. Robinson, J.V. and Shaffer, I. E., U.S. Pat. No. 2,943,954 (July 5, 1960).

24. Smith, J.W., U.S. Pat. No. 2,950,214 (Aug. 23, 1960).

25. Warner, E., U.S. Pat. No. 2,316,202 (April 13, 1943).

26. Warner, E., U.S. Pat. No. 2,360,919.

Chapter 3, Section VI

Surface Sizing and Precoating of Basestock

Charles P. Klass, Klass Associates Inc.

Introduction

Surface sizing or precoating of basestock is a critical unit operation which has significant effects not only on coater runnability but also on the quality of the final coated sheet. Coating holdout is a key objective of surface treatment with chemicals, but improvement of strength, surface uniformity, and optical properties may also be achieved. The choice of surface sizing or precoating chemicals and applicator is dependent on the properties of the paper raw stock, coated sheet quality desired, and the characteristics of the coater and coating formulation to be used. The conventional two-roll size press has been the standard equipment for surface sizing and precoating for many years, but it has limitations. Recently introduced modified size presses provide some advantages. Alternatives to the size press, such as simultaneous two-sided blade coaters and hydrophilic transfer roll coaters, may also provide some advantages. Special care must be taken in handling and drying the surface-sized or precoated web.

Basestock characteristics have dramatic effects not only on coater runnability but also on the quality of the final coated sheet. Basestock characteristics which have direct impact on coated sheet quality include:
- Coating holdout
- Pore structure
- Surface uniformity
- Microcontour
- Optical properties
- Stiffness
- Bulk density
- Internal bond
- Surface strength
- And, in some cases, resistance to vessel segment picking.

Fiber furnish, refining, additives, wet-end chemistry, forming, pressing, and drying all affect these properties. Chemical treatment via surface sizing or precoating is also a critical unit operation in production of a basestock with the desired combination of properties.

The contribution of surface sizing or precoating to basestock quality depends on:
- Sheet properties of the raw stock
- The materials applied
- The application equipment and process conditions used
- Web handling and drying of the surface sized or precoated sheet.

In some cases, raw stock characteristics may prevent the use of a desired surface sizing or precoating technique. For example, a sheet with low wet web tensile strength such as a lightweight groundwood sheet cannot be successfully run through a size press.

Coating Holdout

To provide a basis for comparison of various surface sizing or precoating systems, it is useful to review the basics of coating holdout. Coating holdout, defined as the volume ratio of coating remaining on the surface to that penetrating into base paper pores, can be roughly characterized by the relationship:

$$\text{flow resistance in base paper pores} \geq a + b,$$

Where: a = hydraulic application pressure
b = hydrodynamic metering pressure
and
a/b = constant (1).

The factors which affect flow resistance to coating color in raw stock pores and therefore coating holdout include:

- Structural viscosity of the wet coating film
- Mean pore radius. If the sheet is easily compressible, mean pore radius is further reduced by applying load to the sheet during the metering stage.
- Wettablility expressed in terms of contact angle. Higher contact angle reduces the velocity of capillary water transport.

On absorbent papers with minimal internal sizing, rapid initial wetting of the fibers is the primary cause of poor coating holdout. Surface sizing with a reactive size, which increases the apparent contact angle between the paper surface and the fluid, may be the key to improving holdout (2).

On sized papers, where hydrophobicity is already provided by internal sizing, capillary pore size appears to be the controlling factor. The optimum surface sizing or precoating treatment uses barrier formation or some other mechanism to cover or fill the pores in order to provide holdout. A film former is the treatment of choice.

Applicators vary widely in the amount to hydraulic pressure developed during the application of wet coating (1). An applicator roll inverted blade coater can develop peak hydraulic pressure values of 2.5 bar. The applicator roll nip subjects the wet coating to a wide pressure zone with gradually increasing pressure. The hydraulic load on the wet coating film results in extensive pressure filtration, causing the formation of a concentrated, immobilized coating matrix at the base paper/coating interface. Base paper pores are plugged with immobilized coating. However, large quantities of water are transported from the coating into the basestock. In the subsequent metering stage, deformation and friction forces are exerted onto the sheet, which requires a minimum level of wet web strength. For groundwood papers, adjustment of wet web strength requires rather large percentages of chemical pulp in the furnish, significantly increasing costs.

Surface Sizing and Precoating Chemicals

In choosing a system of surface sizing or precoating chemicals for basestock, it is necessary to take into account not only the raw stock properties but also the coating color and coating applicator to be used.

Starch

Basestock surface sizing operations have typically utilized starch films to fill in the surface voids in the sheet, reducing the pore radius and therefore the rate of penetration of fluids (3). In common paper mill practice, a relatively low-viscosity starch solution of 3-9% solids is used at the size press to achieve pickup in the range of 60-100 pounds of dry starch per ton of paper produced.

Starch is a carbohydrate synthesized in corn, tapioca, potato, and other plants by polymerization of dextrose units. The polymer exists in two forms: a linear structure of about 500 units and a branched structure of several thousand units. The linear chain polymer called "amylose" constitutes 27% of natural corn starch, while the branched polymer called "amylopectin" makes up the remaining 73%. Fractionated starches are available for special uses (4).

Starch is supplied as a white, granular powder which is insoluble in cold water because of the polymerized structure and hydrogen bonding between adjacent chains. However, when an aqueous suspension is heated, the water is able to penetrate the granules and causes them to swell, producing a "gelatinized" solution or paste, depending on concentration. Cooling this hot solution causes thickening, which is called "setback." Starch for surface sizing application is "cooked" using either batch or continuous systems (5).

Unfractionated and unmodified starch, called "pearl starch," is "thick boiling" (i.e., viscous) and has a tendency toward gelling or setback, even without cooling. Setback is avoided by using 100% amylopectin, "waxy" starch which forms a clearer paste and is nongelling; however, a loss of sizing efficiency results because the linear fraction contributes more toward film formation.

Lower viscosity and setback resistance are achieved by using chemically or thermally modified starch (e.g., "oxidized" starch). A "thin boiling," low-viscosity starch can be produced by enzyme conversion, with film formation properties and setback resistance unaffected.

One of the potential problems with starches is their anionic characteristic. This may be undesirable when the starch is reintroduced to the white water system through the use of broke (6). There may be undesirable effects on wet-end chemistry and the mill effluent treatment plant. Some mills have switched to the use of cationic starches for surface sizing. Cationic starch is prepared by reacting starch with a cationic polyelectrolyte. The cationic sites on the starch are attracted to the anionic sites on the fiber, thus making the starch cellulose substantive. This can result in decreased BOD loading on the paper mill effluent treatment system.

Film Formers Other Than Starch

A number of water dispersible film formers are used for surface sizing basestock. Among these are carboxymethyl cellulose, polyvinyl alcohol, and alginate. These materials are higher in cost than starch and their use is limited to higher value added papers.

Addition of the sodium salt of carboxymethyl cellulose (CMC) to size press starch or a precoating formulation improves film forming properties (2). When coating is applied to the surface of CMC treated paper, the CMC rapidly picks up water from the penetrating liquid. This caused the polymer to swell which (a) increases the extent of pore coverage and (b) adds viscosity which aids in slowing both capillary and pressure-induced migration. The use of CMC also increases surface strength.

Polyvinyl alcohol is used as a surface size to form films that have very high tensile strength and a high degree of transparency, flexibility, and oil resistance. The water resistance of these films is fairly good, and it can be improved by addition of a crosslinking resin or an alkali-stabilized colloidal silica. Solutions of polyvinyl alcohol can be mixed with starch and applied via size press, but they tend toward dilatancy, which may cause operating problems at high speeds. The use of polyvinyl alcohol in surface sizes is usually limited to basestocks for specialty coated papers.

Alginate can be used for surface sizing alone or in combination with starch. Alginate forms films which are clear and tough but may be brittle unless plasticized. Alginate films are water-sensitive, and a crosslinking resin may be needed. The primary use of alginate surface size is in grades where a high-density, well-closed surface is required.

Other Specialty Surface Sizes

Several other sizing agents are used on coating basestock. As shown in Fig. 3.81, ketene dimer reactive size and styrene-maleic anhydride copolymer increase the contact angle, even at low levels of pickup. This capability is especially useful to improve the coating holdout of an absorbent, unsized sheet (2).

In the manufacture of some specialty coated papers, additives such as fluorochemical or chromium complexes of long chain fatty acids, e.g., myristic acid and stearic acid, are used in surface sizing to provide special properties to the base sheet (7). An example is the manufacture of wickproof label paper; the fluorochemical must be added to the surface size to protect against edge wicking. These additives are usually used in combination with starch, polyvinyl alcohol, or both.

Pigments

Pigments may be added to the surface size or precoating to provide a combination of physical and optical properties. Pigmented precoating is used to provide a better microcontour by filling valleys and also to aid in coating holdout by plugging the pores (8). Pigments in the precoat also contribute to the brightness and opacity of the finished sheet.

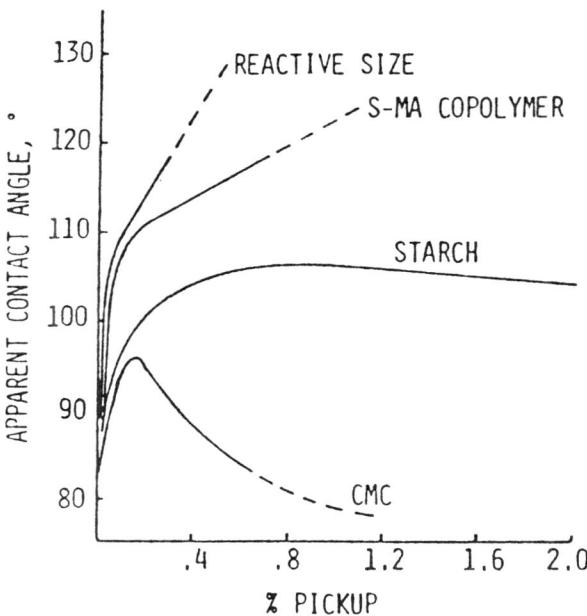

Fig. 3.81. Effect of size press additives on contact angle

Latex

With the exception of some technical specialty papers, latex is usually added only to pigmented surface sizes or precoatings. In precoating, latex is normally used as cobinder with starch and another film former such as CMC.

Size Press

The size press is the most commonly used applicator for surface sizing or precoating of basestock. In a conventional size press, the objective is to flood the entering nip with sizing solution; the paper absorbs some of the solution, and the balance is removed in the nip. The overflow solution is collected in end funnels or a pan below the press and is recirculated to the nip.

Conventional size press configurations are generally categorized as vertical, horizontal, or inclined, as shown in Fig. 3.82. The vertical configuration provides the easiest sheet run, but the pond depth of solution in each nip is unequal. The horizontal size press arrangement solves the problem of unequal absorption by providing identical pond forms on either side of the sheet. The inclined size press is a compromise, developed to avoid the awkward vertical sheet run of the horizontal size press.

Size Press Pickup

Size press operations must be carefully controlled to assure that the sheet picks up the desired amount of chemical uniformly across the machine. At the same time, the amount of water absorption should be minimized to maintain after-dryer requirement at the lowest practical level. The primary factors determining solids pickup at the size press are listed in Table 3.5.

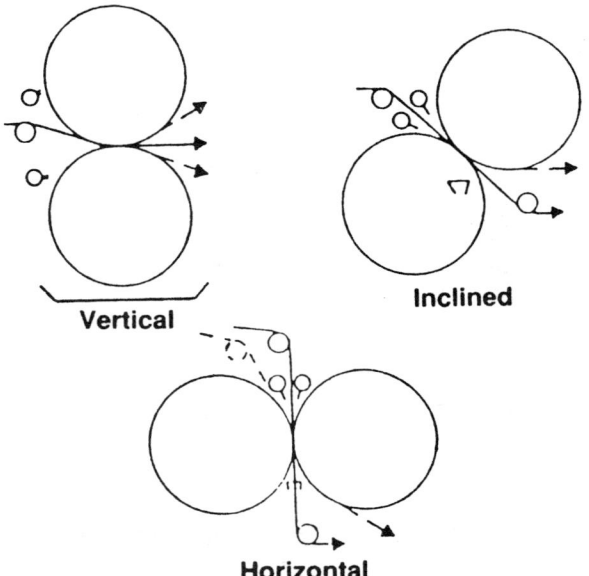

Fig. 3.82. Size press configurations

Table 3.5. Factors influencing size press pickup

I. Sheet characteristics
    A. Paper substrate (basis weight, density, smoothness, capillary structure, void size etc.)
    B. Level of internal sizing
    C. Moisture content

II. Sizing solution
    A. Solids content
    B. Temperature
    C. Viscosity
    D. Composition (type of starch, use of additives)

III. Design and operational
    A.   Machine speed
    B.   Pond depth
    C.   Nip pressure
    D.   Nip width (affected by roll hardness and diameter)

There are two basic mechanisms for incorporating sizing solutions at the size press (9). The first is the ability of the sheet to absorb the size solution. The second is the amount to solution film passing through the nip and the manner in which the paper and the roll surfaces separate. Factors that favor greater absorption include low solution viscosity (higher solution temperature), low machine speed, and low level of internal sizing. The factors favoring greater film thickness include sheet roughness and low nip pressure.

Of the factors affecting pickup, solution viscosity and solids content are the easiest to manipulate. The solids concentration, however, is usually maintained at the highest manageable level consistent with the desired viscosity in order to minimize the amount of water that must be subsequently evaporated.

Conventional Size Press Limitations

In the conventional size press, the solution is delivered to the nip between a pair of oppositely rotating rolls to form a pond of material above the nip. With high-speed roll rotation and high-speed movement of the web, the pond tends to absorb kinetic energy form the moving web and rolls (10). As shown in Fig. 3.83, an excess of liquid is flowing toward the nip, but the nip pressure of the rolls limits the amount of sizing solution passing through the nip, causing the remaining solution to flow back upward. If hydrodynamic forces are too great, this upward velocity becomes high enough to cause the solution to break the surface of the pond and splash out of the nip. This pond turbulence and "nip rejection" results in uneven pickup of solids across the machine.

Size press pond

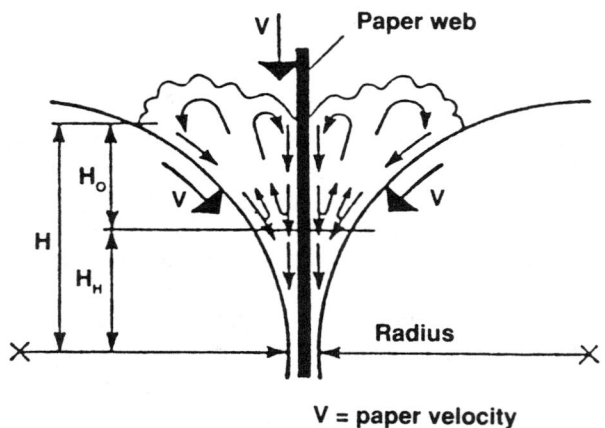

V = paper velocity

Fig. 3.83. Hydrodynamics of the size press pond

Size solution viscosity contributes to the hydrodynamic force in the roll nip. This may make it impractical to use the size press to apply dilatant materials. As machine speeds increase, it may be necessary to dilute the starch solution or go to a more degraded starch to avoid nip rejection. These approaches, in turn, may increase after dryer requirements and also affect sheet properties adversely.

Another problem associated with the size press is film splitting. Although most of the size solution is absorbed into the sheet at the nip, excess solution remains on the surface of the sheet between the paper and the roll. At the exit of the nip, this film is split into two layers with part following the paper and the rest remaining on the roll surface, as shown in Fig. 3.84 (11).

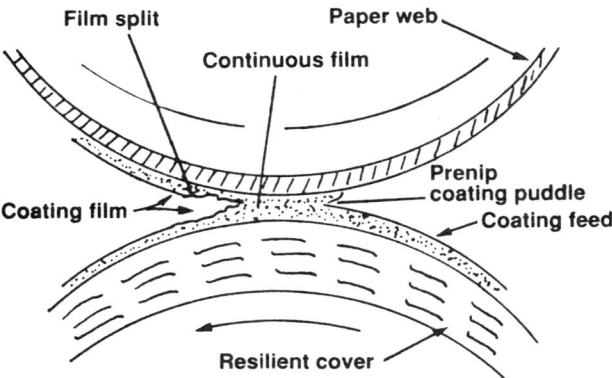

Fig. 3.84. Development of film split pattern

This pattern will appear as heavy rings of the fluid across the roll face (12). The severity of the pattern depends on the rheology of the size, absorption properties of the sheet, speed of the rolls, and volume of material being split. The problem becomes worse with higher speeds, higher viscosity, and higher film thickness (13).

Moisture content of the sheet entering the size press has a significant effect on pickup. Higher moisture content promotes absorption; thus a variable moisture content across the machine causes variations in pickup and sheet properties affected by surface sizing. Thus the ingoing sheet is usually dried down to 1 or 2% or less to ensure an even moisture profile and to keep the sizing agent nearer to the surface. This "over drying" of the sheet is inefficient from an energy standpoint and may place a limit on machine speed.

As illustrated in Fig. 3.85, studies have shown that the level of internal sizing is the greatest factor in size press pickup control (14). In fact, it is essential to have at least a minimal level of internal sizing to even run the sheet through a size press and into the after dryer section without breaking.

The required level of internal sizing to provide runnability is inversely related to basis weight and wet tensile strength. Thus it may be necessary to use high internal sizing levels and expensive wet-strength agents in order to get lightweight papers to run though the size press. Even with these additives, the size press is usually a major source of lost time on machines making lightweight weak papers. Since the commonly used reactive neutral-alkaline internal sizes are slow curing, problems are often encountered with size press breaks in alkaline papermaking (15). Another difficulty with the conventional size press is matching the speeds of the two rolls.

If the roll speeds are not matched precisely, "sheet scuffing" or "scrubbing" can lead to paper surface marking and increased web breaks. The difficulty of precision roll speed matching increases as machine speed goes up.

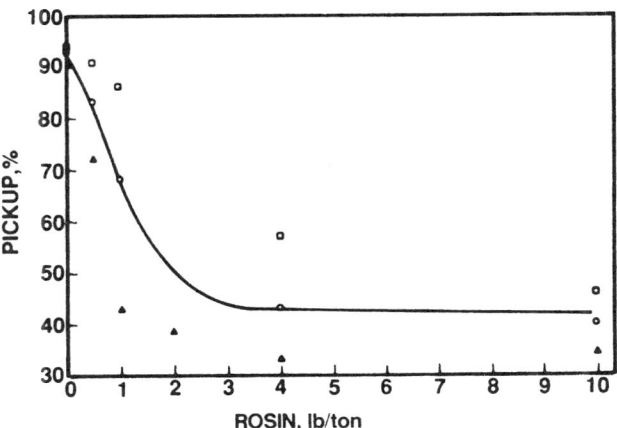

Fig. 3.85. Effect of internal sizing on size press pickup

Size Press Design Improvements

As machine speeds have increased, paper machine designers have increased size press roll diameters to help compensate for greater hydrodynamic forces. As shown in Fig. 3.86, increased roll diameters reduce the proportionate accelerative forces on the sizing solution. Some of the newer uncoated free sheet machines, equipped with 60-in. diameter size press rolls, operate successfully at speeds approaching 3000 ft/min; however, it has been found necessary to use short chain length (degraded) starch and solution concentration in the 2-3% solids range to keep pond turbulence manageable.

Figure 3.87 shows the apron size press, patented as the "high-speed size press" (16). It is designed to prevent pond splashing by isolating the sizing pond from the high-speed surfaces of the rolls and paper. Since less kinetic energy is absorbed by the size solution, circulation velocities are reduced, and the size solution does not have a tendency to jump out of the pond. The baffles are usually made of sheet plastic material. This size press modification is effective in reducing pond turbulence, but there may be problems in keeping the baffles clean and with excessive wear on the dry sides of the baffles. "Icicles" building up on the baffles can create not only streaks but also scratches severe enough to break the sheet. Very few mills have been able to use the apron size press in day-to-day operations.

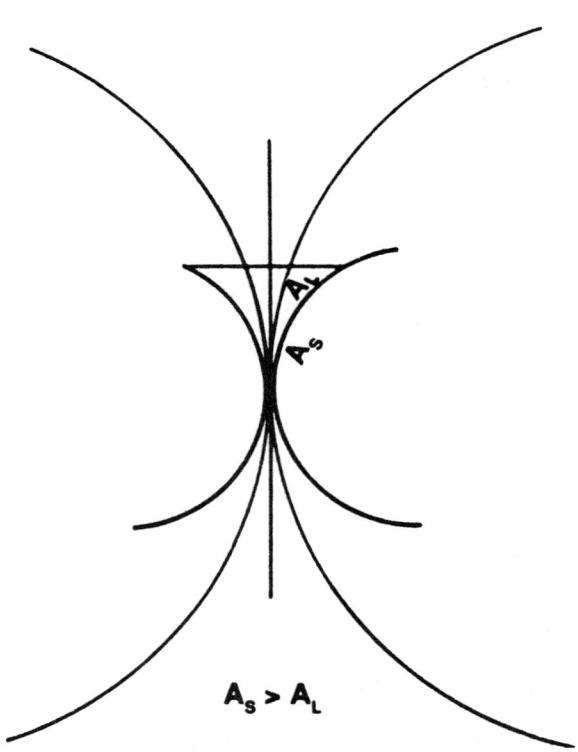

Fig. 3.86. Larger roll diameters decrease relative acceleration distances

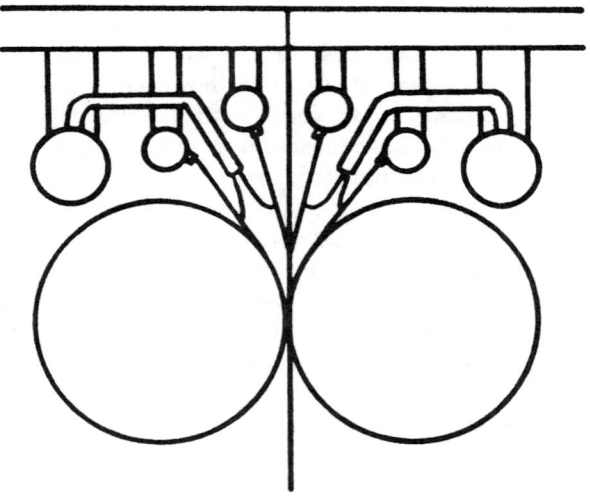

Fig. 3.87. Apron size press

An improvement on the apron size press concept is to combine its use with restricted flow of sizing solution to the pond. The simplest and most reliable device for this use is weir and spreader pan. Restricting the flow over the baffle aprons decreases the loads on them and mini-

mizes the dry-side wear problem. It also aids in reducing the amount of splashing in the size press nip.

Modified Size Press Designs

The problems with pond turbulence motivated development of modified size press designs which eliminated the pond. The gate roll size press, illustrated in Fig. 3.88, has an offset pond which is not in contact with the sheet. This offset pond feeds a metering nip which controls the amount of size going to a second nip. The second nip controls the film uniformity. It is essential that the rolls in the gate roll train be run at different speeds to minimize film split patterning (17).

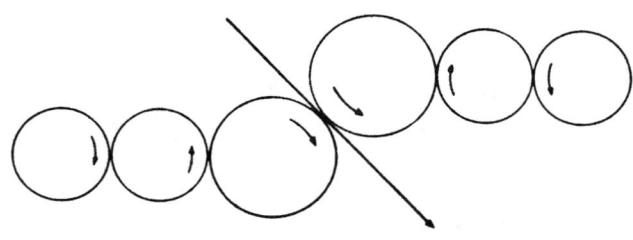

Fig. 3.88. Gate roll size press

With the gate roll size press, sizing solution solids and pickup rate are independent variables. It is possible to increase starch solution concentration to reduce load on the after dryers, and the "print on" application method keeps the starch on the sheet surface. Nevertheless, the problems associated with film split and fiber picking on the out-going side are still present and may be increased due to higher starch solution concentration. The high z-direction shearing forces created as the sheet leaves the nip cause sheet roughening and linting. These factors contribute to the "orange peel" sheet surface defects often associated with the gate roll size press. Substitution of a gate roll size press for a conventional size press involves replacing two rolls with six, contributing to increased initial investment and the potential of significantly higher maintenance costs and downtime for roll changes. It also involves installation and maintenance of six drive units compared to two. It is necessary to crown all six rolls and run them with relatively high nip loads; this causes heat buildup and short roll cover life with a corresponding requirement for a large investment in spare rolls.

The roll maintenance and film split pattern problems of the gate roll size press led to development of the blade/rod metering size press shown in Fig. 3.89 (18). Short-dwell blade coater heads are used to apply and meter the size to the size press rolls. Blade or rod metering permits operation without a pond or with a very limited one.

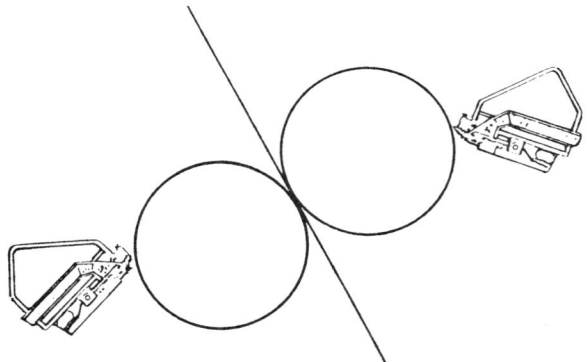

Fig. 3.89. Blade/rod metering size press

With blade metering, the amount of surface size being applied may be controlled and easily changed by varying the blade pressure. This feature makes it practical to control pickup and solids content as independent variables. The ability to control wet film thickness provides the capability to control the degree of penetration of surface size into the sheet, i.e., to locate the size near the surface or to drive it into the sheet with pressure-induced migration.

Use of bent blades for metering provides the capability to control cross-machine profile of the wet film thickness and thus the amount of starch applied. Although it may provide the papermaker with a valuable tool, this profiling capability should be used with discretion. It has the potential to mask profile problems that should be corrected at the wet end or in the main dryer section. In addition, application of varying amounts of starch across the machine can lead to differences in sheet properties.

Rods are also used as metering elements in modified size presses (19). Initially wire wound rods were used. As shown in Fig. 3.90, the application of size is not done with hydraulic pressure as in a conventional coater, but is achieved solely by volumetric application. The wet application of size suspension to the size press roll is a function of the wire diameter. The relationship is linear and independent of speed. Pressure differences across the width do not cause variation in the amount of sizing material. Volumetric metering does not subject the material being applied to the same amount of shear as blade metering, and thus the rheology constraints on the sizing material may not be as critical. Since control of wet film thickness is a function of wire diameter, it is necessary to change to a different rod in order to vary pickup.

There is a concern about the use of wire wound rods on high-speed machines, i.e., the potential for damage or injury

if the wire unwraps or wears through. These concerns have been addressed by a new profile rod construction which has machined grooves similar to those of the wire wound rod as shown in Fig. 3.91. It has higher resistance to mechanical loads, compared to rods with thin wires.

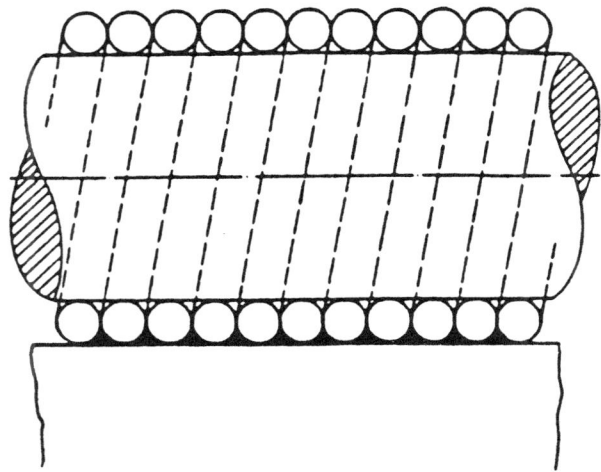

Fig. 3.90. Wire wound metering rod

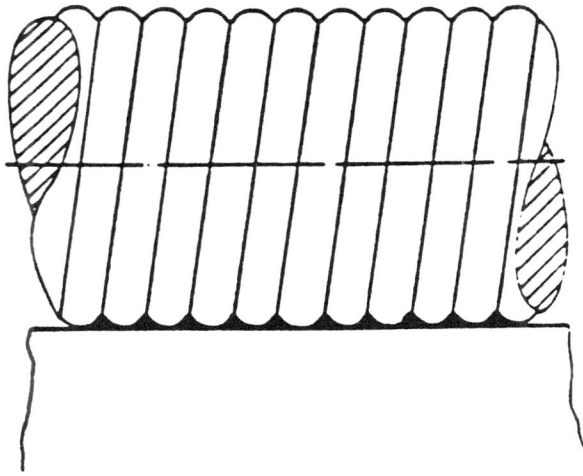

Fig. 3.91. Profile metering rod

In most designs of the blade/rod metering size press, short-dwell coater heads are used to supply the sizing material. Standard short-dwell coater heads require relatively high pumping rates in the range of 2.0-2.5 gal/min per inch of width. This flow rate results in recirculation rates which are considerably higher than those required for a conventional size press. High recirculation rates necessitate larger filtration equipment, pumps, heat exchangers, and tanks in the surface size supply system. One of the machine builders has a novel applicator design which incorporates a sealing blade in the application chamber head of the metering blade (20). The sealing blade is perforated, which allows the sizing liquid to pass through the blade, with the excess amount recirculated. The puddle before the sealing blade forms an air tight seal at the tip of the sealing blade and also lubricates the sealing blade. This feature minimizes the potential for skips from air entering the metering zone. Recirculation rates can be kept in the same range as a conventional size press. Figure 3.92 shows the dynamics of this arrangement.

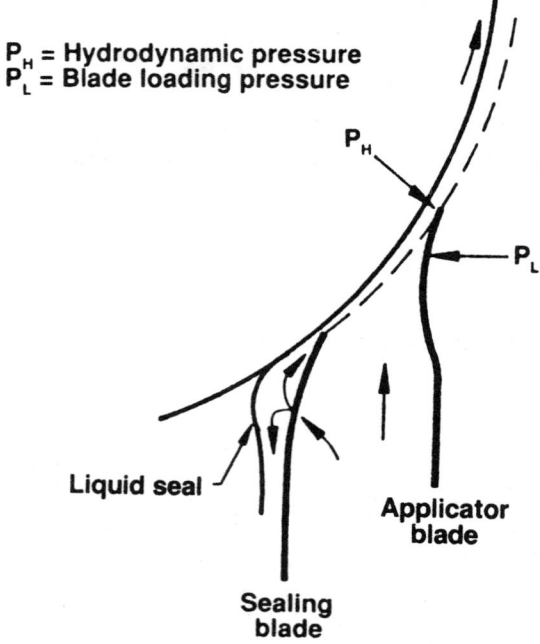

P_H = Hydrodynamic pressure
P_L = Blade loading pressure

Liquid seal

Applicator blade

Sealing blade

Fig. 3.92. Sealing blade arrangement

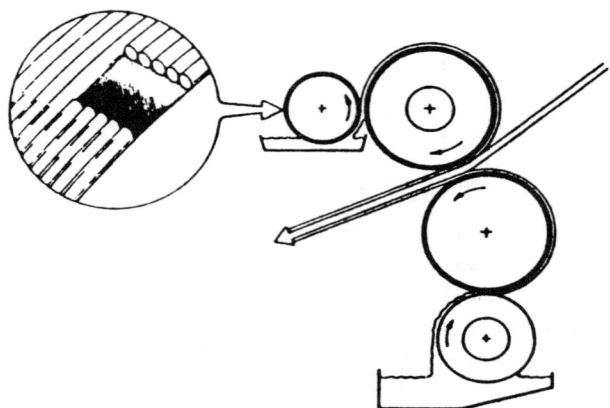

Fig. 3.93. High-solids metering roll size press

Another development which combines some of the advantages of volumetric metering and low recirculation rate is the high-solids metering roll size press illustrated in Fig. 3.93 (21). Large-diameter wire-covered metering rolls are used to apply wet film to the size press rolls. Metering is volumetric, analogous to a wire wound rod metering size press. The shear rate with the high-solids metering roll is not as great as with a blade or rod metering applicator. Thus the rheology requirements for the surface size are not as critical.

Alternatives to the Size Press

The size press is not the only device used for surface sizing and precoating of basestock. A number of alternatives to the conventional size press have been introduced to the paper industry over the past two decades.

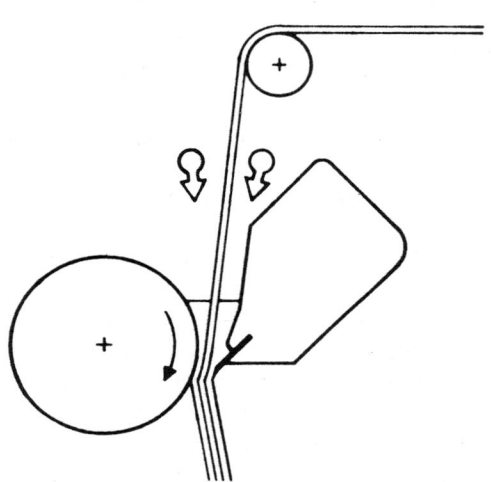

Fig. 3.94. BillBlade coater

The BTG BillBlade coater was invented about 20 years ago as a size press replacement to apply higher coat weights without film-split-related problems such as sheet marking. The principle of the standard BillBlade coater is shown in Fig. 3.94 (22). The paper runs vertically down through a pond formed between a rotating rubber-covered roll and a flexible blade. The thickness of the blade is typically 0.012-0.015 in. The backing roll is similar to a size press roll, but it has a softer rubber cover (70 P&J) and generally runs 2-5% faster than the speed of the web. This speed differential, combined with a take-off angle of the web away from the roll and toward the blade tip, is the key to achieving an even-sided application on both sides and minimizing film split pattern on the roll side. Starch is applied at twice the solids content normally used in a conventional size press. Pigmented coating can be applied at up to 64% solids with viscosities of up to 1000 cps (Brookfield 100 rpm). Coat weights of up to 6 pounds per side can be applied without film-split-related defects such as "orange peel" or "crow's foot."

Valmet has recently introduced a size press alternative under the trade name TwoStream coater, illustrated in Fig. 3.95 (23). The TwoStream coater is similar in principle to the BillBlade coater, but it runs in an upward direction. The upward sheet run provides several advantages, including better visibility of the sheet turn leaving coater and the capability to utilize infrared supplementary drying directly after the coating head without fear of having coating spillage on the emitters. The TwoStream coater uses pressurized ponds as applicators, allowing for precise control of the hydraulic pressure, volumes, and flow velocities of the material being applied to each side of the sheet.

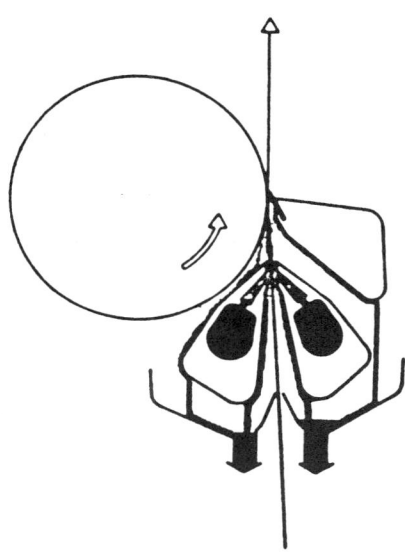

Fig. 3.95. TwoStream coater

Another alternative to the size press is the BTG LAS hydrophilic roll coater, illustrated in Fig. 3.96 (24). The heart of the LAS is the hydrophilic transfer roll, which is chromium plated and treated by a proprietary etching process. The treated roll can carry wet films 3-15 microns thick. Liquid is applied to the hydrophilic transfer roll by a resilient rubber-covered metering roll running in a supply pan. Varying the speed of the metering roll maintains a flooded nip to prevent air intrusion into the starch solution. Pressure in the metering/transfer roll nip is controlled to provide a uniform film on the transfer roll. The hydrophilic transfer roll is run in the same direction as the web to apply the liquid film to the web. A speed differential is maintained between web speed and transfer roll speed to minimize film split pattern on the

sized sheet. A resilient rubber-covered backing roll is used to assure uniform contact of the sheet on the hydrophilic transfer roll. The hydraulic pressure in the LAS coater is very low, making it a good choice for surface sizing or precoating high groundwood content sheets.

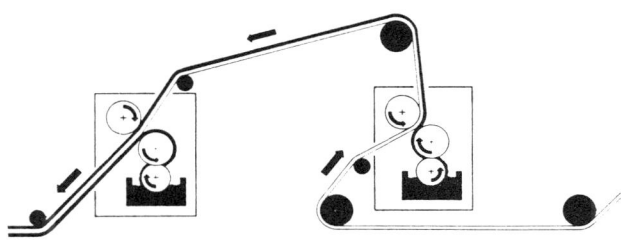

Fig. 3.96. Hydrophilic transfer roll coater

Since the LAS coater applies liquid to only one side of the sheet, two units must be installed to do C2S surface sizing. This requires considerably more space than a size press. As machine speeds increase and basis weight decreases, the cumbersome sheet run between these two units may be a source of runnability problems. At higher speeds, picking and patterning potential from the wet coated sheet contacting the second backing roll may necessitate the installation of infrared dryers between the LAS units and/or chilling the second coater backing roll. Application separately to each of the two sides can be either a cause of or a potential cure for two-sidedness.

Effect of Application Method on Sheet Properties

The key factors which differentiate the effects of various surface size application methods on sheet properties are:

• Pressure-induced migration
• Dwell time at pressure
• Lifting forces on the sheet surface as it exits the nip.

A combination of the first two factors determines the location and uniformity of distribution of starch in and on the sheet. The second factor is the primary determinant of the amount of fiber swelling. The third factor controls integrity of the starch film and disturbance of the sheet surface. As shown in Table 3.6, the conventional size press and each of the major alternatives discussed above represent various combinations of the above factors, resulting in differing effects on sized sheet properties.

Table 3.6. Comparison of key factors for various applicators

	Size press	Gate Roll Blade/Rod Metering	BillBlade TwoStream	LAS
Pressure induced migration	High	Moderate and controllable	Low	Moderate and controllable
Dwell time at pressure	Long	Long	Minimal	Medium
z-direction lifting forces	High	Highest	Lowest	Moderate and controllable

In a size press, liquid is forced into the sheet interior in an enclosed, pressurized nip. When aqueous solutions are applied, penetration of the cellulose fiber cell walls causes swelling and the potential for debonding. When suspensions such as pigmented coatings are applied, the liquid phase is forced into the fibers, causing a filtercake effect and nonuniformity in the coating.

The size press and modified size presses all have rolls which run at the same speed as the sheet. The sized sheet exits via a diverging nip, which creates film split and lifting forces on the sheet surfaces. These forces are higher in a gate roll coater or blade/rod metering size press than in a conventional size press. In the modified size presses, the wet starch film is premetered onto the applicator rolls. Dwell time at pressure is the same as for a conventional size press, but the nip is not full of starch solution. This dewatering causes an increase in viscosity, which adds to the z-direction pull on the sheet as it exits the nip.

Since the severity of film split pattern is dependent on viscosity, it imposes a limitation on the amount of pigmented coating that can be applied via conventional or modified size presses to produce a sheet without objectionable sheet surface marks.

Fiber swelling in the size press contributes to surface roughness and a decrease in smoothness. With alternatives to the size press, use of high concentrations and a starch layer close to the surface results in reduced surface roughness.

A starch layer located closer to the surface does not reduce opacity to the same extent as deeply penetrating starch.

A switch from size press to one of the alternatives can also help to increase stiffness. This effect is related to location of the starch near the surface, which creates a modulus effect similar to an I-beam.

Higher concentrations of starch solution can influence porosity. Alternative starch applicators provide better film forming and thus reduce porosity. Less refining horsepower may be needed to reach a given porosity.

Orientation of the starch closer to the surface, combined with better film forming and elimination of film splitting, results in better ink holdout, improved solvent holdout, and higher ink gloss.

Surface sizing with one of the size press alternatives which minimize the z-direction forces on the sized sheet can reduce linting by providing better bonding of loose fiber onto the surface, thus reducing the potential for streaks and scratches in blade coating.

The size press forces starch into the interior of the sheet, resulting in an increase in internal bond. The lack of penetration of starch into the sheet with a surface applicator means that internal bond improvement cannot be obtained via surface sizing.

Deep penetration of starch into the sheet tends to decrease tear strength. An applicator which keeps starch near the surface does not contribute to the tear strength loss commonly associated with the size press. An applicator with penetration control capability makes the tear strength to surface sizing relationship controllable.

Location of starch nearer the surface increases its effectiveness in providing sheet surface strength; IGT dry pick resistance improves at the same starch pickup.

Effect of Application Method on Machine Speed and Drying

In general, alternatives to the size press apply starch or precoating at about twice the concentration of a size press. They also make it practical to hold pickup in pounds per ream constant as basis weight varies. With a size press, wet pickup usually remains at a constant percentage. This causes pickup in pounds per ream to go up as basis weight increases. On a machine which is afterdryer limited, a switch to high-solids surface sizing can result in significant machine speedup and added value of incremental tonnage.

Web Handling and Drying of the Surface Sized or Precoated Sheet

Surface sizing or precoating rewets the sheet. Absorption of water from the surface size causes fiber swelling and sheet expansion. It also causes a decrease in tensile strength. As water is absorbed by the fibers, the solids and viscosity of the sizing material at the sheet surfaces increases, and the sheet becomes tacky. This tackiness can cause picking and sheet quality problems. All of these factors must be taken into consideration in designing the sheet runout of the size press and through the after dryers.

One of the biggest potential problems is sheet expansion, which can lead to wrinkles and calender cuts. The most commonly used device to compensate for sheet expansion is the curved spreader (bowed) roll. A curved spreader roll in the sheet run between the size press and the first after-dryer will provide a level sheet of uniform tension on the first after-dryer cylinder. The spreader roll should be covered with a nonstick surface.

The draw to the after-dryer section should be precisely controlled. Too tight a draw on the expanding sheet can adversely affect physical strength properties. Too slack a draw may cause wrinkles, cockles, or grainy edges. Modern machines use a load cell mounted on the spreader roll bearing housing to measure sheet tension out of the size press. This tension signal is used as the basis for control of after-dryer section speed.

Another option for transporting and spreading the surface-sized sheet is the air turn (25). The wet sheet is not contacted by a roll but rather rides on a cushion of air. The ratio of the air cushion to air supply pressure provides a reliable measurement of sheet tension for use in after-section drive control.

The after-dryer section usually consists of conventional cylinder dryers. In most cases, the first two after-dryers are run at lower temperatures and covered with a nonstick surface to avoid picking. These first two dryers are usually unfelted. Dryer fabrics used on after sections are usually monofilament fabrics with pin seams to avoid sheet mark. It is desirable to have the capability to control the temperatures of the top and bottom after-dryers separately to control curl.

In some cases, it may be necessary to provide supplementary drying between the size press and the first after-dryer. Infrared or infra-air dryers are usually used in this location because they can provide a large input of energy in the short sheet run distance available. When infrared dryers are used, the first two after-dryers can be run at full temperature. On some machines, air foil dryers are used in the after-size press supplementary dryer position to provide not only supplementary drying but also web guiding, spreading, and tension control.

Bibliography

Literature Cited

1. Baumeister, M., "European Coating Technology," paper presented at the 73rd Annual Meeting, Canadian Pulp & Paper Association, January 30, 1987, p. B241.

2. Adams, A.A., "Effect of size press treatment on coating holdout," *1983 Coating Conference Proceedings,* TAPPI PRESS, Atlanta, p. 23.

3. Smook, G.A., *Handbook for Pulp & Paper Technologists,* Joint Textbook Committee of the Paper Industry, Atlanta, 1982, p. 262.

4. Kirby, K.W., "Starch chemistry and reactivity," *1985 Papermakers Conference Proceedings,* TAPPI PRESS, Atlanta, p. 59.

5. Kearney, R.L., "Starch Binders in Coatings," *1988 Blade Coating Seminar,* TAPPI PRESS, Atlanta, p. 99.

6. Harvey, R.D., "Is there a better size press starch?" *1985 Papermakers Conference Proceedings,* TAPPI PRESS, Atlanta, p. 83.

7. Fournier, L.B., "Chromium complex treatments for paper," *1984 Papermakers Conference Proceedings,* TAPPI PRESS, Atlanta, p. 181.

8. Totty, C., Tappi 50(8):93A (1967).

9. Beals, C.T., Pulp & Paper 54(4):132 (1978).

10. Hansson, J.A. and C.P. Klass, "High speed surface sizing," *1983 Papermakers Conference Proceedings,* TAPPI PRESS, Atlanta, p. 49.

11. Perry, J.A., Pulp & Paper 48(5):144 (1972).

12. Eklund, D.E., Tappi J. 67(5):70 (1984).

13. Beals, C.T., Pulp & Paper 54(4):132 (1978).

14. Dill, D.R., Tappi 57(1):97 (1974).

15. Pollart, K.A. and D.R. Dill, "Size press sizing additives on neutral and alkaline base sheets," TAPPI/PIRA Seminar Notes: New Sizing Methods and Effects on Fibers, Fillers and Dyes, TAPPI PRESS, Atlanta, 1982, p. 1.

16. Justus, E.J., "High speed size press," U.S. Pat. No. 4,340,623, (July 20, 1982).

17. Alheid, R.J., "Operation and applications of the gate roll size press," *1978 Papermakers Conference Proceedings,* TAPPI PRESS, Atlanta, p. 83.

18. Wight, E.W., "Operation and applications of the blade metering size press," *1988 Engineering Conference Proceedings,* TAPPI PRESS, Atlanta, p. 549.

19. Sollinger, H.P., "Speedsizer," *1988 Engineering Conference Proceedings,* TAPPI PRESS, Atlanta, p. 553.

20. Finch, K.W., "Sym-sizer: the film transfer size press," *1989 Papermakers Conference Proceedings,* TAPPI PRESS, Atlanta, p. 117.

21. Pazckowski, M. and T. Engevik, "A high speed metering system for BillBlade and LAS coaters," *1990 Conference Proceedings,* TAPPI PRESS, Atlanta, p. 165.

22. U.S. Pat. No. 3,489,592 (1970).

23. Rautiainen, P., "The Twinstream, Up-stream and Twostream coaters—new tools for coating," *1983 Coating Conference Proceedings,* TAPPI PRESS, Atlanta, p. 41.

24. Klass, C.P., Pima 70(9):19 (1989).

25. DeSanti, E.A., Southern Pulp Paper 49(5):16 (1986).

Chapter 3, Section VII

Coated Board

Terry Kellogg, J.M. Huber Corp., Clay Division

Recycled Board

Basestock

Recycled board has a furnish consisting of a mixture of various grades of wastepaper. This diversity in available fiber, combined with the multi-ply construction of the cylinder machine, allows recycled board manufacturers to place each grade of wastepaper fiber where it best suits quality and economic needs.

Typically, the top liner consists of white grades of wastepaper (SBS, ledger, manila) so that the whiteness of these fibers can provide much of the final sheet brightness, thereby lessening the brightness and opacity demands on the coating.

Recycled board is also manufactured with dark body basestock, which has a brightness much closer to unbleached fiber. In this case, the top liner, like the back liner, will be a cleaner version of the filler ply which is made up of a combination of news, corrugated, and box cuttings. The filler ply can handle contaminated waste better than either the top liner or bottom liner. The various contaminants, such as inks, coatings, waxes, and glues, generally do not cause quality problems if kept away from the coated surface.

Coating surface defects such as "stickies" (transfer of thermoplastic materials from the back liner to the coated surface), non-adherence of the coating to ink, and functional coating contaminants in the top liner are caused by insufficient cleaning of the furnish.

Fillers are not added to the basestock because clays and coating particles from the wastepaper serve to fill in the basestock structure, which along with the short, soft, well-refined recycled fibers tends to result in a compressible, dimensionally stable, low-strength sheet. The smoothness and holdout of the basestock prior to coating is enhanced by a machine calender stack which may use starch boxes, water boxes, or both to achieve the finished basestock for coating.

Basestock characteristics of recycled board include stiffness ratios which are much higher in machine direction (MD) than cross direction (CD), high internal bond strength, low tear strength, and a top liner brightness which could be between 15 and 65 (TAPPI brightness) depending on the furnish used. For basestock with higher brightness top liners, formation must be very uniform or the mottled basestock appearance will show through the coating, resulting in a mottled appearance and possibly a mottled print depending on the severity of the problem. Recycled basestock is smooth and fairly uniform in density and porosity due to the high degree of refining imposed on the fibers.

Coating Application and Metering

Basecoat

The basecoat typically utilizes a rod metering system. Many different rod metering systems are manufactured, although two basic types are used by the majority of recycled board mills: the free-standing rod or the pressurized backing-roll rod coaters.

Free standing rod metering systems generally have single-roll applicators which apply an excess of coating to the basestock followed by a rotating rod metering element. Lack of coat weight control and coating uniformity along with the low coating solids necessary to operate the free standing rod metering systems has led to the next generation of rod coaters, the pressurized backing-roll systems.

These systems allow the rod to be loaded against a rubber backing roll which supports the sheet allowing for more precise coat weight control, better coat weight uniformity, and higher solids. Coating application to the web can be accomplished either roll applicator or short-dwell application methods. The trend in coating recycled board has gone toward the pressurized backing-roll rod system with short-dwell application, thereby taking advantage of higher solids to achieve a smoother surface.

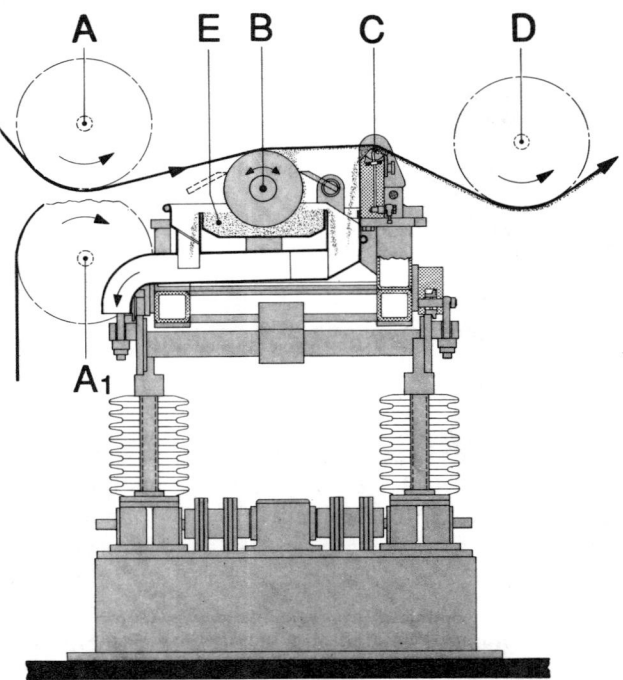

A   Infeed guide roll
B   Applicator roll
C   Metering bar
D   Guide roll at exit
E   Coating pan

Fig. 3.97. Metering bar coating head (courtesy of Jagenberg Inc.)

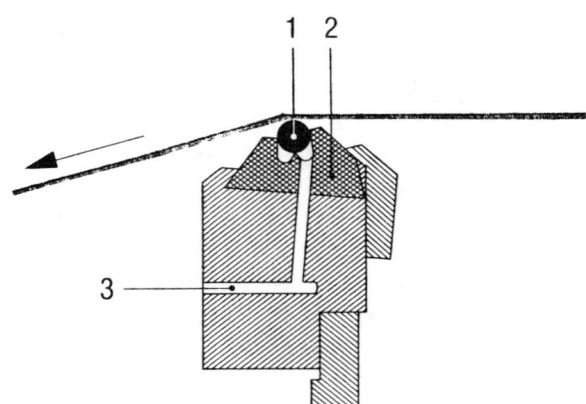

1   Metering bar
2   Metering bar bed
3   Flushing water infeed

Fig. 3.98. Metering bar bed (courtesy of Jagenberg Inc.)

Fig. 3.99. Pressurized backing-roll rod metering systems: flooded nip application (courtesy of Beloit Corp.)

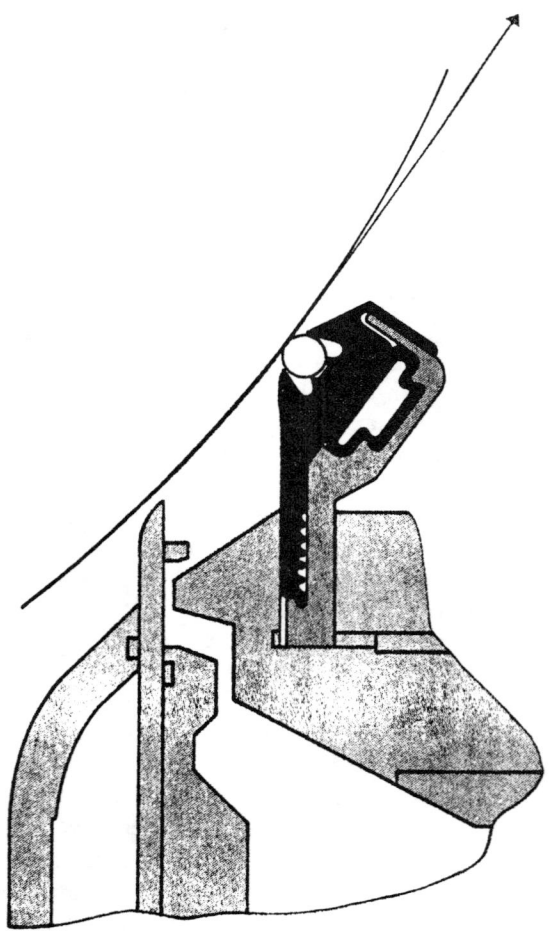

Fig. 3.100. Pressurized backing-roll rod metering systems: short-dwell application (courtesy of Beloit Corp.)

Blades are seldom used as a basecoat system for coating recycled board. The high pressure at the blade tip could force contaminants out of the sheet and into the coating, resulting in scratches and streaks. Although a few mills utilize a blade basecoat, generally they are reserved for a third coating station.

Topcoat

A typical topcoat utilizes an air knife metering system whereby an excess of coating is applied to the web by means of a single or double-roll application system which is subsequently metered off to a given coat weight by an air jet. Because of the contouring effects of the air knife, it is used to apply low solids, high-opacity, high-brightness, uniform coatings to the rod coated basestock. Drawbacks to the air knife system include coating waste, foam, a porous coating structure, and speed constraints.

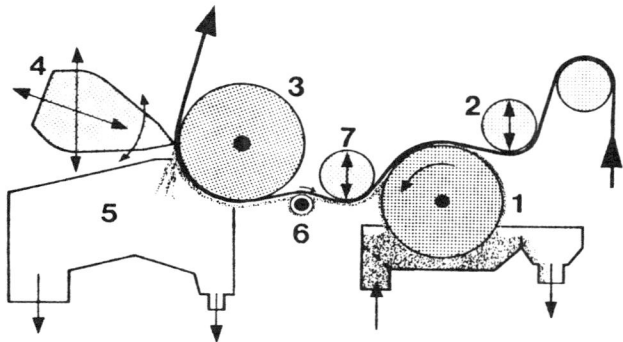

1	Applicator roll
2	Deflecting roll
3	Air knife backing roll
4	Air knife
5	Injector catch pan
6	Presmoothing roll
7	Deflecting roll

Fig. 3.101. Air knife application and metering system (courtesy of Jagenberg Inc.)

To get away from these limitations, some recycled board mills have gone toward using a rod topcoat. The rod-on-rod sequence requires a very smooth basestock to minimize mottle, but can result in a very smooth and dense coated surface with excellent ink holdout.

Third Coat

The trend in the United States and Europe is toward a triple-coating sequence as gravure and offset printability requirements become more demanding. Sequences currently in use include rod/rod/rod and rod/air knife/blade. Although triple coating adds more complexity to the coating system, custom-designed coating structures can be made to suit end-use needs while optimizing coating formulations from a cost and opacity viewpoint.

Coating Formulation

Recycled board requires a low-brightness, high-clay, high-solids basecoat and a high-opacity, high-brightness, uniform topcoat. If a third coat is used, it is usually formulated to be dense in structure, either of the same brightness as the topcoat or transparent to see through to the opaque and bright topcoat. On dark body stocks, high levels of titanium dioxide (TiO_2) are used in the topcoat for opacity. Cobinder systems are generally used in both the basecoat and topcoat, usually consisting of a synthetic latex or latex blend with protein or a synthetic soluble cobinder such as carboxymethyl cellulose (CMC) or sodium alginate. Gluability, smoothness, gloss, opacity, and high surface strength are important coated board characteristics.

Drying and Finishing

The basecoat and topcoat applications are usually wet-on-wet with no inter-station drying due to space constraints. Recent trends have been toward adding a bank of infrared (IR) preheaters to dry the basecoat prior to topcoat application for increased smoothness development. Low-solids topcoat applications may be followed by a bank of IR preheaters to immobilize the coating, and then a zoned high-velocity hot-air tunnel dryer to remove the bulk of the moisture.

Finishing consists of either a dry calender stack or a gloss calender, depending on quality needs. Calendering leads to de-opacification of the coating structure, or calender blackening, which must be minimized by proper coating formulation if calendering is necessary for gloss and smoothness development.

End-use Considerations

The primary end-use for recycled board is as folding boxboard for food packaging applications. The dimensionally stable, internally strong, easily scored, and folded multi-ply sheet is a primary substrate for sheet-fed offset printing. As the trend from sheet-fed to web-fed offset continues in the folding boxboard market, recycled board has adapted well to the increased speed and surface strength requirements of this new printing process.

The improving smoothness, gloss, compressibility, and ink holdout characteristics of recycled board has also made it a substrate capable of high-quality gravure printing. The trend toward triple coating will further improve the coated surface of recycled board for even the most demanding high-quality gravure printed graphics.

Unbleached Kraft Board

Basestock

High-yield sulfate pulp consisting of 50-90% softwood and 10-50% hardwood is used to manufacture unbleached kraft basestock. A small amount of recycled virgin unbleached kraft may also be added, although due to the high-tensile and tear-strength characteristics desired, recycled use is closely controlled.

The softwood fibers provide strength, and the hardwood fibers provide smoothness and compressibility. To optimize these properties on the fourdrinier, a very high percentage of softwood pulp is used as the primary basestock, and the hardwood is concentrated in a top ply by the addition of a secondary headbox or top former to the machine. The shorter hardwood fibers provide a smoother, less porous surface for coating.

Fillers, such as clay, can be added to the high hardwood top ply or at the size press in the form of a pigmented starch solution. A size press, machine calender stack, or both may also be used to apply starch or water to the basestock prior to coating to enhance smoothness and holdout. Yankee dryers have also been used to promote basestock smoothness prior to coating.

Unbleached kraft basestock is characterized by high strength in the form of high tear and tensile. Smoothness, compressibility, and porosity are controlled by the amount of hardwood incorporated into the sheet. Brightness of the basestock is very low, requiring an optically efficient coating to obtain brightness, opacity, and uniformity.

Coating Application and Metering

Basecoat

The basecoat typically utilizes a trailing blade or bent blade metering element with either a short-dwell or roll applicator system. The trailing blade is used to achieve maximum smoothness development for subsequent coating applications. The trend in application is from roll applicator to short dwell, as mills try to take advantage of the increased holdout attainable with the short dwell to increase smoothness. A combination "wet-on-wet" system is also available but not in widespread use.

Topcoat

Traditionally, an air knife has been used as the topcoat on unbleached kraft board. The contouring effects make it ideal for applying high-opacity, high-brightness, very uniform coatings to the blade basecoat. The main problem in utilizing an air knife for unbleached kraft board involves speed limitations. Air-knife metering of traditional pigmented coating formulations is currently limited to 1500 ft/min, focusing future efforts at increas-

ing productivity around the elimination of the air knife completely.

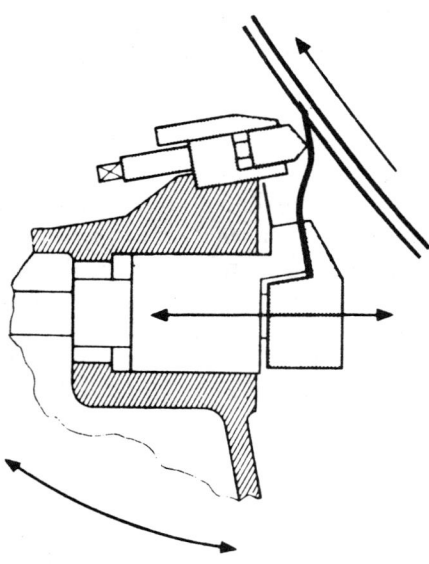

Fig. 3.102. Combi-blade with rigid blade (courtesy of Jagenberg Inc.)

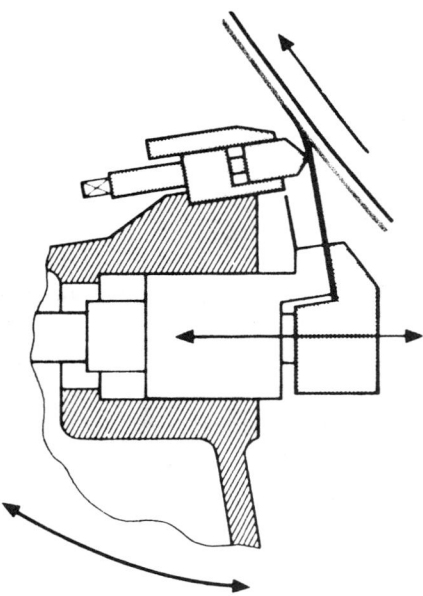

Fig. 3.103. Combi-blade with bent blade (courtesy of Jagenberg Inc.)

An alternate metering system for a second-down coating application in a triple-coating sequence is a blade coater. Although the need to apply more coating may require a bent blade, a trailing blade may also be used. Either a roll applicator, short-dwell, or a combination application system can be used depending on the surface characteristics desired.

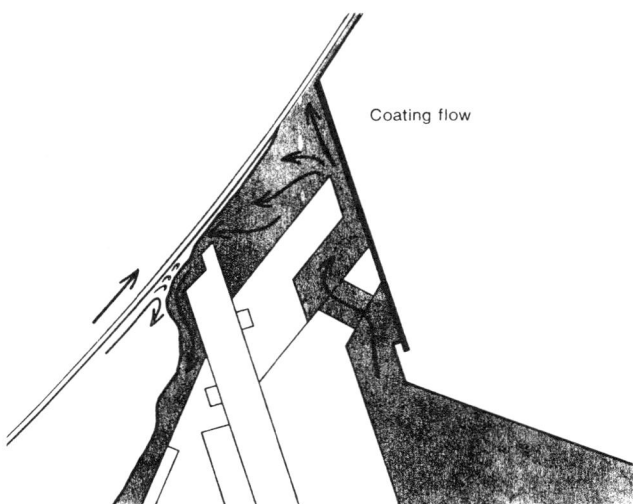

Coating flow

Fig. 3.104. Short dwell blade application system (courtesy of Beloit Corp.)

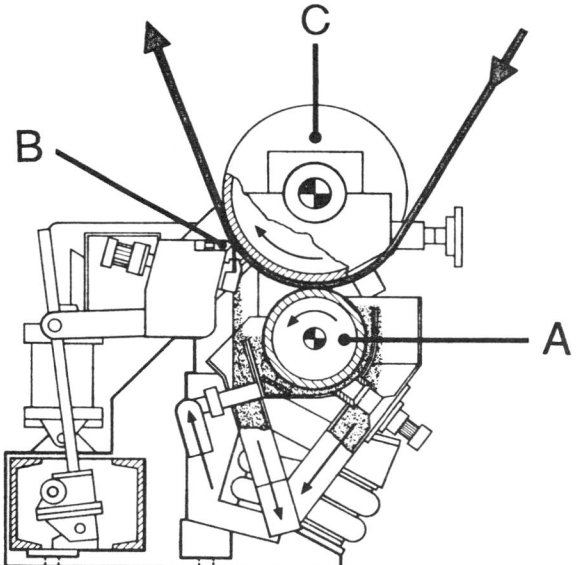

A   Applicator roll
B   Metering unit
C   Backing roll

Fig. 3.105. Combi-blade coating head (courtesy of Jagenberg Inc.)

Third Coat

As seen with recycled board, the trend in unbleached kraft board is toward a triple coating sequence to further enhance gravure and offset printability. Sequences in use include blade/blade/air knife and blade/air knife/rod. It is interesting to note that in both cases, the air knife was retained in the coating sequence due to its unmatched ability to apply a high-opacity, uniform coating to a rough basestock even though it is speed limiting.

Coating Formulation

Unbleached kraft board requires a high-solids, very low brightness basecoat which is usually dyed-back to minimize the mottle obtained by the leveling effects of the blade metering system.

If a blade is used as a second coating station, the formulation must begin to build opacity while continuing to increase smoothness for the third coating application. If an air knife is used, a very high-opacity, high-brightness, high-TiO_2 formulation must be utilized to hide the nonuniform blade basecoat and dark basestock.

A rod third coat formulation is usually dense in structure and somewhat transparent if applied over an air knife top coat to show through the opacity and brightness created by the air knife formulation. If an air knife is used as a third coat, it must be a high-opacity, high-brightness, high TiO_2 formulation to provide hiding, brightness, and uniformity.

Cobinder systems are generally used, usually consisting of a strong synthetic latex or latex blend with a protein, casein or synthetic-soluble cobinder such as polyvinyl alcohol, CMC, or sodium alginate. Coated board requirements include wet and dry gluability, gloss, smoothness, opacity, and high surface strength as the coating must be strong not to be the weak link on a very high-strength basestock.

Drying and Finishing

IR dryers are used as inter-station dryers and as preheaters for the lower viscosity and solids air knife coatings. Air-knife IR preheaters are usually followed by a zoned high-velocity hot-air tunnel dryer to remove the bulk of the moisture.

Finishing may include a gloss calender, a brush polisher (see Chapter 7) or not be done at all depending on the properties required. The gloss calender produces both gloss and smoothness at the expense of densification and de-opacification of the coating, whereas the brush polisher merely glosses the sheet without sacrificing opacity and density. As always, the coating formulation must be modified to work in conjunction with these finishing processes for optimum results.

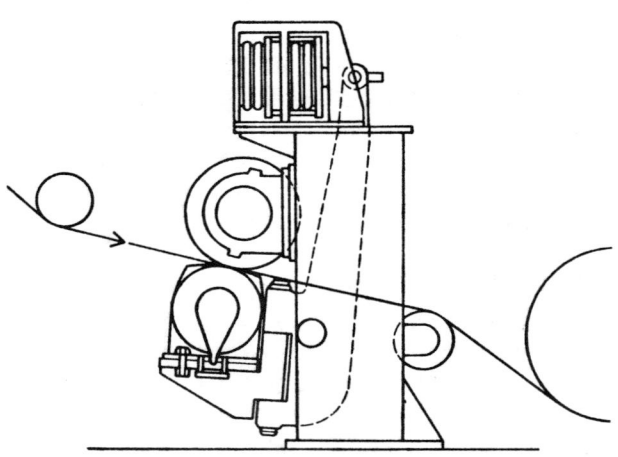

Fig. 3.106. Gloss calender (courtesy of Beloit Corp.)

End-use Considerations

The primary end use for unbleached kraft board is as carrier board for beverage packaging applications. High wet strength and dry strength (tear, tensile, stiffness) of the coated board make it ideal for packaging heavy wet cans and bottles. Although Flexo printing has not placed severe demands on coating, the trend toward more detailed half-tone gravure printing has made a push for increased smoothness and compressibility a major factor driving the unbleached kraft industry toward using higher hardwood levels in the top ply and going toward the triple-coating sequences. A trend toward utilizing coated unbleached kraft in the offset folding boxboard market has also initiated substrate changes in the direction of improving compressibility or scorability, coating strength, and ink or varnish holdout.

Solid Bleached Sulfate Board

Basestock

A typical solid bleached basestock fiber mix is between 75-98% hardwood and 2-25% softwood. The low-yield sulfate pulp is subsequently bleached, resulting in the bright, white basestock characteristic of solid bleached board. It is this basestock brightness which shows through the coating, resulting in a very bright, opaque, uniform surface.

Single-ply fourdrinier construction with a strong emphasis on good formation, smoothness, and low-density typifies solid bleached basestock. The high hardwood content throughout the sheet contributes smoothness and compressibility as does the bleaching process, which tends to weaken and soften the fibers for increased smoothness, flexibility, and compressibility.

High-brightness clays or TiO_2 are occasionally used as fillers and can be added in the wet end or at the size press in the form of a pigmented starch solution. Because fillers tend to increase sheet density, their use is limited. Either a size press, machine calender, or a combination of both may also be used to apply starch or water to the basestock prior to coating to enhance smoothness and holdout.

Solid bleached basestock characteristics include softness, compressibility, high brightness, low density, low porosity, excellent smoothness, high surface strength, and high internal bond strength. Because of the high hardwood to softwood ratio used to obtain optimum surface properties, strength characteristics such as tear and tensile are low.

Coating Application and Metering

Basecoat

The basecoat utilizes a trailing blade or bent-blade metering element with either a roll applicator or a short-dwell application system. Combination "wet-on-wet" systems are also available but not in widespread use.

Topcoat

Typically, the topcoat station consists of a second trailing blade or bent-blade metering element with a short-dwell application system. Although a few air-knife metering systems are still in use, the majority of solid bleached board mills have made the change to the blade metering system to take advantage of the extraordinary smoothness a blade-on-blade sequence has to offer.

Third Coat

The third coating station on a solid bleached board machine refers to a blade metering element on the backside of the sheet, used to improve wax holdout for board going into freezer applications.

Top-side third-coating stations have not been the trend in the solid bleached board industry, perhaps due to the excellent gravure smoothness attained with the current blade-on-blade coating sequence.

Coating Formulation

Because of the very high brightness basestock, opacity development is not necessary on solid bleached board resulting in very limited use of TiO_2. Instead, a combination of No. 2 or No. 3 clay and calcium carbonate ($CaCO_3$) is used in the basecoat to keep the sheet open for the clay-containing topcoat, which may also utilize plastic pigment for gloss.

The blade third backside coating formulation used for wax holdout is usually very similar to the blade basecoat, consisting of clay and $CaCO_3$.

Cobinder systems are generally used. These are typically high-synthetic formulations consisting of a latex or latex blend with a small percentage of protein, starch, or synthetic-soluble cobinder such as CMC or sodium alginate. Coated board requirements include smoothness, ink or varnish holdout, gluability, high surface strength, and gloss.

Drying and Finishing

IR dryers are used as inter-station dryers and as preheaters for the lower viscosity and solids air-knife coatings. Final drying may consist of a zoned high-velocity air-tunnel dryer to remove any remaining moisture.

Finishing typically consists of a machine calender stack followed by a gloss calender. Since calendering produces sheet densification, only the minimum amount of calendering necessary to achieve smoothness and gloss quality targets is performed.

End-use Considerations

The end-use areas for solid bleached board fall into two categories: printing grades and folding cartons.

Printing grades include applications such as book covers, greeting cards, and bristols. Folding carton applications include food packaging, especially in freezer applications where the bright, white basestock, and excellent holdout obtained by the blade coatings on both the top and back sides make it ideal for wax coating and polyethylene extrusion.

Although solid bleached board has always had a quality presence in the offset printing area, the excellent smoothness, gloss, compressibility, uniformity, brightness, and ink holdout characteristics have made it a primary substrate for high-quality gravure packaging applications. The increasing demand for even higher-quality gravure graphics has been the major factor behind the trend toward the use of increasing percentages of hardwood in the basestock and the blade-on-blade coating sequence.

Chapter 4

Drying

William L. Bracken, Bracken Resources Inc.
Robert A. Daane, Consultant
Ernest A. DeSanti, The Black Clawson Co.
James Y. Hung, Hung International
Donald W. Lawton, Advanced Systems Inc.
John F. Munce, IR Application Consultant
Robert E. Peiffer, Beloit Corp.
E. C. Porter Jr., Repap Wisconsin Inc.
Marion R. Ricks Jr., Impact Systems, Inc.
Michael Wunderlich, Infrared Consultant

Introduction

This chapter on drying is comprised of a number of individual sections describing the equipment used today in the drying of coated paper for publication grades. Each section was prepared by Coating Process Committee members whose names have been recognized.

Fig. 4.1. Off-machine coater (courtesy of Repap Wisconsin Inc.)

The success of present drying systems has been a result of a long history of progression in drying technology and knowledge. While there have been few revolutionary developments in drying equipment used by the papermaker, modern materials and electronics have resulted in many improvements in drying design and control. With the demand for coated publication grade papers on the increase in the past decade, combined with worldwide competitive pressure, the need for increased productivity and quality has led to focusing more attention toward the drying process, and this in turn has led to more rapid improvement in equipment technology with existing dryer designs. In particular, radiation and air impingement systems have seen significant developments in design for use on coating machines and are now widely used and accepted.

For the production drying requirements in the coating of publication grade papers, five major categories of dryer designs, listed below, are the preferred drying methods used today and are the topics of discussion for this chapter.

1. Steam-heated cylinder dryer
2. High-velocity air cap dryer
3. Air impingement or tunnel dryer
4. Air flotation dryer
5. Infrared dryer.

Of the common drying systems in use today, all thermally remove the excess solvent, in this case water, from the liquid coating solution applied to the sheet. The process of thermal drying occurs when coated paper is subjected to the simultaneous transfer of heat and mass. The drying rate is governed by the speed at which heat and mass transfer can be applied and maintained. Heat transfer is the transport of energy resulting from a temperature gradient in a system or a temperature difference between two systems.

In the analysis of drying, three basic mechanisms of heat transfer are identified: conduction, convection, and radiation.

1. Conduction involves the transmission of heat between molecules through a stationary solid, liquid, or gas. Steam cylinders in direct contact with the sheet.

2. Convection involves the process of heat transfer between a surface and a liquid or gas in motion. Impingement air from an air impingement hood or flotation dryer.

3. Radiation involves the transfer of heat in the form of electromagnetic waves. Infrared heaters.

Conduction and convection are major contributors to the total heat transfer at low temperatures (below 800°F),

whereas radiation is the important factor at very high temperatures.

As importantly, one generally identifies the three phases in the drying process of coated paper: pre-heat phase, steady-state evaporation phase, and falling rate phase.

1. Pre-heat Phase: involves the energy required to heat the paper, coating, and water to the equilibrium drying temperature. While most of the energy absorbed is used for pre-heating the mass, some evaporation does occur, but is usually considered minimal in this phase.

2. Steady-state Evaporation Phase: involves the evaporation of water proportional to the amount of heat transferred at the equilibrium drying temperature. The majority of evaporation is accomplished in this phase.

3. Falling Rate Phase: as the paper and coating dry, the evaporation rate begins to fall, and the energy from the dryer causes the paper temperature to rise.

For most applications today, dryer types are likely to be combined, taking advantage of the benefits that each has to offer and applying the above mechanisms that best fit the drying process and product. The sections in the chapter will review the general design of each dryer type and how the system can be applied in the drying equation.

Steam Cylinder Dryers

Steam cylinder dryers have been the workhorse of the paper industry as a result of their use in the early days of paper production. This is not an accident, but rather due to the many inherent advantages that cylinders dryers can provide:

1. Supports the wet sheet during drying, eliminating long draws and reducing tension requirements
2. Sheet is dried on a flat surface, reducing paper stress, cockle and curl
3. Contact drying is a very effective method of transferring the heat required for drying.
4. The conversion of steam latent heat to drying energy is reasonably efficient resulting in an economical drying method where steam is available.
5. Ease of threading and broke handling.

Drying coated paper with hot cylinders is accomplished by conduction heat transfer from the heat source to the paper. The most common heat source for cylinder dryers is saturated steam (Fig. 4.2), although some off-machine coaters in converting plants may use a circulating hot fluid, such as oil, as a heat source. Since all paper mills have steam available, steam is by far the most widely used and most efficient heat source.

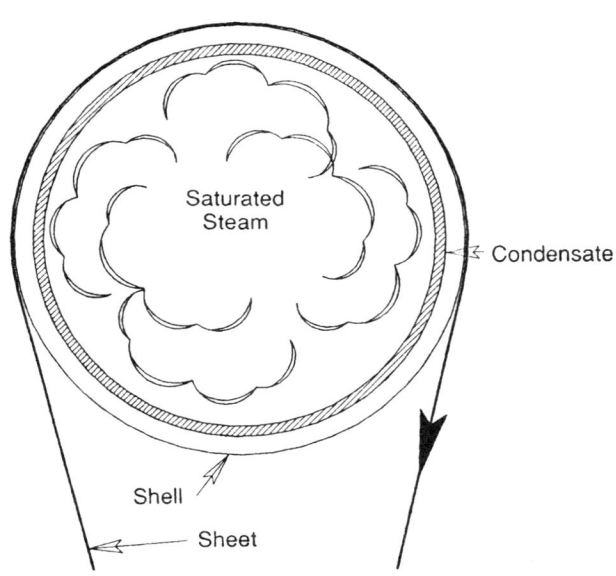

Fig. 4.2. Cylinder dryer cross section (courtesy Beloit Corp.)

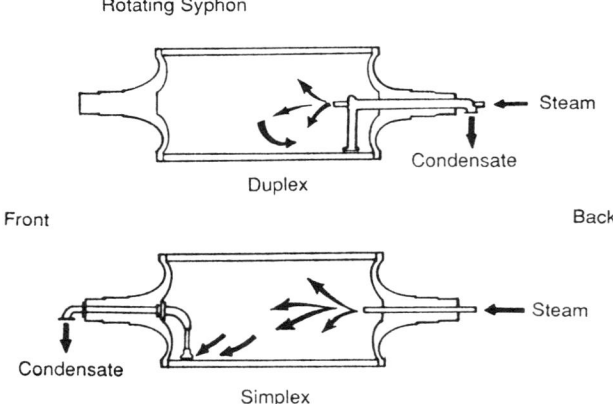

Fig. 4.3. Dryer steam and condensate flow (courtesy Beloit Corp.)

Saturated steam is introduced inside the cylinder through rotary unions (Fig. 4.3). Energy is utilized from the condensing of the steam on the inside of the dryer shell (latent heat of vaporization), and heat is transferred from the steam to the web by conduction through the condensate layer, the dryer shell, and to the coated paper. The drying rate is largely dependent on the rate at which heat from the condensing steam can be transferred to the sheet. This rate, referred to as the heat transfer coefficient, can be calculated as follows for cast-iron steam cylinders.

$$H(oa) = \left[\frac{1}{H(c)} + \left\{ \frac{WW}{360°} \right\} \times \left\{ \frac{1}{H(s)} + \frac{1}{H(m)} \right\} \right]^{-1} \quad (1)$$

Where: H(oa) = Over-all heat transfer
   H(c) = Condensate heat transfer
   WW   = Web wrap
   H(s) = Surface contact heat transfer
   H(m) = Metal shell heat transfer

While the inherent advantages of contact drying are obvious for the production of plain papers, the use of drying cylinders for coated paper requires that the coated side of the sheet be sufficiently dry to prevent sticking or picking of the coating on the hot dryer surface. On most coated paper grades, this requires some type of noncontact drying prior to the steam cylinders. This can be radiant (IR) or convection (hot air) or a combination of both. As a general rule, for a groundwood containing sheet, 20-25% noncontact drying may be sufficient. For a woodfree sheet, 40-45% noncontact drying may be required.

On the first cylinder dryer the coated side of the sheet contacts, it is advisable to have a nonstick surface on the cylinder (Fig. 4.4). However, surface coatings or sleeves tend to reduce the overall heat transfer of the drying cylinder, and should be accounted for in the drying analysis.

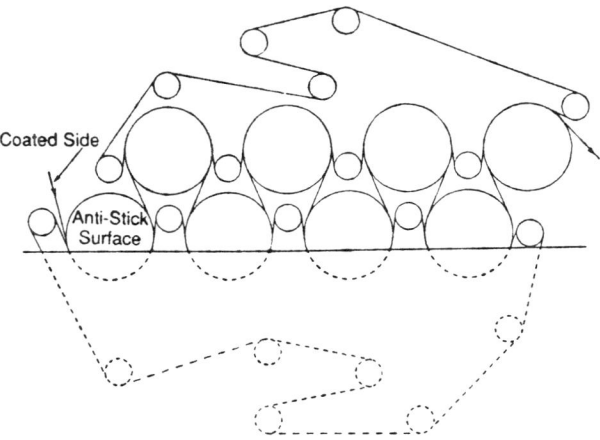

Fig. 4.4. Steam cylinder dryer arrangement (courtesy Beloit Corp.)

Cylinder dryers should be felted to maximize sheet contact and to ensure uniform heat transfer across the sheet. The felted cylinders also increase sheet contact friction to eliminate web slippage and improve web tension stability through the section.

To control drying from side to side and for curl control, the first two steam cylinder dryers of the section are normally designed with individual steam control (Fig. 4.5). The remaining cylinders should have separate top and bottom steam control. A typical steam control system is shown in Fig. 4.6.

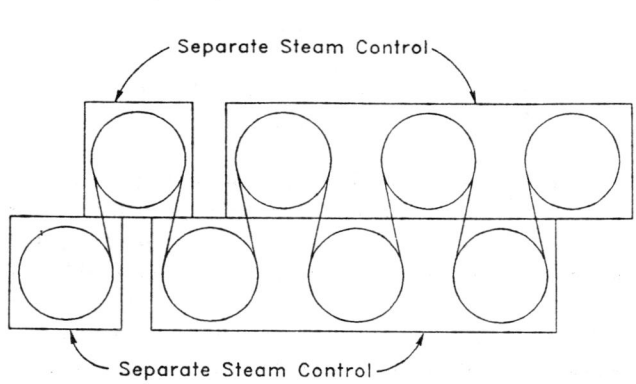

Separate Steam Control

Separate Steam Control

Fig. 4.5. Dryer temperature control arrangement (courtesy Beloit Corp.)

Typical evaporation rates for steam cylinder drying of coated paper grades can vary with sheet and coating formulations along with the stage the drying process is in.

Since the cylinder dryers for coated paper usually follow some noncontact predrying, evaporation rates can vary from 0.5 to 2.5 pounds of water per hour per square foot.

When sizing a cylinder dryer section for after-coater drying, the total amount of water to be evaporated using the cylinders must be determined. The following is a simple method to calculate the water load.

$$W(tot) = \left\{ \frac{C(w) \times B}{S} - (C(w) \times B) \right\} \times P \qquad (2)$$

Where:
W(tot)= Total amount of water to be evaporated (lbs/hr/ft width)
C(w) = Coat weight on the reel (lbs/ream)
B = Percent bone dry of coating on reel
S = Percent solids of coating entering cylinder dryers
P = Production rate (reams/hr/ft width)

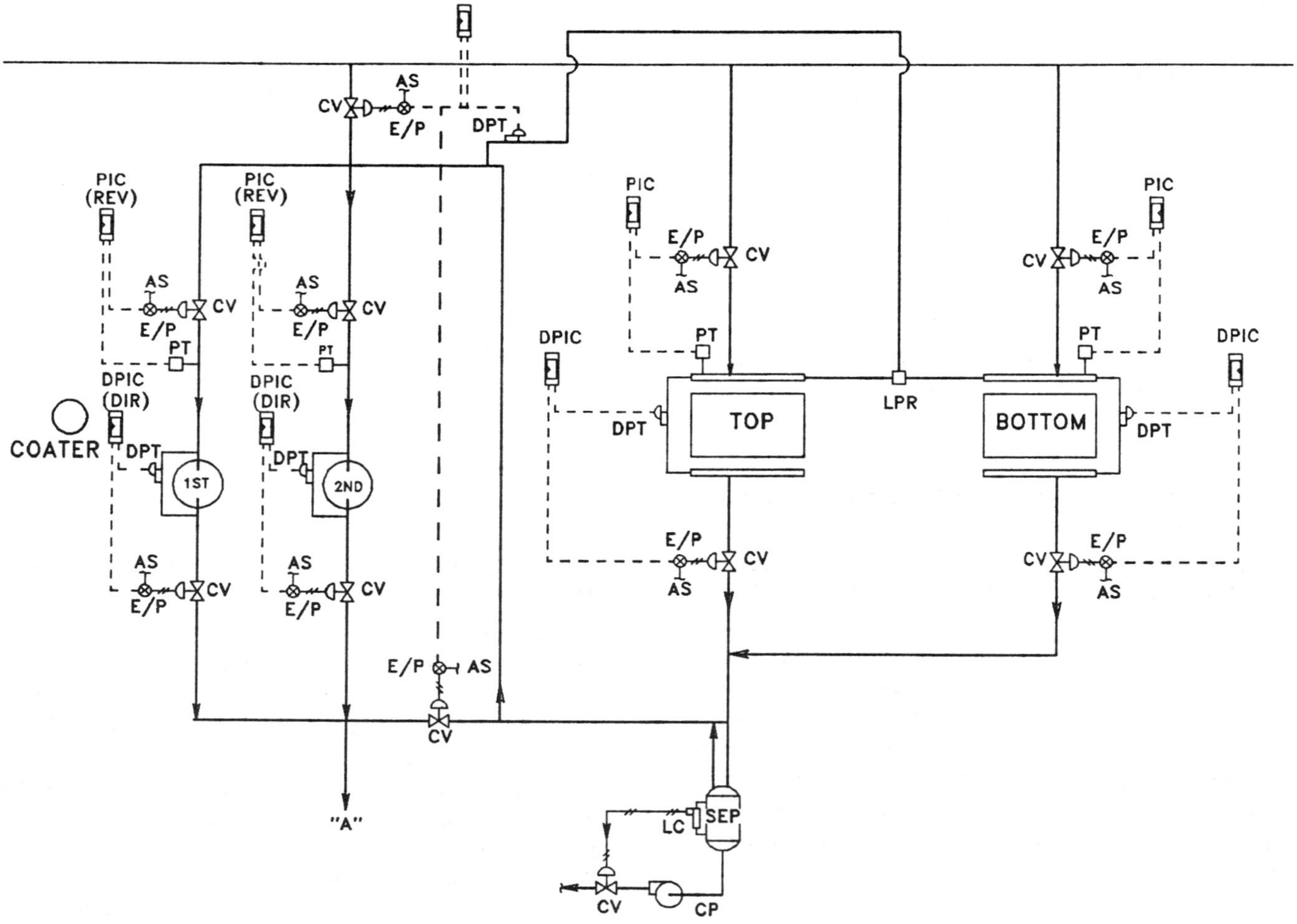

Fig. 4.6. Dryer steam control arrangement (courtesy Beloit Corp.)

From the calculation for the total amount of water to be evaporated, the number of dryers can be determined, based on the average evaporation rate for the section.

$$N = \frac{W(tot)}{EV \times D \times \Pi} \qquad (3)$$

Where:
N = Number of cylinder dryers
EV = Evaporation rate (lbs/hr/sq ft)
D = Cylinder diameter (ft)
Π = Pi

Air Impingement Dryer

In a paper presented in 1960, Gardner described a drying concept which was highly efficient in terms of space requirements. This process was developed for use in drying tissue on a Yankee dryer where space was limited by the circumference of the steam cylinder dryer. This concept was applied to the process of coated paper drying in the quest for ways to increase speed on the production coater.

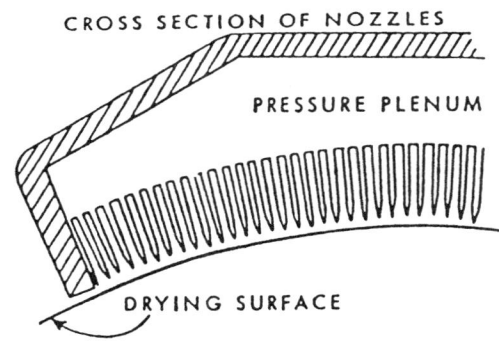

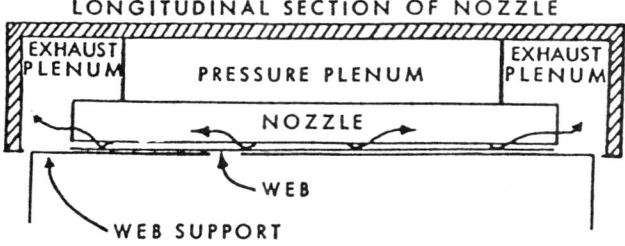

Fig. 4.7. Gardner dryer design

The Gardner dryer had a series of slot orifice air nozzles arranged circumferentially around a steam cylinder (Fig. 4.7), enclosed in an insulated hood. The nozzles were as long as the sheet was wide, and were spaced ¾-1 in. apart. The height of the nozzle from the sheet was typically ⅛-½ in. The width of the nozzle slot orifice was typically 0.025 in. or less. Nozzle air velocities

of 10,000-20,000 ft/min and temperatures of 300-600°F were commonly used. It was claimed that "under very special conditions, as with a very thin web in close contact with a small high-pressure steam cylinder shrouded with a high-velocity dryer, drying rates of over 60 lb/hr/ft² had been achieved." However, the narrow slot, combined with the close proximity of the nozzle to the web, resulted in operating and maintenance problems for this particular dryer design when used with coated paper production equipment.

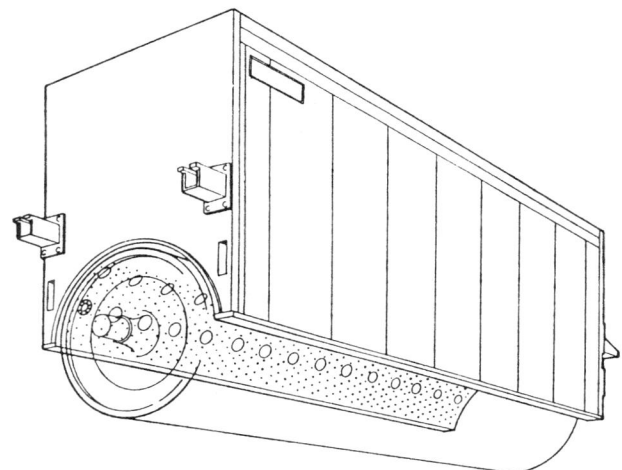

Fig. 4.8. Radial Aircap™ (courtesy Beloit Corp.)

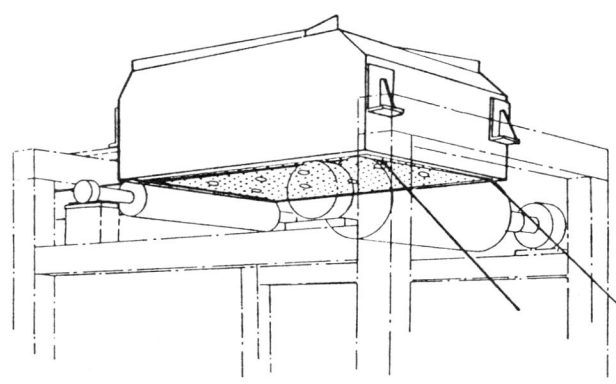

Fig. 4.9. Flat Aircap™ (courtesy Beloit Corp.)

Air impingement cylinder hoods, other than the Gardner design, using round holes rather than slots (Fig. 4.8), were also widely used. Daane and Han in 1961 evaluated the design of air impingement drying, concluding "that the performance of round jets is better than that of two-dimensional jets (slots), other conditions being equal." By performance, it meant the "practical objective of an air impingement system is to achieve a high heat transfer rate with a minimum expenditure of air blowing power." An important advantage to the round nozzle design was improved operation due to greater nozzle to sheet clearance combined with a nozzle plate that was less prone to damage, warp, or plugging.

A variation of the cylinder hood was the use of a flat hood air impingement dryer which was used primarily for heavy-weight paper, such as coated board. A driven sheet support roll was placed prior to the air cap to support the sheet entering the hood, and also exiting the hood (Fig. 4.9). The air cap was placed above the sheet, approximately 1-2 in. above the coated side. The impinging air on the sheet was then drawn back into the hood where it was recirculated and exhausted.

Evaporation rates obtained with these high-velocity air cap cylinder dryers can be quite high. The typical design ranged from 7.0-30 lb/hr/ft². The graph below (Fig. 4.10) compares the heat transfer as air temperature is varied. For the curve shown on the graph, 3/8-in. diameter holes with a 1-in. nozzle to sheet gap, and an open area (hole pattern) of 2% was used.

Air Cap Heat Transfer

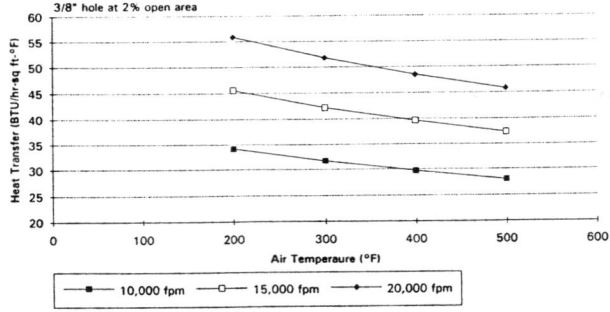

Fig. 4.10. Heat transfer for air impingement cylinder hood

Unfortunately, very few coating applications can take advantage of the high evaporation rates obtainable with cylinder hoods. In many cases, print mottle induced by binder migration will result if the evaporation rate is pushed too high. A specialized case of this is the so-called "railroad tracking" condition which consists of parallel mottled streaks running in the machine direction. These streaks are commonly associated with cylinder air impingement drying, most likely due to uneven sheet contact to the cylinder, and hence, nonuniform heat transfer between the steam cylinder and the paper. The added heat transfer in areas of good contact apparently leads to

high evaporation rates when combined with impinged air. This effect is clearly worse at lower basis weights where the heat is conducted through the sheet quickly. In such cases, the steam cylinder must be operated at very low temperature, thereby reducing the total drying potential of the section.

Because of the potential drying quality problems, the air cap was never fully utilized to its potential with coated paper. The high-velocity cylinder air cap has not proven to be an accepted method for drying of the lighter basis weight and higher-quality papers. Coated paperboard, on the other hand, does present a useful application of this drying method, and many systems are still in use today.

The hood and heating system design has a long history of development and a more complete description of operation can be found in many excellent articles on the design of high-velocity hoods.

As for heating mediums used, it is common at air temperatures below 400°F, for steam to be used as the primary air heating method. But it is necessary to use direct-fired gas or a gas-heated oil recirculation system for operation at air temperatures of 400°F or higher. For coating applications, it is rare that operating air temperatures will exceed 600°F. Because of these operating temperatures, and even at lower temperatures, the hood enclosing the impingement air and recirculated air must be designed to keep the hot air from escaping into the machine room and the hood insulated to keep the outer skin at a safe temperature, generally about 120°F maximum.

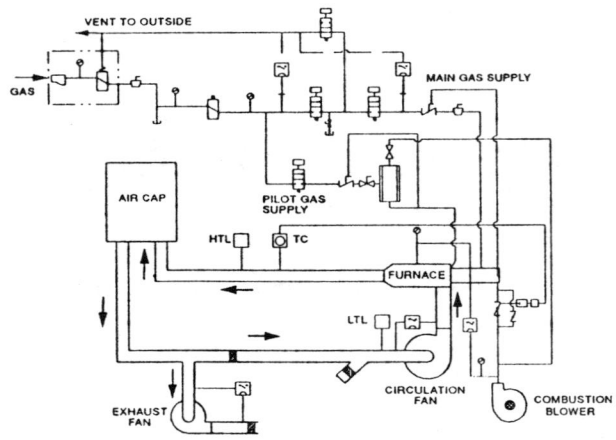

Fig. 4.11. Cylinder hood air system schematic (courtesy Beloit Corp.)

Depending on the size and air volume requirements of the air cap, a single- or two-zone air supply and heating system will be furnished. Preferably the heating and supply air system will be located out of the machine room or in a separate well-ventilated room. A typical diagram showing hood and flow diagram is shown below.

Figure 4.11 shows a schematic of an air recirculation and gas-fired heating system for a typical air cap dryer. The steam cylinder and steam supply to the steam-heated cylinder is not depicted.

Tunnel Dryer

The arched air impingement or tunnel dryer was never fully utilized for publication grade coated paper, except for a few applications in coated paperboard. The tunnel dryer has been used extensively for specialty converting applications, which was typically narrow webs (less than 100 in.) and slower speeds (below 2000 ft/min). However, it is worth mentioning this dryer design, since its history in the developments of drying systems has had significant impact on the drying of coated papers.

Fig. 4.12. Round-hole nozzle design (courtesy TEC Systems)

While several nozzle designs were used, the round-hole type (Fig. 4.12) and the slot-nozzle type (Fig. 4.13) were by far the most used. The slot nozzle, in most cases, was preferred for providing a continuous air jet across the web width. It was believed that a slot jet provided more uniform heat transfer, although not necessarily proven. The nozzles were placed above the web, varying in clearance from 1 in. to 2 in. for most applications. In some special applications, nozzles were also placed beneath the web or opposed, but web instability usually

prevented this type of operation except for applications with high-tension or heavy-weight web applications.

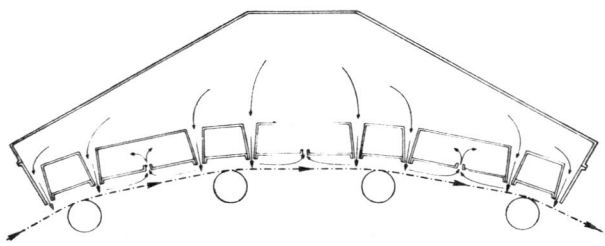

Fig. 4.13. Slot-nozzle design (courtesy of Black Clawson)

Slot designs ranged from 1/8 in. to 1/2 in. in slot width, with the slots generally spaced from 4 in. to 12 in. apart, depending on the intended application and web support method. Typical heat transfer rates are shown in the graph in Fig. 4.14 for a 1/4-in. slot orifice on 6-in. centers.

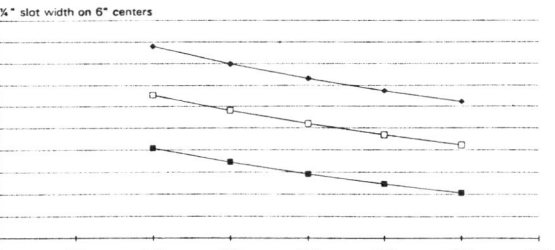

Fig. 4.14. Tunnel dryer heat transfer

Early tunnel dryers supported the web on bars carried through the dryer length on conveyor chains. For higher-speed applications, rolls were used to support the web in an arched arrangement (Fig. 4.15), with the rolls driven for the lighter-weight web applications. For sensitive webs and coatings, the slot nozzles are usually positioned above each web support roll. This provides reasonable heat transfer without causing the web to sag or flutter excessively between rolls.

When lightweight webs are coated at higher speeds, a conveying fabric, called a felt, is used to support the web

(Fig. 4.16). With the support of a strong stiff felt for the web to lay on through the dryer, fewer support rolls are needed, and are usually spaced approximately 4-5 ft apart. With a felted dryer, heat transfer can be improved by spacing the nozzles closer together, typically to about 6 in. on centers. Dryer lengths of over 100 ft were common with this type of web support and conveying system. With these longer dryers, the conveying fabric must be made to pass over automatic guiding and tensioning rolls, usually on the return run under the dryer enclosure.

Fig. 4.15. Idler roll support with housing in retracted position (courtesy of Black Clawson)

The air-heating systems used with tunnel dryers are very similar to the cylinder air cap designs, as mentioned in the previous section, with exception that exhaust volumes were typically increased to compensate for lower recirculating air humidity and, in some cases, maintaining lower solvent explosive limits.

Air Flotation Drying

The first off-machine coater to be designed with all flotation air drying was No. 5 OMC (off-machine coater) at Blandin Paper Co., Grand Rapids, MN. An outline of the machine is shown in Fig. 4.17. Other machines had used the flotation dryer design, but either as supplementary drying or to replace existing cylinder dryers, air cap dryers, or both. The Blandin machine had all the coater drying done with air flotation without the use of dryer cylinders or other auxiliary drying equipment.

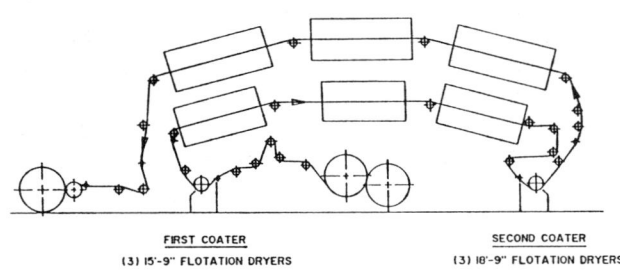

FIRST COATER
(3) 15'-9" FLOTATION DRYERS

SECOND COATER
(3) 18'-9" FLOTATION DRYERS

SPEED = 4000 FPM   SHEET WIDTH = 181 in.

Fig. 4.17. First off-machine coater with drying by flotation dryers (courtesy of Advanced Systems Inc.)

Richardson had described the results of a research program that would develop a floater dryer to dry both sides of the sheet without contact. The program resulted in the use of an airfoil type nozzle as shown in Fig. 4.18.

It was shown, at that time, that among other advantages of flotation drying, the quality of drying with a two-side floater was measurably better than the conventional impingement dryers. This was later confirmed by additional work by Johns. Later Jaeger and Gregory reported on coating machines that were retrofitted with flotation dryers to replace the older impingement dryers for the purpose of increasing quality as well as production.

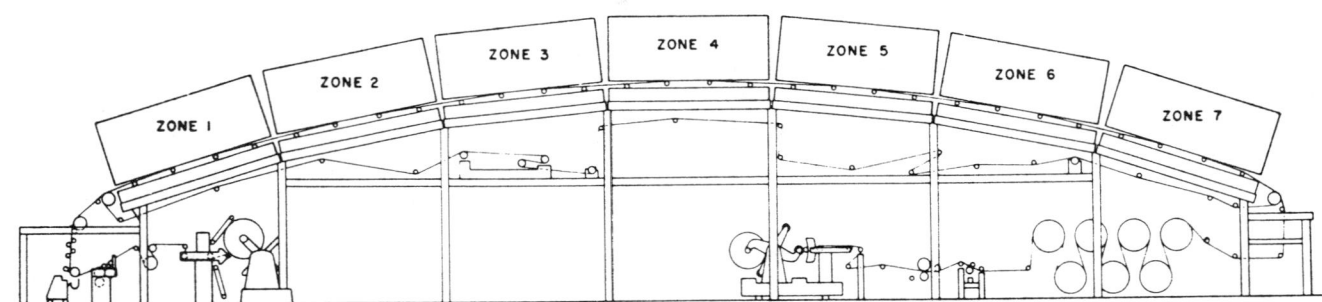

Fig. 4.16. Felted tunnel dryer arrangement (courtesy of Black Clawson)

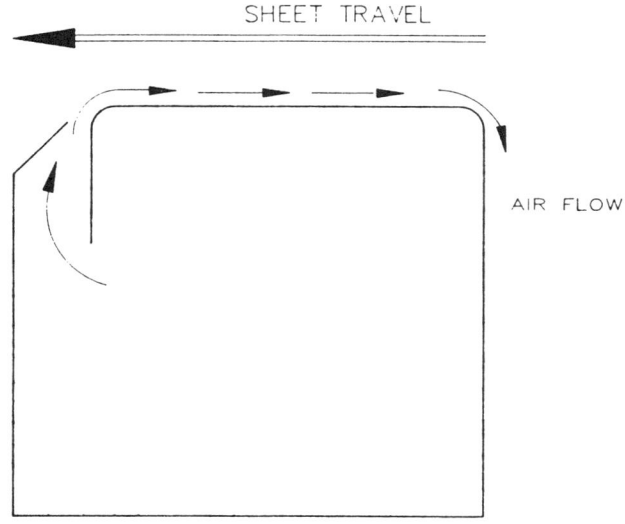

Fig. 4.18. Air foil-type flotation nozzle (courtesy of Advanced Systems Inc.)

The airfoil nozzle design consists of a single slot with curved orifice to a flat support area. Air exiting the slot will tend to follow the curve portion of the orifice and along the flat support area of the airfoil. When a web is placed as close to the airfoil as the jet thickness, the air flow "engages" the web, so to speak. The air-jet flow will tend to maintain the web position above the airfoil with a clearance that is roughly equal to double the slot jet thickness. When the web is too far from the airfoil surface, a negative pressure is developed which draws it closer. And when the web is too close to the airfoil surface, a positive pressure develops which pushes the web further away. Under proper conditions, the forces developed will maintain a web in the equilibrium position. With this type of airfoil design, it is possible to float the web with air bars on one side of the web only, top or bottom.

About the same time that the single slot airfoil was being successfully used for coated publication grade papers, a two-slot Coanda design was being predominantly used for the drying of offset printing inks and specialty coatings (Fig. 4.19). This air bar design consisted of a two-slot air bar with the slots separated by a flat support surface. In this case, air exiting the slots is made to converge toward the center of the support surface, creating a pressure pad along the air bar top. With this design, the web is always subjected to a positive force pushing away from the air bar surface. This required that the air bars be staggered on both sides of the web for stable operation. The advantages of this design are greater web to air bar operating clearance combined with a sine-wave effect on

the web for enhanced web stability. Also higher heat-transfer coefficients are obtainable with this type of air bar.

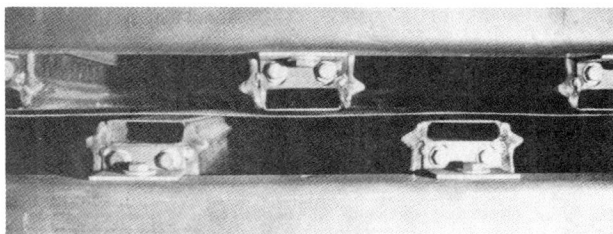

Fig. 4.19. Two-slot Coanda flotation air bars (courtesy of TEC Systems)

Numerous variations to the single-slot and the two-slot design have been developed, but all utilize the principles of flotation as described above. Variations of the two-slot direct, impingement-type air bar (Fig. 4.20) report more effective heat transfer potential and stronger pressure pad for better web support through the dryer than the initial Coanda design.

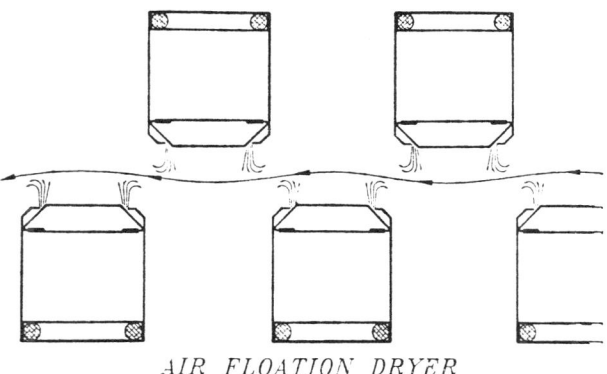

AIR FLOATION DRYER

Fig. 4.20. Sinewave flotation with staggered two-slot floater air bar (courtesy of Advanced Systems Inc.)

For both off-machine and on-machine installations, the dryers are usually arranged in series called zones. Each zone consists of an enclosure of specified effective drying length, sometimes with driven support rolls between zones. In a few applications, the last zone is followed by a steering roll to maintain the sheet on center, particularly if the total combined dryer lengths are long with a relatively narrow sheet width.

As with cylinder dryers, air cap dryers, or both, the hood enclosing the air bars must be designed to keep the hot air from escaping into the machine room and the hood outer skin at a safe temperature, approximately 120°F maximum.

Materials of construction are normally aluminized steel, stainless steel (type 304), CorTen "A," or a combination of each. The type of construction typically consists of an inner tub of heavy gauge backed by a carbon steel frame, 3-4 in. of high-temperature insulation, and clad with a sheet metal skin.

The enclosures vary in size dependent on the width of the sheet, but the length of the zones usually do not exceed about 20 ft due to air-handling equipment size, the physical handling of the enclosure, or both.

For access into the enclosure, two types of retraction are common: clam shell and screw jack. The clam shell type is a pivoting design, usually from the fan side or both as a double pivot, when full open access is required. The screw-jack type consists of screw jacks on each of the enclosure corners, moving one half of the enclosure uniformly, which in most cases opens about 18-20 in.

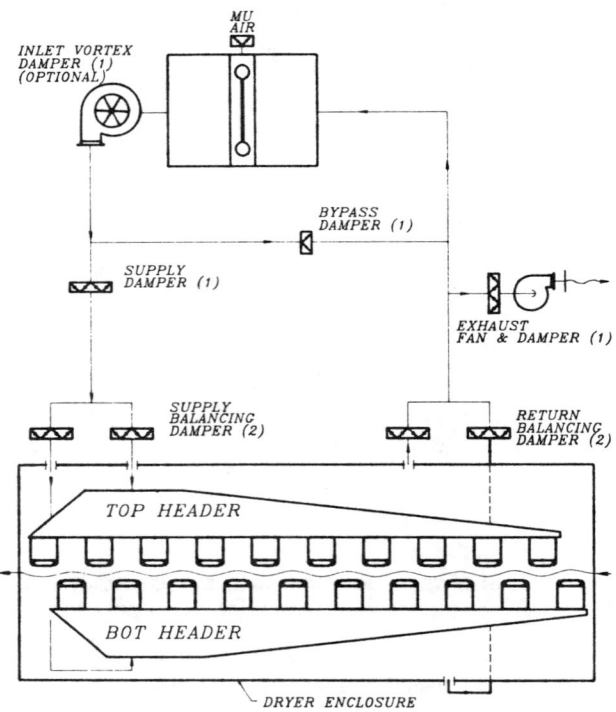

Fig. 4.21. Draw-through air supply system (courtesy of Advanced Systems Inc.)

Flotation dryers use two types of air supply systems, termed a "draw-thru" (Fig. 4.21) and "blow-thru" (Fig. 4.22). The terminology refers to the location of the heating plenum in relation to the supply fan, with the "draw-thru" being on the suction side of the fan, and the "blow-thru" being on the positive side of the fan. The air systems are designed using either steam- or direct-fired gas or oil for heating. Operating temperatures of 300-400°F are typical, with maximum design commonly being 600°F, making gas the preferred heat source. Because of the lower operating temperatures associated with the drying of coated paper, the preferred design of the air systems for flotation dryers is toward the use of a "draw-thru" arrangement, as the fan will always deliver a constant air volume regardless of varying air density keeping the airfoil velocity constant.

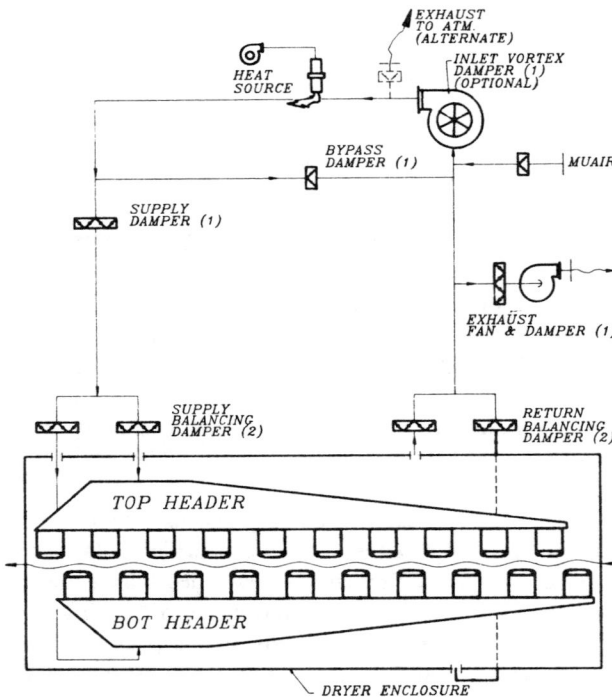

Fig. 4.22. Blow-through air supply system (courtesy of Advanced Systems Inc.)

With most air drying systems, the outlet nozzle velocity is defined as low, medium, and high. With low velocity being up to 5000 ft/min, medium velocity up to 10,000 ft/min, and high velocity up to 15,000 ft/min.

The exhaust air is taken from the recirculating air prior to the heating plenum with makeup air drawn into the heating plenum supplied from a relatively stable temperature and moisture area to keep from oversizing burners or steam coils. Exhaust volume is sized to keep the enclosure negative and the operating dewpoint less than 0.1 pounds vapor to pounds of dry air.

The air system apparatus is usually located on the back side of the enclosure, isolated from the machine room where possible, with connecting ducts to and from the enclosure.

For dryer controls, it is desirable to adjust process air temperature, nozzle outlet velocity, and in some cases exhaust volume.

- Control of air temperature can be by use of a standard temperature controller which is operated by remote setpoint or from a moisture measurement and control system.
- Control of air velocity is by use of remote set pneumatic or electronic damper actuator. Digital pressure indicating gauges are recommended, but not necessary. It is common to have manually set dampers for balancing top and bottom air velocity, with velocity control being accomplished by adjusting an inlet damper to the fan.
- Control of exhaust air, while most often left as a manual operation, is simply done by remote operation of the fan outlet damper. Restrictions on minimum exhaust should be set to maintain a negative exhaust pressure in the box.

Basically the operator controls drying by adjusting air temperature, and it is common practice to have this under computer control. air-impingement velocity can also be adjusted, if kept within the limits of flotation requirements. But it is not common practice to adjust velocity for drying, and if used is usually done manually.

For the purpose of sizing, the application of flotation drying is normally used with air bars on both sides of the sheet with uniform velocity and temperature on both sides. Based on this typical dryer setup, the graph for heat transfer can be used as a general guide to dryer performance (Fig. 4.23).

Web Turning with Air Flotation

A unique application of the flotation dryer is its use as an air turn. Special applications after a coater or size press that apply a wet solution to both sides simultaneously such as a Billblade or similar type equipment most often require the need for noncontact drying of the coated sheet.

The most successful design uses a two-slot pressure-type air bar arranged in a circular array, as shown in Fig. 4.24 to provide a continuous air cushion for support of the sheet. Because the two-slot air bar produces a positive pressure along the face of the air bar, sufficient forces can be generated to handle typical machine tensions.

Fig. 4.24. Flotation air bars arranged in a circular array for an air turn (courtesy of Spooner Industries Ltd.)

With this type of air bar, the supporting cushion pressure (Pc) is proportional to the air bar supply pressure (Ps). For an air bar operating with a nominal 1/4-in. clearance, the support pressure is approximately 40% of the supply pressure. The relationship of support pressure to tension is expressed in the equation:

$$\text{Pressure} = \frac{\text{Tension}}{\text{Radius}} \qquad (4)$$

Float Dryer Heat Transfer

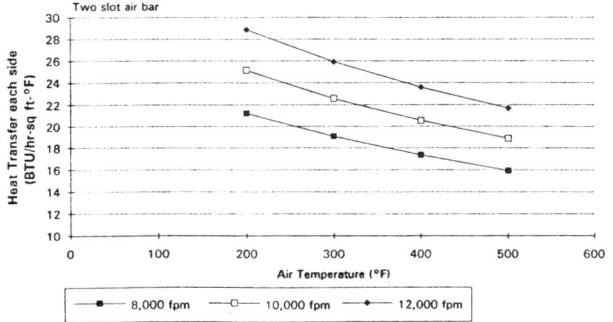

Fig. 4.23. Flotation dryer heat transfer for each side of sheet

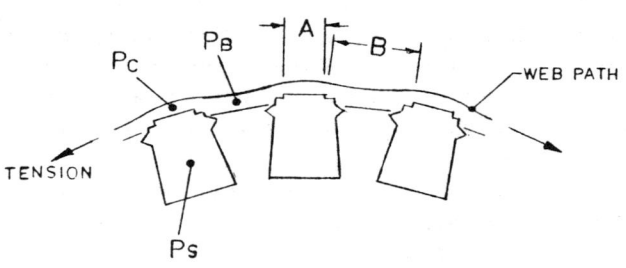

Fig. 4.25. Pressure area under the sheet to float the web with air turn

In actual operation, the support pressure under the web through the air turn is not continuous, as shown in Fig. 4.25. Therefore, the various pressure areas must be equated to an average pressure, which is expressed as:

$$(A \times Pc) + (B \times Pb) = (A + B) \times \text{Tension} \div \text{Radius} \quad \text{(5)}$$

Where:
A = Air bar face
B = Spacing between air bars
Pc = Cushion pressure or support pressure developed from the air bar
Pb = Back pressure or support pressure controlled between the air bars

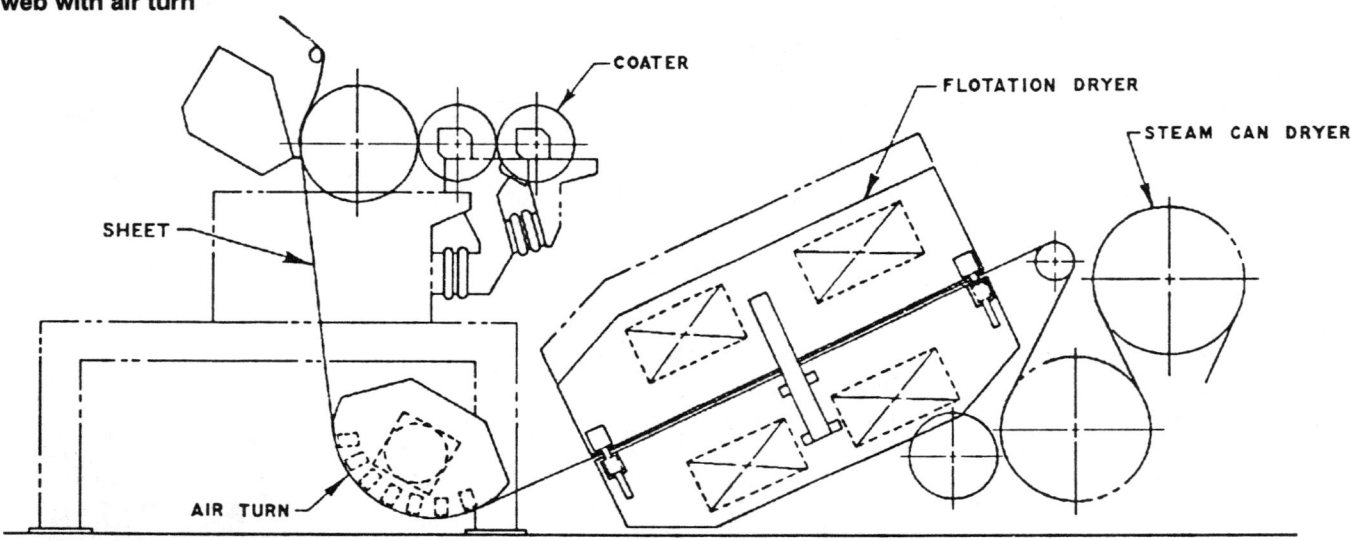

Fig. 4.26. Billblade coater followed by an air turn and float dryer (courtesy of TEC Systems)

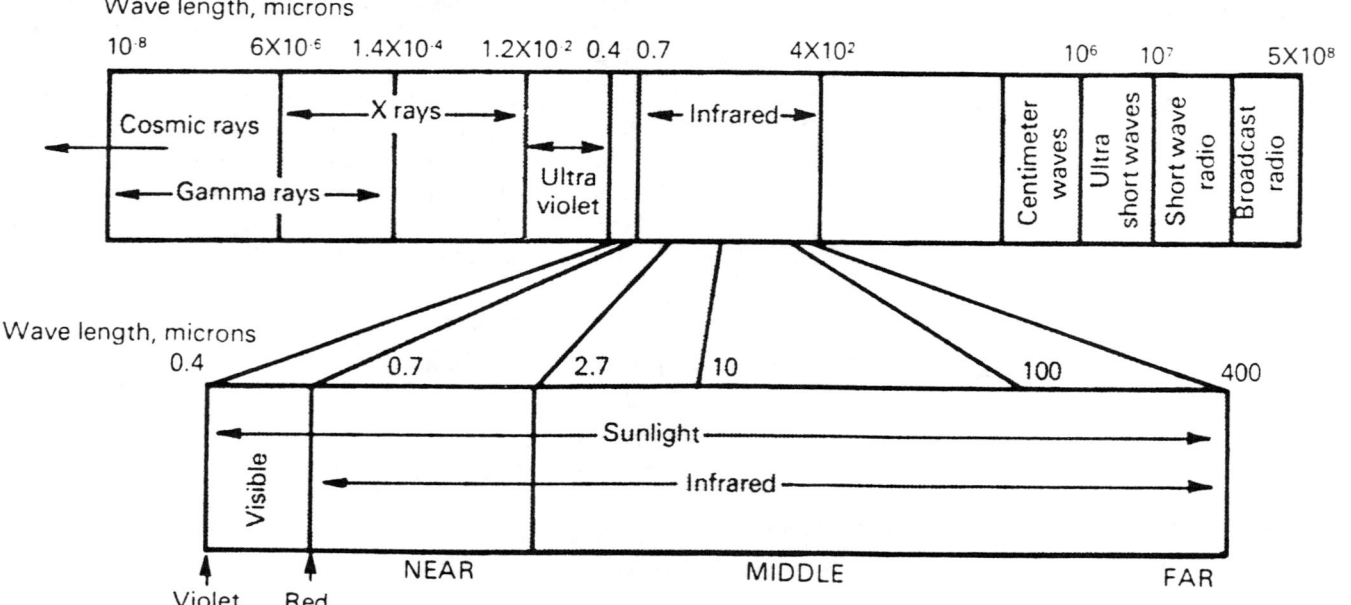

Fig. 4.27. Electromagnetic spectrum

Since the air bar used in the air turn is the same air bar that is used in a float dryer, heated air can be supplied in the process of turning the sheet, adding versatility for drying. Air turns have been installed as individual units as well as incorporated within the dryer enclosure. Figure 4.26 is an example of an air turn being used, following a Billblade coater.

IR Drying

It was not until the 1950s that infrared (IR) began to be used to any extent in the paper industry. With the advantage of the high energy available per unit area, the use of IR was conveniently used for the drying of coatings on paperboard, where space constraints were many times a factor as coating stations were added to existing machines. The more recent and wide acceptance of the use of infrared energy for the drying of coated publication grades has led to many developments and improvements in materials and controls, making IR a practical and safe method of drying for this industry.

Infrared energy is radiated from any object or material that is warmer than its environment. IR radiation, whether generated from an electric or a gas system, is emitted by atomic excitation of any substance. The IR energy travels in the form of electromagnetic waves at the speed of light, even through a vacuum, until it strikes another substance where it may be absorbed, reflected, or transmitted. Electromagnetic waves occur over the range of wavelengths, from 10^{-16} μm (cosmic wave) to 10^6 μm (electric power waves), with IR energy falling in the "invisible light" range between 0.7 μm to 100 μm (Fig. 4.27). The useful wavelengths for industrial applications are from 1-10 microns, because absorption of many materials in the 3-8 micron range is especially effective.

The theory of radiant energy for energy transfer is described by three fundamental laws:

1. Plank's distribution law: distribution of radiant energy over a band or wavelengths for a given temperature

$$E_{b\lambda} = \frac{A \times \lambda^{-5}}{e^{\left[\frac{B}{\lambda T}\right]} - 1} \tag{6}$$

Where: $E_{b\lambda}$ = the emission from a black body
surface at wavelength
(btu/hr/sq ft/μ) or (watt/sq m/μ)
  A = 1.187×10^{-8} (btu $\times$ μ^4/hr/sq ft) or
3.741×10^{-16} (watt $\times$ sq m)
  λ = wavelength (μ)
  B = 2.5896×10^4 (R $\times$ μ) or
1.439×10^{-2} (K $\times$ m)
  T = absolute temperature in °R or °K

2. Stefan-Boltzmann law: radiant energy density is dependent on absolute temperature of the source

$$Q = \sigma \times T^4 \tag{7}$$

Where: Q = emitted heat quantity from a black
body surface at absolute temperature T,
(btu/hr/sq ft) or (watt/sq m)
  σ = Stefan-Boltzmann constant,
0.1714×10^{-8} (btu/hr/sq ft/R^4) or
5.6697×10^{-8} (watt/sq m/K^4)
  T = absolute temperature in °R or °K
(°R = 460 + °F) or (°K = 273 + °C)

3. Wien's displacement law: wavelength or most energetic radiation.

$$\lambda_{max} = \frac{N}{T} \tag{8}$$

Where: λ_{max} = wavelength for the maximum emission
in micron
  N = 2998.8 for absolute temperature °R or
5215.6 for absolute temperature °K
  T = absolute temperature in °R or °K

At any particular temperature, radiation leaves a surface through a wide spectrum, determined jointly by the surface characteristics and by surface temperature. At higher temperatures, the amount of energy emitted at shorter wavelengths increases; whereas, as the temperature is lowered, the amount of energy emitted at longer wavelengths is increased. The approximate peak wavelength λ(max) for the shorter and medium waves are:

$$\begin{aligned} \lambda(max) &= \text{1.15 micron for T > 3600°F} \\ &= \text{2.5 micron for 1400°F < T > 2000°F} \end{aligned} \tag{9}$$

The electric and gas IR heaters typically used for coated paper fit into three temperature ranges:

1. 650-1000°F (gas-fired and electric IR)
2. 1000-2000°F (gas-fired and electric IR)
3. 3000-4000°F (electric IR).

The operating temperature and efficiency depends on the design of the IR unit as well as the moisture content, base sheet furnish, sheet thickness, and coating materials. While it is not possible to make an absolute statement as to the optimum wavelength for all coated grades, the absorption of infrared radiation in water and paper appears to be coupled in the 2.8 μm wavelength as shown in Fig. 4.28.

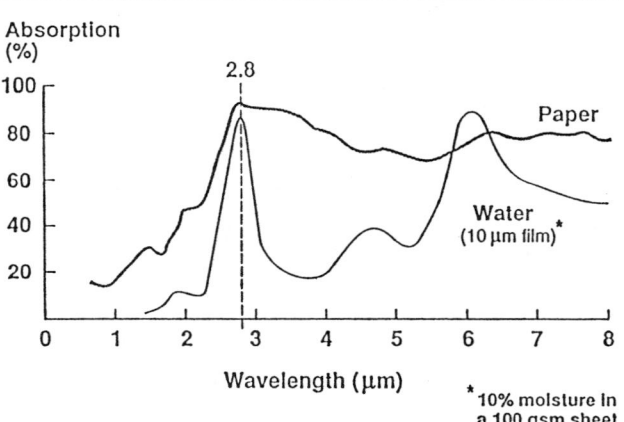

Fig. 4.28. The absorption of infrared radiation in water and paper

Electric Infrared

There are three basic physical configurations of electric IR emitters used in web process applications, which can be classified as line, point, and area sources.

1. The line source configuration is constructed of a fine tungsten wire surrounded by halogen gas housed in a quartz tube about 1/2 in. in diameter. The quartz tube is generally from 6-25 in. long (Fig. 4.29). The tube or tubes are then supported in a case which also holds a reflector and protective quartz cover. The tungsten filament can reach temperatures as high as 4100°F and can operate at a peak wavelength range of 0.8 µm to 2.0 µm. Because of the high filament temperature and energy developed by this type of system, a cooling air circulation fan is required, interlocked with the power, to prevent damage or reduced lamp life.

Fig. 4.29. Quartz tube module showing reflector tile lamp and quartz reflective plate (courtesy of Impact Systems)

Electrical IR is used for drying of coatings and other applications where it is felt that high energy drying is necessary to set the coating and binders and coating picking. However, high intensity IR is most widely used for moisture profiling (Fig. 4.30) on the paper machine and coater. In some cases, combining both gas and electric IR has been used to obtain the benefits of both for base sheet drying and for profiling.

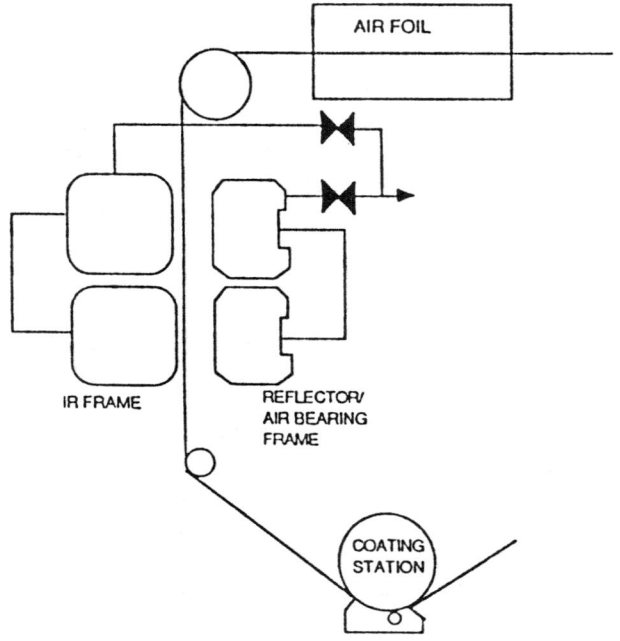

Fig. 4.30. Location after the coater for electric IR dryer section (courtesy of Impact Systems)

2. The point source configuration is least common to this industry and is not generally used for the drying of coated paper. This type of IR emitter looks and functions much like an ordinary light bulb. The bulb employs tungsten filaments and operates at wavelengths in the 0.8 µm to 3.0 µm range.

3. The area source configuration (Fig. 4.31) is more commonly used in "specialty" coated applications where lower temperature drying control is likely to be critical to the process. These units use a "secondary emission principle" where an imbedded heating element heats a flat surface which in turn becomes the radiant heat source. The types of materials used as the area source of radiation can be Pyrex, ceramic, quartz, glass cloth, or other ceramic-type materials. The performance requirements of these materials are that they exhibit both good thermal conductivity and high emissivity and are good electrical insulators. Using a combination of fused quartz emitter plate, woven quartz cloth, or black ceramic coatings, temperatures as high as 1800°F are obtainable, allowing operation in the 2 µm to 11 µm range.

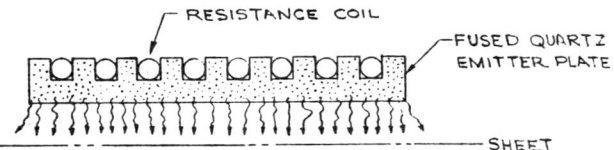

Fig. 4.31. Area source electric IR emitter

The control system for electric IR usually consists of a power unit which houses the controls for each individual zone or SCRs and firing circuits, an operator station, computer station, and process measurement link (Fig. 4.32). One advantage with electric IR is the ability to precisely control the power to the filaments. Therefore, by selective design of filament sections, cross-web profiling can be closely controlled. Because of this, high-intensity electric IR systems have been predominantly used at the dry end of a paper machine, coater, and calender sections to provide final moisture and caliper profile.

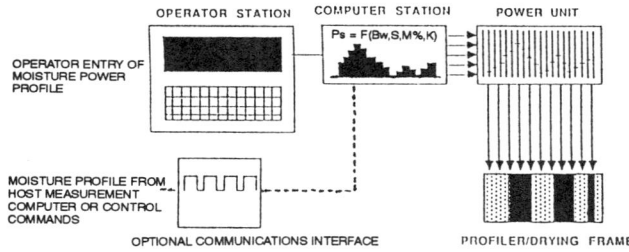

Fig. 4.32. Basic electric IR control system (courtesy of Impact Systems)

Gas Infrared

Gas-generated infrared systems are the most widely used for the drying of coated paper, primarily due to its lower cost of operation in areas where gas is readily available. In the temperature range that most competitive gas-infrared systems operate today, the highest wavelength output is in the 1.8 μm to 2.5 μm range. Gas-fired generators of IR can be described more readily by classifying them by energy output potential, although several different physical configurations are used.

1. Low-energy types are radiant tube burners where the gas/air mixture is fired into a tube. The tube surface heats up, becoming the source of radiation. Since the source temperature operates in a range of 600-800°F, the radiation energy is only about 1000-3000 Btu per square foot of radiant surface (Fig. 4.33).

Catalytic burners that use a catalyst to oxidize gas at low temperatures also fall into this energy output range. Neither system is used to any extent in the drying of coated publication grade paper.

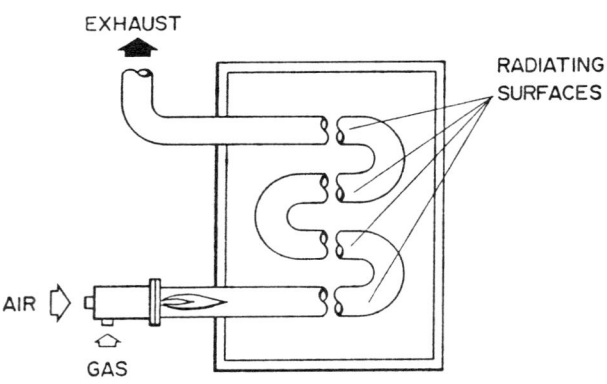

Fig. 4.33. Radiant tube-type gas emitter (courtesy of John Munce, IR application consultant)

2. The most common type of gas IR systems used in the drying of coated paper is surface combustion burners which utilize a metal or porous refractory for the radiating service on which an air and/or gas mixture burns creating the radiator. Materials used are high-temperature stainless steel, ceramic, and combinations of metal screen and ceramic. These burner types emit medium- to high-intensity radiant energy, ranging from 1400°F to 2200°F, and several types are listed below.

Figure 4.34 is an emitter which incorporates a matrix of stainless steel plates with a nozzle centered immediately behind each plate. A wire mesh holds the emitter plates in place, maintaining the integrity of the emitter surface. The nozzle directs the flame to the center of each emitter plate, causing it to heat, incandesce, and radiate infrared energy.

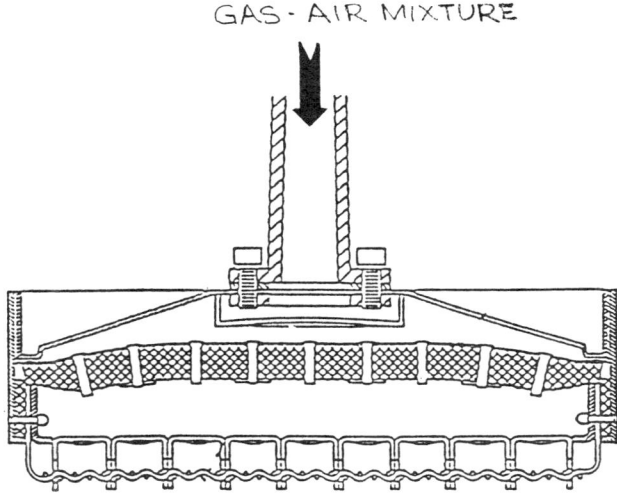

Fig. 4.34. Gas IR emitter using a matrix of stainless steel plates (courtesy of Marsden, Inc.)

Figure 4.35 shows the use of ceramic emitter material, used in various shapes and forms, which consists of a cast ceramic block with a number of specially sized nozzles cast into the block during the molding process. The gas/air mixture flows through these nozzles to the emitter surface, ignites, and causes the ceramic surface to incandesce and radiate infrared energy.

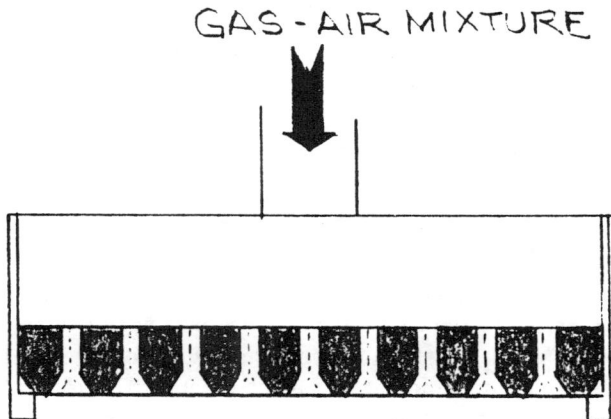

Fig. 4.35. Gas IR emitter using cast ceramic block (courtesy of Marsden, Inc.)

Another type of ceramic block, shown in Fig. 4.36, is where a porous ceramic material is used to serve as nozzles. The gas and air mixture passes through the multitude of pores. Permanently bonded to the high-porosity primary substance is a secondary ceramic surface structure of lower porosity. The gas/air mixture ignites at the interface of the two different porosity materials, causing the low-porosity material to incandesce and radiate infrared energy.

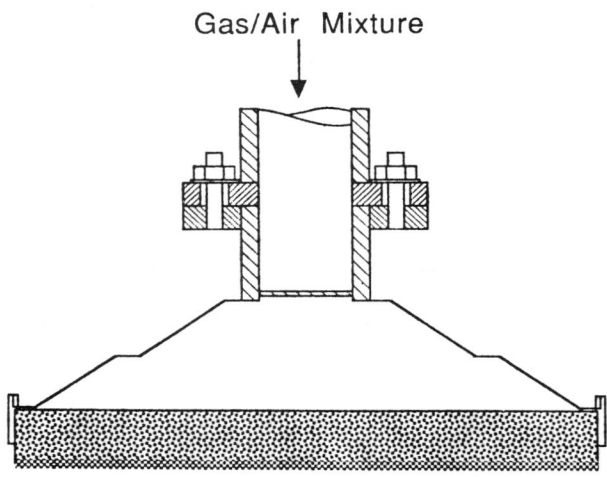

Fig. 4.36. Emitter using a porous ceramic block (courtesy of Marsden, Inc.)

A variation of the cast ceramic block is the formed ceramic composite matrix. Unlike the porous ceramic block, this material is manufactured by vacu-molding a ceramic fiber composite slurry, creating a ceramic mat as shown in Fig. 4.37.

The stoichiometric air/gas mixture burns within 1-2 mm of the surface of the matrix, heating the surface, causing it to incandesce and radiate infrared energy.

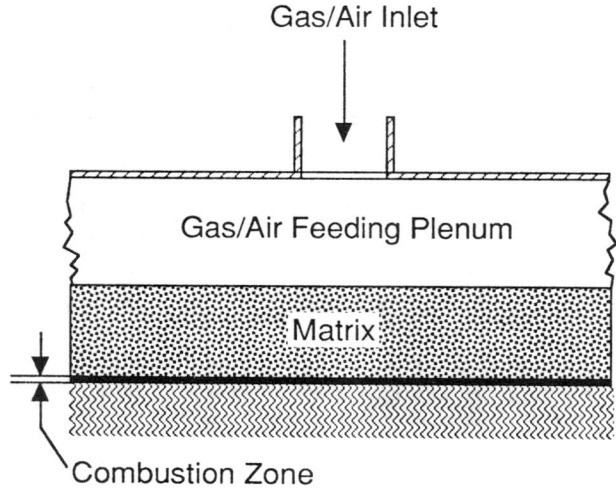

Fig. 4.37. Emitter using a ceramic composite matrix (courtesy of Marsden, Inc.)

Figure 4.38 shows yet another type of ceramic block which has the flame directed via a flame-spreading nozzle against the face of the ceramic material. This impinging flame causes the ceramic surface to heat and radiate.

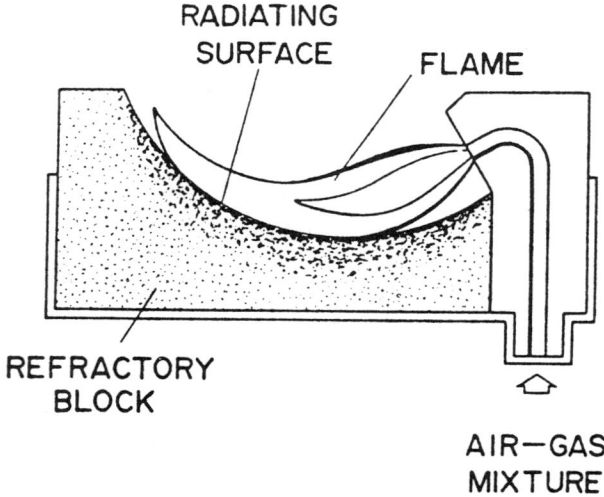

Fig. 4.38. Gas IR emitter flame-impingement-type (courtesy of John Munce, IR application consultant)

When used as a means of drying coated paper, gas IR systems are typically arranged in rows, usually double rows. Each burner consumes approximately 1/2 ft². The burners are usually butted cross machine from edge to edge of the sheet (Fig. 4.39). An air flow system is commonly supplied with ducting to and from the units to remove exhaust gases and remove evaporated water. This air flow breaks up the laminar layer of air and vapor traversing with the sheet, decreasing the surface vapor pressure, and facilitating water removal. The air flow also keeps the sheet from touching the burner frame and is used to cool the unit on shutdown.

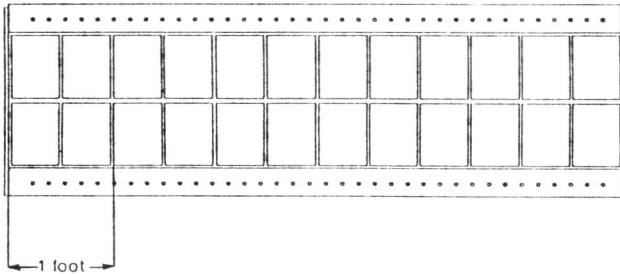

Fig. 4.39. Typical emitter arrangement for gas IR dryer showing a dual-row hood (courtesy of Marsden, Inc.)

Since gas IR systems must rely on the mixture of gas and air for effective operation, all emitters need a stoichiometric gas/air mixture delivered to the individual emitters for combustion. Two types of systems that are in use today are premix and proximity mix.

In early gas infrared systems, premixing proved to be a versatile and reliable system, which mixed gas and air before entering the combustion manifold. Because the proper mixture of gas and air exists before the IR hood, opening and closing a manual valve ahead of the combustion manifold was a simple method to control emitter temperature. However, the proper ratio of gas and air relies on several variables such as air density, pressure, humidity, and temperature, as well as other variables which were ever changing within the mill environment. And if not adjusted for each condition, performance and efficiencies suffer.

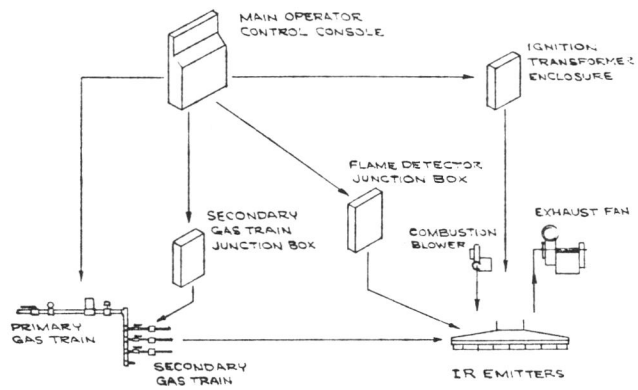

Fig. 4.40. Typical gas IR dryer control system (courtesy of Repap Wisconsin Inc.)

Proximity mixed systems introduce raw gas and combustion via separate headers to individual emitters with

Fig. 4.41. Off-machine coater (courtesy of Repap Wisconsin Inc.)

the correct ratio of gas and air relying on nozzle size and pressure. Since the gas nozzle on each emitter mixing tube is small, any blockage of the nozzle opening results in decreased temperature at the emitter. To keep the nozzles from plugging, strict adherence to maintenance must be followed.

Figure 4.40 shows a typical IR system which consists of a flame-detection system, ignition transformer, primary gas train, secondary gas train, combustion blower, and operator panel.

Application

Applications of a drying system for coated paper grades on a modern off-machine or on-machine coating line will typically include the use of IR, flotation, and cylinders. The task in the application of drying is to determine the optimum combination, length, and features of the variety of dryer designs available. By combining experience with the backup of pilot tests, the assurance of success on the production machine is greatly improved.

However, approximation of drying requirements for any new project is generally required before testing is begun or completed. A simple method of estimating the removal of water from the coating is always helpful in developing a drying scheme. The purpose of this section is to provide a simple but practical method of estimating the average drying rates for the equipment selected.

The most common method of communicating drying is the use of drying rates, most typically stated in terms of pounds of water evaporated per unit width (or area) per hour:

Example: lb/hr-ft width or lb/hr-ft^2

To those unfamiliar with drying technology, dryer equipment, and dryer specifications, drying rates are most often associated with average evaporation over the entire length of the dryer system being specified. Caution is advised, as the actual drying rates from point to point throughout the dryer length can vary significantly from the average.

The drying process is often described in three phases which are defined as follows:

Phase 1. Sensible heat load, where the sheet temperature increases and drying begins

Phase 2. Steady-state evaporation, where energy is consumed for free water evaporation while the sheet temperature remains relatively constant

Phase 3. Falling rate, where sheet temperature increases as "free" water becomes scarce and the evaporation rate begins to fall.

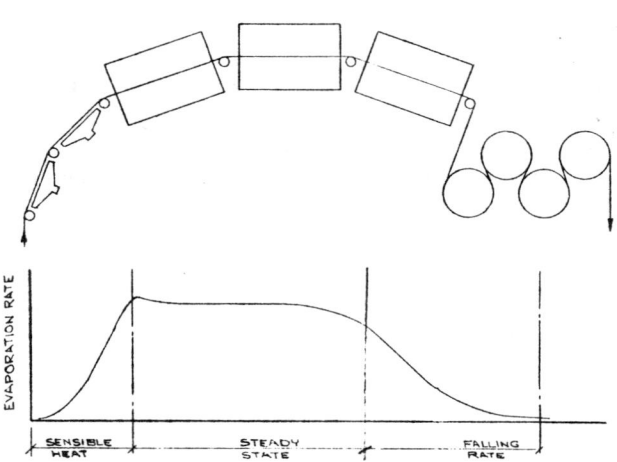

Fig. 4.42. The three phases in the drying process with typical dryer arrangement

Applying the above drying phase theory with the drying systems available typically results in the use of IR pre-heat dryers, followed by air flotation dryers, and finishing with cylinder dryers which, in practice, approximates the drying process as defined above.

1. IR dryers have the advantage of compactness of design and high-energy output, which allows the dryers to be located close to the coating head to initiate drying immediately upon coating.

2. Flotation dryers have the advantage of noncontact drying with exceptional evaporation rate control until the coating is sufficiently dry to allow contact by rolls or cylinder dryers.

3. Cylinder dryers have the advantage of contact drying for improved sheet appearance and web support for tension control.

To begin to layout a drying arrangement, it is important to get an understanding of what the drying scheme is to be. Many papers have been written in recent years that have attempted to predict the optimum drying rates and the proper equipment to achieve those rates, mainly to prevent coating binder migration and mottle. Much work has been done to show that the drying process can play a significant part in improving or degrading the paper coating appearance and performance.

However, the ideal percentage of the drying load that each of the various drying components must contribute is not easily defined and will usually vary with coating formulation and grade. It is common practice to have about 20-30% of the drying with IR, followed by 60-70% with air flotation, and the remainder with cylinder dryers.

To introduce a method for estimating the drying for each phase, we need to set up an example to approximate the drying lengths required of each section. The first step is to define the specifications that are binding. For this example, we will define these requirements as follows:

Basis weight (WT)	68 lb/rm paper @ 5% moisture
Ream size (RM)	3300 ft^2
Coat weight (CW)	7 lbs
Coating solids (CS)	62 %
Speed (S)	2200 ft/min
IR drying load (IR)	reduce coating from 62% to 72% dry.
Air float drying load (AF)	reduce coating from 72% to 90% dry.
Cylinder drying load (CD)	reduce coating from 90% to 95% dry.

With but a few basic equations, we can determine most of the information we need to approximate the various drying lengths for each dryer section. They are:

1. Calculate the water load required for each section.

$$EV = \frac{CW}{RM} \times S \times 60 \times \left\{ R1 - R2 \right\} \qquad (10)$$

Where:
$\quad$ EV $=$ Water evaporated per foot of width (lbs/hr)
$\quad$ CW $=$ Coat weight (lbs)
$\quad$ RM $=$ Ream size (sq ft)
$\quad$ S $=$ Speed (ft/min)
$\quad$ R1 $=$ Ratio of water to solids entering the dryer
$\quad$ R2 $=$ Ratio of water to solids exiting the dryer

2. Calculate the heat load required.

$$Q = \frac{WT}{RM} \times S \times 60 \times SH \times \Delta T \qquad (11)$$

Where:
$\quad$ Q $=$ Energy per ft of width (btu/hr-ft-width)
$\quad$ WT $=$ Basis weight dry (lbs)
$\quad$ SH $=$ Specific heat of substance (BTU/lb°F)
$\quad$ ΔT $=$ Temperature rise (°F)

Estimating the IR Dryer Section

It is generally accepted that the use of IR drying in the early phase of the drying process has the advantage of: (a) placement of drying equipment close to the coater, due to its compactness of design, and (b) the high-energy output can rapidly heat the web to begin the drying process in the shortest possible period.

For this example, a gas-fired IR unit will be estimated. We are given that the IR unit will operate with a sheet input energy of 20,000 Btu/hr-ft^2. We are also given that the sheet temperature exiting the IR is approximately 170°F. In reality, it is not a simple task to calculate this temperature for IR dryers. For lighter-weight paper grades, a practical range to use is between 160°F and 190°F.

Since the IR unit is located in the initial stages of the drying arrangement, the pre-heat phase, we will need to determine the sensible heat load as well as estimate the evaporation load expected.

Step 1. Calculate the water to be evaporated to reduce the coating solids from 62% to 72% dry with the IR dryer.

$$EV = \frac{7}{3300} \times 2200 \times 60 \times \left\{ \frac{38}{62} - \frac{28}{72} \right\} \qquad (12)$$

$$= \boxed{62.7 \text{ lb/hr-ft-wd}}$$

Step 2. Calculate the sensible heat load.

$$Paper = \frac{68 - \left\{ 68 \times \left\{ \frac{5}{95} \right\} \right\}}{3300} \times 2200 \times 60 \times .35 \times 100 \qquad (13)$$

$$= \boxed{90,190 \text{ btu/hr-ft-wd}}$$

$$Moisture = \frac{68 \times \left\{ \frac{5}{95} \right\}}{3300} \times 2200 \times 60 \times 1.0 \times 100$$

$$= \boxed{14,315 \text{ btu/hr-ft-wd}}$$

$$Coating = \frac{7}{3300} \times 2200 \times 60 \times .5 \times 100$$

$$= \boxed{14,000 \text{ btu/hr-ft-wd}}$$

$$Water = \frac{\left\{ \frac{7}{.62} \right\} - 7}{3300} \times 2200 \times 60 \times 1.0 \times 100$$

$$= \boxed{17,160 \text{ btu/hr-ft-wd}}$$

$$EV \text{ load} = 62.7 \times 1000 \text{ }^{btu}/_{lb}$$

$$= \boxed{62,700 \text{ btu/hr-ft-wd}}$$

$$Q \text{ (total)} = Q_{(paper)} + Q_{(moist)} + Q_{(coating)} + Q_{(water)} + Q_{(EV)}$$

$$= \boxed{198,365 \text{ btu/hr-ft-wd}}$$

Step 3. Estimate the length of IR needed.

$$Length = \frac{198,365 \text{ }^{btu}/_{hr-ft-wd}}{20,000 \text{ }^{btu}/_{hr-sqft}} \qquad (14)$$

$$= \boxed{\text{use 10 ft of IR dryer}}$$

Estimating the Flotation Dryer Section

The advantage of the flotation dryer is that noncontact drying provides the means to effectively remove the water from the sensitive coating surface until the coating is sufficiently dry so that further contact will not mark the coating. Generally, this requires that the moisture in the coating be dried below 10%.

For this example, we are given that the average dryer evaporation rate will be 5 lb/hr-ft², and the moisture out of the flotation dryer is to be approximately 8%. Since the IR section has initiated the drying and the web is at evaporation temperature, all the available flotation dryer length will be used for evaporation.

Step 4. Calculate the water to be evaporated to reduce the coating solids from 72% to 8% dry.

$$EV = \frac{7}{3300} \times 2200 \times 60 \times \left\{ \frac{28}{72} - \frac{8}{92} \right\} \qquad (15)$$

$$= \boxed{84.5 \text{ lb/hr-ft-wd}}$$

Step 5. Estimate the dryer length needed.

$$\text{Length} = \frac{84.5 \text{ lb/hr-ft-wd}}{5 \text{ lb/hr-sqft}} \qquad (16)$$

$$= \boxed{\text{use 20 ft of flotation dryer}}$$

Step 6. The graph shown in Fig. 4.43 can be used to estimate the operating conditions within the dryer.

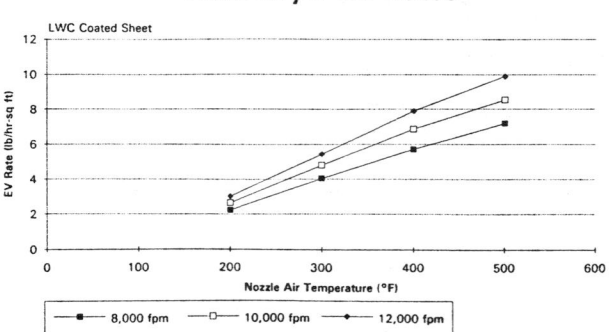

Float Dryer EV Rates

Fig. 4.43. Flotation dryer average EV rate float nozzles both sides of sheet

Estimating the Steam Cylinder dryers

The advantage of steam cylinder drying is that the sheet is held in contact to a smooth surface in the final moisture removal stage or falling rate phase. This is generally accepted as the portion of the drying that most affects sheet appearance and curl. Also the higher heat transfer, due to direct contact with the hot cylinder surface, provides the most effective method to remove the moisture from the sheet that is no longer free water.

For this portion of the drying, we are given an evaporation rate of 0.5 lb/hr-ft², with 4-ft diameter drying cylinders, and are to obtain a final sheet moisture of 4%.

Step 7. Calculate the water to be evaporated using the cylinder dryers to reduce the coating solids from 8% to 4% dry.

$$EV_{\text{COATING}} = \frac{7}{3300} \times 2200 \times 60 \times \left\{ \frac{8}{92} - \frac{4}{96} \right\} \qquad (17)$$

$$= \boxed{12.7 \text{ lb/hr-ft-wd}}$$

$$EV_{\text{PAPER}} = \frac{68}{3300} \times 2200 \times 60 \times \left\{ \frac{5}{95} - \frac{4}{96} \right\}$$

$$= \boxed{29.8 \text{ lb/hr-ft-wd}}$$

$$EV_{\text{COATING + PAPER}} = \boxed{42.5 \text{ lb/hr-ft-wd}}$$

Step 8. Estimate the number of cylinders required.

$$\text{Length per Cylinder} = \Pi \times 4 \qquad (18)$$

$$= 12.5 \text{ ft per cylinder}$$

$$\text{Number of Cylinders} = \frac{42.5 \text{ lb/hr-ft-wd}}{12.5 \text{ ft} \times 0.5 \text{ lb/hr-sqft}}$$

$$= \boxed{\text{use 7 cylinders minimum}}$$

Step 9. The graph shown in Fig. 4.44 can be used to estimate the cylinder operation conditions.

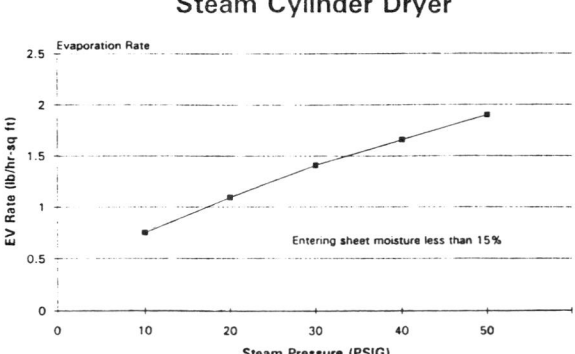

Fig. 4.44. Average evaporation rate for steam cylinder section following flotation dryer

The proper application of a drying system for the drying of coated paper is not a simple task. For the optimum drying setup, only coating and drying tests can determine what temperatures, evaporation rates, and equipment arrangement are to be used. This is usually determined after evaluations based on achieving a coating quality standard. The remainder of this section on drying will attempt to briefly address one of the more difficult and controversial issues of drying, which is in the area of defining and testing for binder migration.

Binder Migration

Binder migration, which is also known as K & N mottle and binder specks, appears to manifest itself primarily in the printing of half-tone colors. And if the coated sheet exhibits this problem, it will be worse in the light blue screens. The problem has been associated with drying, and is apparently caused from the coating binder being nonuniformly dispersed on the surface of the sheet. Ultimately, when the coated paper is printed, the ink is not absorbed uniformly, and the high concentration of binder will show up as a unacceptable printed area.

Testing for binder migration is normally handled by the use of absorbent inks that are applied for a specific period of time, with the excess ink wiped off and the inked area observed or tested by a meter and compared to the meter reading on the background area.

Of the several inks that are used for this test, the most common are:

1. K & N ink—a light purple ink
2. Croda red ink—a red ink
3. Porometrique Noir—a black ink.

It is the opinion of this author that preference is toward the light purple K & N ink, because the K & N is a more surface property ink. The Croda red ink tends to show "too much." The Croda shows the binder migration, the coating lay, and the effect of the base sheet on the ink absorption.

As an example, if a mill is having problems with formation, size addition, and binder migration in the Croda test, it would be difficult to separate the effect of the binder migration from the base sheet formation. The K & N on the other hand, being more of a surface test, would indicate more clearly the binder migration, while the contribution of the base sheet formation would be minimized.

Chapter 5

Process Control

Sanford I. Shapiro, Honeywell Inc.

Introduction

Benefits for the coating mill from implementation of process control technologies include improved runnability and productivity and a higher-quality product at reduced cost. Discussion applies to both on-machine and off-machine coaters for paper and paperboard.

To meet product specifications, proper instrumentation and control are required for the basestock, coating preparation, and supercalender, as well as the coating stations and dryers. On-line product quality sensors available include weight, moisture, ash, opacity, gloss, brightness, color, formation, smoothness, and caliper. At the coater, closed-loop controls include coat weight and moisture, both of average and cross-direction profile. Optimizing controls include coordinated speed change, throughput maximization, grade change, and for off-machine coaters, start-up sequence control. Distributed Control Systems (DCS) are being used for cost-effective coordination of all instrumentation and control functions in the mill from startup through shutdown, as well as the continuous operations. Integration of the machine drive, PLCs, scanning sensors, machine-direction and cross-direction controls, and lab data reporting is accomplished via "Single Window" technology. The resulting greater efficiency benefits operators, technicians, engineers, and management of the coating mill.

Process Control

What is process control? A general definition is "The regulation or manipulation of variables to manage a processing operation within the required quantity and quality specifications." Involved are sensors and other instrumentation, valves, actuators, motors, computers of many types, user interfaces including printers and video display units, and obviously, the logic to make it all work. Excellent textbooks are available for reference on process control theory and general application (1) and, more specifically, on basic pulp and paper applications (2).

Control System Technology

Process control operates at four hierarchical levels:

I. Device level: valve, motor, etc.
II. Loop level: pressure, level, temperature, flow rate, etc.
III. Supervisory level: coat weight, moisture, grade change, etc.
IV. Management level: order scheduling, etc.

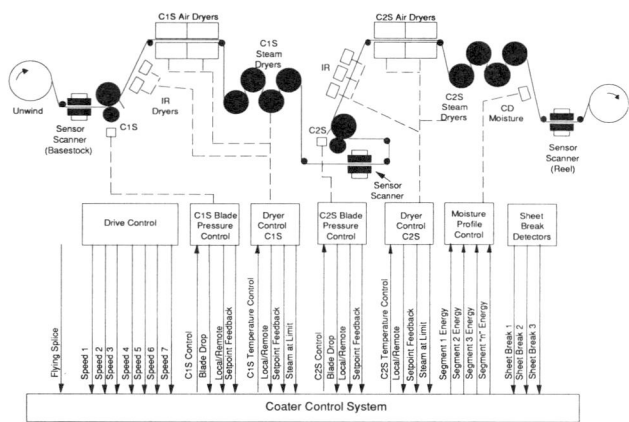

Fig. 5.1. Coater control system instrumentation

Instrumentation for a typical modern off-machine coater is shown in Fig. 5.1. Except for the flying splice, it is similar for on-machine operations. Most coaters today still use single-loop controllers for process variables. There is a trend toward integrated Distributed Control Systems which can coordinate process control for all functions from startup through shutdown, as well as the continuous operations. DCS installations are more common with on-machine coating operations since these systems were adopted earlier in the paper machine area. The modern DCS can handle loop, supervisory, and sequencing (e.g., startup) controls, as well as logic (e.g., safety interlocks).

Operator stations with color video displays are provided for communication with the control system. Printed reports document coater operation and product quality. All information may be continuously stored on hard disk for rapid retrieval and display at any time in the future.

Single-window Integration

In most coating operations, the measurements and controls used for the coating kitchen, coating stations, dryers, supercalender, and quality inspection are independent systems. These are referred to as "islands of automation." There are many potential benefits to be gained from integrating these independent islands. This is already being done with paper machines to obtain a single-window operator interface (3). The term refers to the ability of mill personnel to obtain information and coordinate all controls from a single system's video screens. It involves interfacing via the DCS the previously unconnected islands of automation available for the coating complex:

- Drive controls
- Programmable Logic Controllers
- Scanning sensors
- MD and CD controls
- Speed and draw indication
- Hole and defect detectors
- Lab data reporting.

A typical "Single Window" architecture is shown in Fig. 5.2.

Basestock

Basestock uniformity and adherence to specification are essential to the coating process and final coated product. Basestock quality must be controlled tightly at the paper machine, from stock blending through drying. New basestock process controls are continuously being applied to meet the coater's productivity and quality requirements. For example, if basestock formation is poor, coat weight and moisture variations increase, leading to print mottling. Among the technologies now used at the paper machine, for basestock quality, are on-line formation sensors for monitoring fiber distribution uniformity (4).

Single-window technology makes available to the coater operator detailed information concerning the basestock being coated. This has value as an aid for setup or troubleshooting at the coater. In the future, it should be possible to do manual or closed-loop modification of basestock furnish, forming, and drying based on feedback from on-line quality sensors at the coater windup.

Coating Preparation

For coating prep operations, both batch and continuous, process controls handle metering, mixing, routing, and monitoring. Sensors monitor temperature, pH, viscosity, percent solids, levels, and flows. Measurement of mass flow requires either the combined use of both magnetic flowmeters and density meters, or the newer Coriolis-force mass flowmeters which provide measurements of mass flow directly.

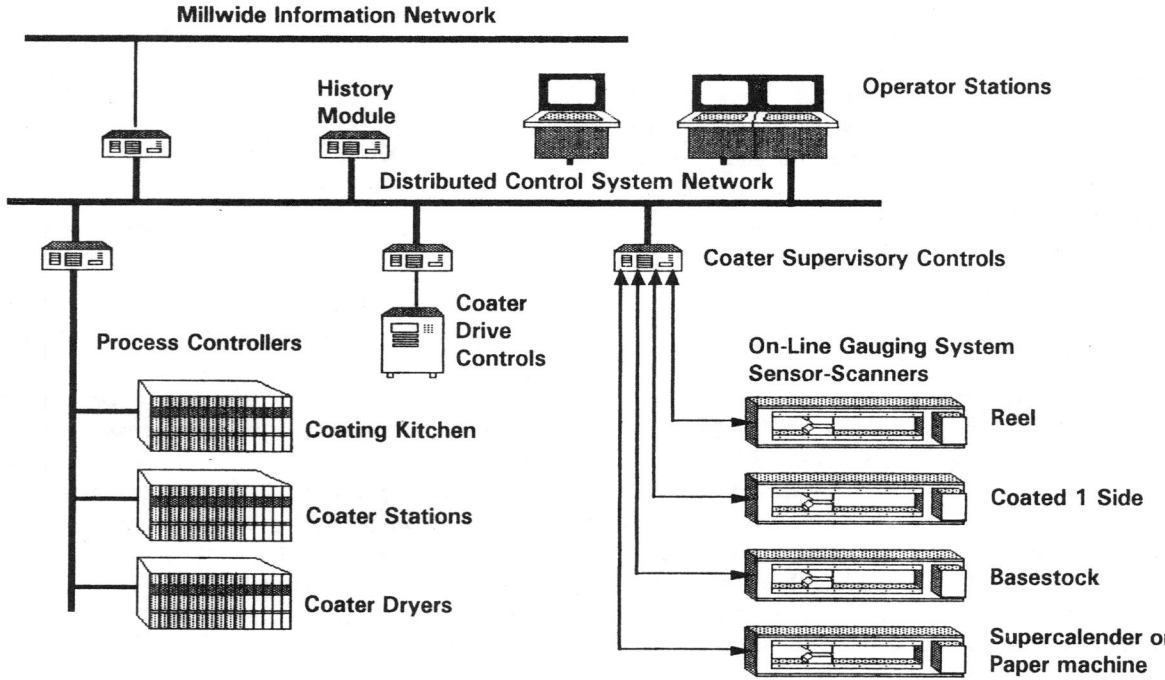

Fig. 5.2. "Single Window" integration of coating controls

Distributed Control Systems are now being applied in batch and continuous coating preparation operations. A system architecture is outlined in Fig. 5.3. One DCS can handle several mixers simultaneously and an almost unlimited number of recipes.

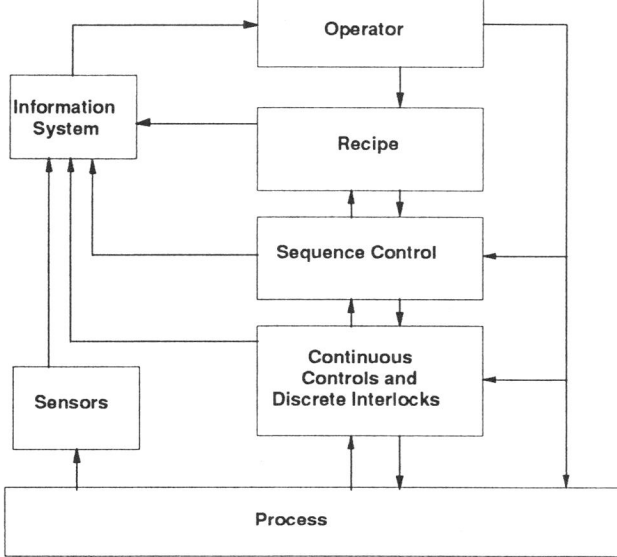

Fig. 5.3. Coating preparation: control system architecture

When an operator enters the identifier for the next recipe to be made, the DCS calls the formulation from memory for verification and modification if necessary. The DCS schedules each ingredient to start in the proper sequence, and controls timing, temperatures, consistency, and flows. There are alarms and interlocks in the system. For example, if a tank level reaches a specified low or high point, a message is flashed to the operator. Depending on the situation, automatic action may be taken by the DCS to start or stop a pump or open or close a valve. Alternatively, the operator may be given, via video message, suggested options for manual action.

As materials are used, the computer calculates and records changes in inventory for the DCS's management information system. By including actual production rate off the coater via single-window integration and its global database, immediate and continuous information is available on coating materials used per ton or per area of product.

Coating Solids Control

Any change in percent solids of the coating supplied to the applicator will result in a change in coat weight. This variation in solids can be significant, particularly where excess coating is recycled. Sensors such as the Coriolis meter are available for monitoring solids content. A simple dilution control in the coating supply line can provide stable solids and resulting coat weight.

The Coater

Machine Drives

Essential to efficient operation of any coater is proper instrumentation and operation of the drive system (5). Improper web tension causes breaks and reduces quality uniformity. Upgrades of older coaters replace line shafts and manual draw controls with independent dc motors and drive systems which react quickly and precisely. For tension control, segmented web-driven tension rolls are seen as an improvement over swing rolls and load-cell supported rolls. Together with the multi-motor dc drive, they provide fast and accurate response to tension changes and good control to target setting. The web is stressed less, reducing causes of breaks.

On some coaters, each roll is driven by an independent motor; on others, groups of rolls and their motors are ganged together and operated from one motor controller. Digitally controlled drive equipment can be easily integrated into automation networks and addressed from the control room. Machine drives can be integrated into the Distributed Control System for operator convenience and overall coordination efficiency.

Coater Station

Operations at the coating station are varied and can be complex in terms of process control requirements. Flexible blades are less predictable in response than rigid blades. Coaters can have one to four or more stations, often of different types (blade, roll, rod, etc.). To meet the process needs for each station, there is available measurement, display, and control of variables including coating supply pressure, temperature and solids, blade (or air knife) angle and pressure, and the resulting coat weight.

Product Sensors

Coaters now are almost all equipped, at the reel and before each station, with on-line sensors which mechanically move across the web. (See Fig. 5.1.) These provide continuous machine-direction and cross-machine information. Measurements available include weight, moisture, ash, caliper, formation, color, opacity, brightness, gloss, smoothness, and roll hardness. Because the web is moving rapidly in the machine direction (MD) while the sensors are moving in the cross-machine direction (CD), the web is actually sampled in a diagonal strip. The ability to continuously see the entire web width (100% inspection) is a technology still developing.

A few on-machine sensors operate in a stationary non-traversing mode, yet continuously "see" the entire width of the moving web. These include the defect detectors for holes, dirt, and blemishes of various types.

Coat Weight Measurement

Differential Beta

Coat weight, in most mills, is calculated via the difference between basis weight measurements before and after the coating station. (See Figure 5.1.) Almost all of these sensors pass a narrow beam of beta radiation, from encapsulated radioisotopes, through the web to measure mass. Radioisotopes used include Krypton-85 and Promethium-147.

In most installations, a traversing moisture gauge is used in line with each weight gauge to document the differential moisture contents at the two locations. As there is some degree of error in any sensor, using four to produce one dry coat weight calculation is not the most desirable approach. In addition, because the two sets of sensors do not see the same spot on the web, the new coat calculation will be in error by the difference in basestock weight of the two spots. At the higher basis weights, particularly coated boards, the variation in basestock weight between the two spots seen can be greater than the coating weight being applied, producing a negative coat weight calculation. Signal averaging, or smoothing, is commonly used to minimize these errors, but this degrades control results. One of the better techniques available is sensor traversing coordination to enable the two sets of sensors to actually see the same spot on the web. Without this same-spot synchronization of the moving sensors, the accuracy of coat weight calculated will be affected by base sheet variability.

Sensors based on other technologies are available to measure coat weight directly (6), and are described below.

X-ray Fluorescence

An X-ray source, either high-voltage electronic tube or Fe-55 radio isotope, causes certain elements in the coating to emit X-rays characteristic of that element. This works well with titanium and calcium, for example. The magnitude of the specific fluorescence is measured and calibrated in units of coat weight. Since the element(s) must be a known proportion of the coating, the measurement is affected by any variations in composition of the coating. A unique advantage of this sensor, being a single-sided reflection design, is the ability to measure coatings applied to opposite sides of the web simultaneously (7).

X-ray Transmission

Ash sensors operate by looking at the transmission of X-radiation through the web, rather than its fluorescence. Suppliers state that by using an electronic tube instead of a radioisotope as the radiation source, they are able to tune the X-ray spectrum generated and obtain equal sensitivity to titanium, calcium, and clay. The sensor sees total ash in the web, both coating and basestock. Some board mills are using this sensor calibrated to measure coat weight directly. It appears to work satisfactorily as long as the ash in the basesheet is consistent or known.

Differential X-ray

To get around the problem of variable ash in the basestock, some mills are using X-ray gauges before and after the coating, doing a subtraction to calculate ash applied, and calibrating it as coat weight. For on-machine coaters, the basestock ash sensor can also be used for filler monitoring and closed-loop control as well as for determining the coat weight.

Moisture and Other Sensors

Moisture is routinely measured after each coating station to control dryer sections and for quality management. Infrared absorption is the predominant technique, and many model variations are offered for specific circumstances.

Other sensors available for coated web quality include caliper, color, opacity, brightness, gloss, and smoothness. Details about these can be obtained from the various suppliers.

Coat Weight Control

Closed-loop control of coat weight average (8) is operational at many mills. Uniformity is being achieved to the point where coat weight average with less than 0.1% deviation from target is reported by mills.

Control of each coating application can be to either a net coat weight target or to total coated-sheet weight (basestock plus coating), depending on mill preference. In either case, control is on a calculated dry weight basis. After each scan of the sensors across the sheet, measurements are compared to target.

When controlling to a total coated-sheet weight, the amount of coating applied will vary to compensate for variations in basestock weight. If the sheet is to be coated on both sides (wire and felt), it is usually desirable to maintain a constant ratio applied to each side rather than allow one side to take up full compensation for basestock variations. On startup, the operator enters targets for total coated weight and the desired side-to-side coating ratio. After each scan of the sensors cross-sheet, the computer calculates the need for corrections at each coater station and adjusts actuators as necessary.

The primary actuator used most often is blade pressure. Secondary actuators used, depending on the specific coater, include blade angle, coating solids, and where applicable, coater speed, air-knife pressure and angle, or roll speed. The limitation to the application of closed-loop coat weight control is usually the mechanical capability (movement resolution and repeatability) of the coat weight metering device actuator.

Dual actuator logic may be used to maintain the primary actuator within the optimum control range. A typical combination is blade pressure as the fine adjustment primary actuator, and angle as the coarse or mid-ranging control. For example, as a rigid blade wears, pressure may have to be increased beyond a workable value. The mid-control logic makes small changes to the blade angle, thus maintaining the blade pressure within the optimum range. Adjustment of coating slurry solids for coat weight control has the disadvantage of slower response times. However, it has been used successfully as a mid-ranging adjustment with blade pressure as the primary control.

The use of speed changes for coat weight control is limited to off-machine coaters and then only under specific conditions.

Drying

Process control of the dryer sections is essential for product quality and fuel conservation. Coated web warming and drying must be carefully controlled for several reasons, including minimizing binder migration and for curl control.

For air dryers of the various types, the control system can automatically adjust fans, dampers, and temperature in each section. With conventional steam dryers, control may be of the individual can's pressure or temperature, but usually is grouped into top and bottom sections. For condensate removal, differential (inlet vs. outlet) pressure is controlled, as well as temperature and vacuum in the condensate tanks, temperature of cooling water, and pump operation.

Dual actuator control often has advantages. For example, temperature in an air dryer section may be used as the fine adjustment actuator because of its quick response, with steam pressure as the coarse secondary control, thus keeping the air section in its optimum energy usage range.

Cross-direction Control

CD Coat Weight

A number of on-machine coaters are doing CD coat weight control using breaker or calender stack actuators. The basic principle is modifying the smoothness, porosity, or density of the basestock, which affects the coating pickup. The smoother, denser, or less porous the web, the less pickup. Calender nip action is adjusted by use of a row of narrow cross-machine actuators, such as hot or cold air showers or induction heaters, acting on the metal rolls.

For direct control of coat weight on blade coaters, mechanical adjusting rods, similar to slice screws used on paper machine headboxes, can be installed to modify blade pressure at intervals across the web. This application requires blade holders and coater stations designed to support these CD mechanical actuators.

CD Moisture

Infrared (IR) dryers are available with a large number of narrow emitters placed side-by-side, cross-machine, for control of the moisture profile. Currently IR actuators are on the market that are only 2.5-in. (63 mm) wide in the cross-machine direction. Since moisture sensors are able to accurately measure and display streaks even narrower, the width of drying modules will be reduced even further in the future.

Optimizing Controls

Several applications of optimizing controls are widely used on paper machines and are being increasingly applied to coaters. These are discussed below.

Coordinated Speed Change

Upon entry of a new speed setpoint, the system smoothly ramps all related actuators so as to maintain coat weight, moisture, and other specified variables within specification, and complete the change in the shortest possible time. Among the benefits are a reduction in breaks and off-speed rejects.

Speed Optimization Control

The purpose of this program is to maximize coater production by gradually increasing operating speed, in coordination with the other quality-maintaining controls, until some limiting condition is reached. Production rate is thus held at the maximum under existing conditions.

Startup Control

For off-machine coaters, startup control may be used to bring the coater from threading speed up to optimum production speed as rapidly as possible after a shutdown or sheet break. Coater production is optimized by slowly increasing speed, in coordination with weight and moisture control, until a limiting condition is reached. Possible limits include blade pressure, steam pressure or flow, dryer air temperature, or a specified maximum speed. Any analog or digital input signal may be monitored for this purpose. When the specified limit of a control actuator is reached, the program switches to an alternate control mode. For example, if a dryer pressure limit is reached, the system switches to coordinated speed changes to hold moisture on target while continuing to run at maximum steam pressure. Production rate is thus held at the maximum possible under existing conditions.

The process control system automates as many functions as possible. A typical example would operate as follows:

A. When the coater is threaded, the operator pushes a button to begin the startup.

B. The coater is smoothly accelerated to a pre-set coating start speed. Weight and moisture sensors are positioned at a fixed location on the sheet, rather than traversing.

C. When the coater dryers are at the proper preset condition, the coating is applied. An optional stabilization period begins, with coater speed held steady. Closed-loop control brings coat weight and moisture to target. After the stabilization period, speed is increased in coordination with weight and moisture control. Use of a single-point position for the sensor package allows rapid control actions compared to the longer intervals based on cross-machine averaging.

D. The coordinated speed increase is continued until an interim-target speed is achieved. The sensors start scanning and control is based on cross-sheet averages.

E. After a short stabilization period, coater speed is gradually increased until standard operating speed is reached.

Automatic Grade Change

Upon operator entry of the new grade identifier and activation of the start change control, adjustments are automatically made, with the proper coordination, to speed, weight, moisture, etc., so as to move production to the new specifications in the shortest possible time. Automatic grade change control is applicable to both on- and off-machine coaters.

Target Optimization

While closely monitoring quality variability of the coated web, control setpoints for weight, moisture, or both are carefully modified up or down so as to keep these quality parameters as close as possible to the economic optimum, while remaining safely within customer specifications.

Supercalender

Most new supercalenders are instrumented with traversing sensors, typically for gloss but increasingly also for moisture and caliper. Steam showers can be used to control the gloss development, cross-machine as well as average. Actuators with a series of high-resolution steam jets put a small amount of warm moisture on the web prior to the nips. The technique modifies the coating just enough to improve finishing. This method of gloss optimization control can be applied independently to each side of the web.

Summary

Control technology is now advancing almost faster than industry can implement. As the coating industry continuously moves to use and take advantage of the best of the available process control technologies, resulting benefits include:

- Product quality more consistent
- Productivity increased
- Raw material savings increased
- Production flexibility improved
- Information more timely and accurate
- Better operator decision tools
- Better management decision tools
- Labor utilization improved
- Cost accounting improved
- Strategic competitive advantages gained.

Bibliography

Literature Cited

1. Shinskey, F.G., *Process Control Systems, 3rd Edn.,* McGraw-Hill Book Co., New York, 1988.

2. Lavigne, J.R., *Instrumentation Applications for the Pulp and Paper Industry,* Miller Freeman Publications, San Francisco, 1979.

3. Shapiro, S.I., "Single Window for the Paper Mill," 1989 Engineering Conference Proceedings, TAPPI PRESS, Atlanta.

4. Shapiro, S.I., "On-Line Formation Sensors," 1989 Papermakers Conference Proceedings, TAPPI PRESS, Atlanta.

5. Kheboian, G.I., "Off-machine Coater Drive Systems: Back to Basics," 1989 Engineering Conference Proceedings, TAPPI PRESS, Atlanta.

6. Rutledge, W.C., "On-line Coat Weight Measurement," 1988 CPPA Control Conference.

7. Boos, J.S., *Paper Age* 105(8):34 (1989).

8. Shapiro, S.I., "Computer Control of Blade Coaters," *1988 Blade Coating Seminar Notes,* TAPPI PRESS, Atlanta.

Chapter 6

Web Handling and Off-machine Coater Drives

Gerald I. Kheboian, Kheboian & Associates
Editor, Robert J. Alheid, Beloit Corp.

Introduction

This chapter stresses the fundamentals necessary to begin to understanding how a coating line works and what it does. Therefore, we will define terms, describe the equipment, and show how to select motor horsepower and clutch/brake sizes. For purposes of this discussion, a coating machine consists of an unwind, pull rolls or draw rolls, coating stations, dryers, and a rewind.

Web Handling

The web transport system includes the motor, the gearbox, the power supply of the motor, the control system for the power supply, the roll that drives the web, the cover of the roll, the drive interface between the roll and web, and the frictional and physical character of the web itself. Web transport also includes the tracking of defect-free and wrinkle-free web on the center line of the machine. The web transport system is not the motor alone. The web transport system must include the web itself.

The transport of web will also be influenced by the characteristics of the individual sections. For instance, the air-handling system in the air dryers must be designed so it helps rather than impedes the movement of the web through the machine. Other sections must be designed to ensure that they maximize the straight tracking of the web through the machine without distorting coating and without adding wrinkles, scratches, or web tears.

A web is any paper, film, or foil that is delivered in a continuous flexible strip to the coating machine for coating, laminating, or other processing. (See Fig. 6.1.) The web width will directly affect the selection of horsepower and brake/clutch size. The web materials define how much tension will be run for each product. For this discussion, the web width is expressed in inches and the

web thickness is expressed in mils (thousandths of an inch).

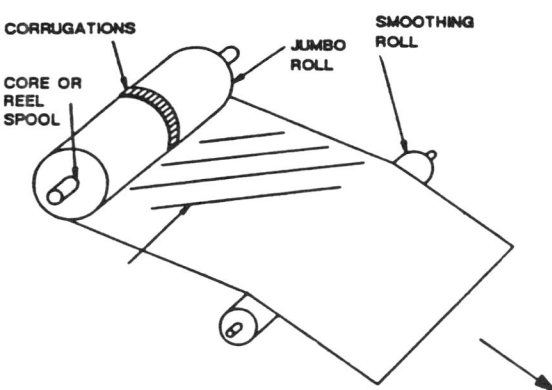

WHAT IS WEB?
ANY PAPER OR PLASTIC DELIVERED TO THE MACHINE IN A CONTINUOUS, FLEXIBLE STRIP FOR COATING, LAMINATING OR ANY OTHER PROCESS.

Fig. 6.1. What is a web?

Each coating mill must understand what defines the parameters of an acceptable web and set up "practical" specifications for the process and the supplier (within or outside of the mill) of the basestock. Practical specifications represent a commitment by the supplier and coating mill to establish web parameters that will run satisfactorily on the machine without making the production of the basestock impossible or economically infeasible.

As an example of how a poor web has an impact on quality consider a web which comes off the paper machine with crushed quarter points from the press section and that is almost bone dry off the reel and is brittle. This web may be loaded with edge cracks which can break down at the coater at the first nip or first coating head. In this case, if edge cracks don't cause a web break, then the crushed web probably will. Wrinkles that form at crushed quarter points and then turn into creases will break in the machine.

The responsibility for determining basestock specifications falls on the user. Basestock problems are often masked when coater operators adjust the coater to make a salable product. They may misalign rolls, put rolls out of level, and change process speeds, tensions, and air handling to get rid of wrinkles to satisfy the production schedule. Such techniques will be used until the machine is so out of specification that production stops. Usually, at this time a task force is formed, the machine will be blamed, all rolls will be realigned and leveled, and the wrinkle problem will still exist, because the problem is not in the coater but in the basestock.

Tension through a coating machine is the force which is applied to the web between adjacent machine sections. This force tends to stretch or elongate the web and is expressed as some unit of web width. Those units are defined as:

- Pounds per linear inch, PLI, P/LI
- Ounces per linear inch, OLI, O/LI
- Grams per linear inch, GLI, G/LI
- Kilograms per meter, KGM, KG/M.

It is important to know and record the operating tension values of each product throughout the coating machine. These values can be recorded on any computer-operated data collection system. It is important for the operators, process engineers, and equipment engineers to know these values so they can determine if the machine and web is performing normally. Tension information is also valuable in process design, when looking at new equipment, and in troubleshooting machine problems. A knowledge of how each product runs may show how a machine can be rebuilt to operate at higher process speeds without increasing the size of existing drive equipment.

Sheet tension on a coating machine should be just high enough to allow good tracking of the web. Table 6.1 (end of chapter) gives some typical tension values for a variety of web materials and web thicknesses. Lower tensions tend to reduce sheet breaks that are due to sheet defects such as web wrinkles, holes, and edge cracks. Excessively low tension will allow the web to wander in the machine and will contribute to telescoping that occurs in loosely wound rolls.

In addition to expressing tension as a unit of web width, tension in a coating machine is also expressed as total tension. Total tension is the product of unit tension times the web width. For instance, 2 PLI x 68 in. = 136 pounds total tension. The product of the web width times the tension per linear inch represents the amount of work that the drive motor or brake must do. This value is used to calculate brake, clutch, or motor size. The range of tension for any machine will enter into the design of the tension control system components and the design of the rolls which support the web. At the end of this chapter, Table 6.2 lists some helpful formulas for drives.

Unwind

The unwind's primary role is to introduce the paper to the coater. Before an unwind is selected, it is important to know if the web is involved in a continuous process or an intermittent process. If the process demands a continuous stream of web to the process, then a multiple arbor unwind is necessary. A multiple arbor unwind must have the ability to splice the web while the machine and process are in full-speed operation.

Web Control

Make no mistake, the two real "bosses" on the coater are the unwind and rewinder. Transients caused by splicing at the unwind and transferring on the rewinder can be felt throughout the machine. Tension will vary as the roll diameter changes if the torque is maintained constant. As an example, we will use a 30-in. diameter roll with a 6-in. core. Unwind tension will change from 2 PLI to: 30 in./6 in. x 2 PLI = 10 PLI. At the same time, the rewind tension will change from 2 PLI to: 6 in./30 in. x 2 PLI = 0.4 PLI.

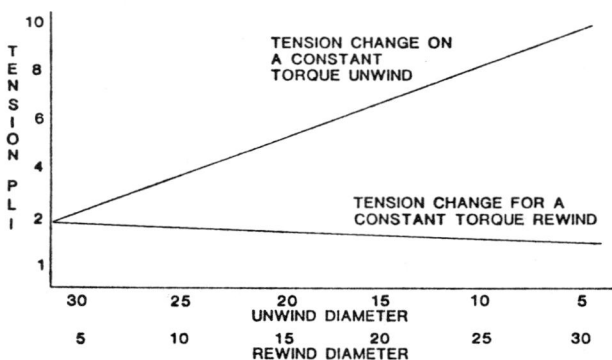

Fig. 6.2. Tension vs. diameter

If the unwind tension is reduced to allow poorly wound rolls to be processed when the rewinder tension is increased to wind perfect rolls, then the web will frequently break away from the lead section and the web will accelerate to a speed higher than the lead section. This condition is difficult to detect unless all sections are monitored or a coating defect or coating scratch attracts the operator's attention.

Center Unwind

The center unwind comes in two configurations: the transfer unwind and turret unwind. In the transfer unwind, a jumbo roll is loaded into primary arms and runs until half the original diameter is left on the spool or core. The roll is then transferred to a pair of secondary arms where it runs until there is almost no web length left on the spool. The arms are then moved into a splicing position and a new jumbo roll in the primary arms is spliced onto the expiring web. The empty spool is ejected from the secondary arms and is held in readiness for the next transfer cycle.

Transfer unwinds are very complex because the drives on each arm must be synchronized for both the transfer cycle and the splice cycle. The transfer cycle is particularly critical because the drives of both arms will be clutched to the spool for a short period of time before control of the web is transferred from the primary arm drive to the secondary arm drive. Precise speed matching is required during the transfer period to avoid a transient which could cause a web break, or worse, damage to the machine. Precise speed matching is also required for the splice cycle if a web break is to be avoided.

In the turret unwind, the jumbo roll is loaded into one of the spindles of the turret and remains there until almost no paper is left on the core. A splice is then made to a roll which has been loaded into the other spindle, and the empty core is ejected. A new jumbo roll is inserted into the first spindle and is held in readiness until the splice. The turret of unwind requires the same precision of splice pattern sensing as the transfer unwind and can be automated with equally good results.

Drive Motor on Unwind Stands

If the unwind uses a mechanical braking system, the web must supply the power to accelerate the jumbo roll without breaking the web. A brake is used on an unwind if the machine runs at low speeds, if the unwind makes zero speed splices, and if the web can provide the force that is required to accelerate the undriven jumbo roll to line speed after a splice. If the web cannot transmit the force without deforming or breaking the web, disrupting the process, or if the acceleration time is too long, then the unwind must be motor-driven. In addition to helping the jumbo roll to accelerate to line speed, it also helps control web tension when the web is up to speed. Clearly, tension control of some type will be required to ensure that all sections accelerate and decelerate together and that tension transients will be small during the acceleration and deceleration process.

On modern high-speed coaters designed for speeds up to 5000 ft/min, motor-driven unwinds are necessary. The machine operators prefer to accelerate the jumbo roll to splice speed and quickly make the splice. If a jumbo roll is left running at splice speed for an extended time, the hold-down tabs on the splice pattern may fracture before a splice is attempted, causing a splice failure and machine shutdown.

Roll Diameter and Weight

Roll diameters together with roll weights are used to determine horsepowers or brake size based on the inertia of the roll. Consider an unwinding jumbo roll or a calender roll as having high inertia loads. They may not track the line reference as closely as a low inertia draw roll or a coating roll. The horsepower (hp) to inertia ratios for all coater sections are not equal, and the section with the lowest hp to inertia ratio will determine how fast the whole machine will accelerate or decelerate.

The maximum roll diameter is needed to design support arms that allow adequate clearance between the maximum roll diameter and other parts of the machine and process. Structural members are designed to support roll weight with minimum deflection.

Core Diameter

The paper web comes to the coater wound on a core. In the paper industry, these cores are called reel spools. Cores can be made of cardboard, metal, or plastic; 6-in. inside diameter (ID) cores are 6 5/8-in. outside diameter (OD), and 3-in. ID cores are 3 1/2-in. OD. The core size, together with the largest diameter wound on the core, will determine the size of the brake and motor horsepower.

To get around any contamination from cores or from particles which come on the web, web-cleaning devices have been installed to remove dust and dirt from the web. These cleaning devices are located just before the coating station. Web cleaners usually consist of a device to neutralize static electricity, followed by a method of removing the contaminants from the web.

Build Down Ratio

Machinery builders and drive builders prefer to use large-diameter cores in order to keep the ratio between the maximum roll diameter and core as small as possible. This ratio is called the build down ratio. Increasing the core diameter does not require a proportional increase in maximum roll diameter to store the same feet of web.

Machinery builders have charts which show the amount of web that can be stored on a given core diameter and various maximum wound roll diameters. For instance, a transfer unwind starts with a 96-in. diameter jumbo roll wound on a 6-in. core. The roll is transferred from the primary arms to the secondary arms at 50 in., representing a ratio of 96:50 or 1.9 to 1. The secondary arms must now process the roll from 50 in. down to the 6-in. core. This represents a ratio of 50:6 or 8.33 to 1.

A small build down ratio simplifies the selection of drives or brakes, reduces the size of drive motors and brakes, and reduces the need for extraordinary control measures to satisfy large build down ratio. If brakes are used on the unwind stand, small build down ratios may reduce the brake torque range to a point where separate cooling may not be needed. Most wide high-speed unwinds use drive systems which control the field of the drive motor over the build down range. This is the constant horsepower range of the motor. Drive motors are designed with low base speeds and are very expensive. If the total build down is less than 6:1, it reduces the need to use special armature voltage control of the motor in order to bridge the build down ratio.

Center unwinds on small slow-speed machines often use armature control and standard motors to drive the unwind. Standard motors are used because they are less expensive than the low base-speed motors. The break-even point for standard motor versus low base speed motors used to be about 25 hp. Motor horsepower (hp) is directly related to the build down ratio. If the build down ratio is high enough, it may be necessary to use a more expensive drive system.

Flying Splice

Splices can be performed on the fly through a flying splice (Fig. 6.3). The diameter at which the transfer is made must guarantee that enough web is left on the spool to ensure that the operator has enough time to prepare and load a new roll into the primary arms, and be in position for a flying splice, all before the expiring roll is depleted. When transfers are made, at half the original roll diameter, less than half the web footage remains on the spool. Heavyweight webs running at high speeds will expire in a very short time. Therefore, heavyweight webs are transferred at fairly large diameters to allow the operator enough time to prepare for the flying splice.

Splices performed on the fly are usually lap splices. Splices can also be performed with the unwind at zero speed and the machine in full operation if a web storage device (festoon or accumulator) is used. The splice configuration either can be lapped by overlapping the two webs with an adhesive between or can be butted by placing the two web ends together with an adhesive tape

bridging and unifying the two webs. Some vendors are now offering flying butt splicers.

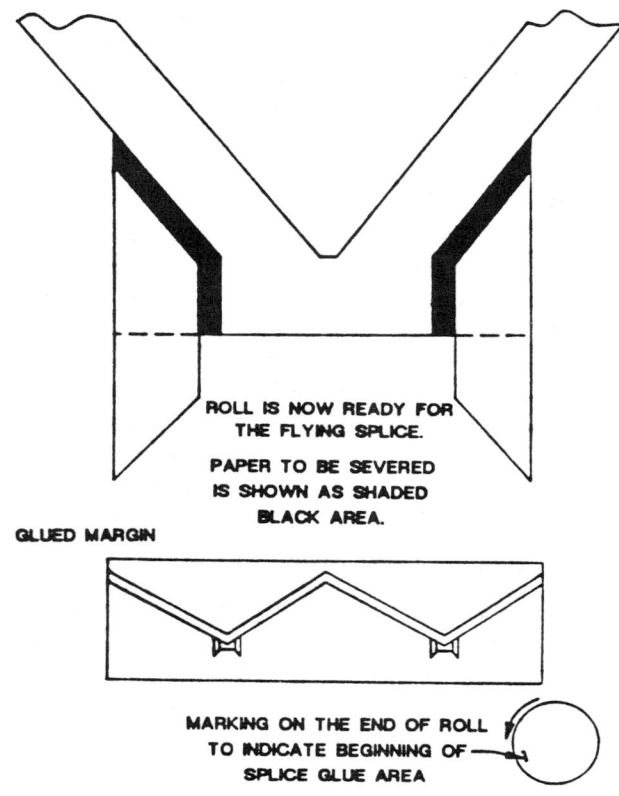

Fig. 6.3. Flying splice

The splice cycle requires precise position-sensing of the splice pattern to initiate the firing of the paste roll and knife and ensure successful splicing with a minimum tail length. Many vendors offer systems which completely automate all the functions of the unwind including loading of the roll into the primary arms.

Rewinder

At the end of the drive system is the rewind. If the process is an intermittent process, then a simple reel with a single pair of arms is all that is necessary to wind the jumbo roll. If the process is a continuous process, then two pairs of arms are needed, just like the unwind, to ensure continuous operation.

Surface Rewinds and Surface/Centerwind Rewinds

Surface/centerwind rewinders allow the operator to transfer the web onto the core in the surface mode and then finish winding the roll in the centerwind mode. The surface start allows a good start at the core and then allows the operator to centerwind the remainder of the roll with a controlled taper tension profile. This type of windup is frequently used for pressure-sensitive coatings where a good start is critical, but the pressure of the surface wind can damage pressure-sensitive coatings.

The surface wound reel consists of a driven reel drum, a pair of primary arms, and a pair of secondary arms. The web is threaded through the machine and is fixed to the reel spool which is located in the reel secondary arms. The spool is loaded against the driven reel drum. Power is transmitted to the winding roll from the driven reel drum. The reel drum drive incorporates a tension controller for the operator to control the quality of the wound roll. The secondary arm loading is another operator tool which can enhance the quantity of the wound roll.

The winding roll is wound to full diameter in the secondary arms until the web is transferred to a spool held in the primary arms. The primary arms hold the winding roll against the drum while the roll in the secondary arms is unloaded. When the secondary arms are returned to the winding position, the primary arms deposit the winding roll into the secondary arms and then the primary arms retract to a ready position.

The reel core accelerator should be sized adequately to drive over-greased reel spools up to machine speed during the startup phase of the drive. Over-greased spools and undersized motors do not get up to line speed and decelerate rapidly as they are indexed down to the drum. The resulting speed mismatch can severely crease or shred the web causing a web break.

Center Rewinds

Center windups need special control when a transfer is made. During a transfer, the paste roll will push the web against the incoming core. As the web is pushed against the core, tension increases. At the moment the paste roll contacts the core, it isolates the tension sensor from the winding roll. The tension sensor senses the increase in tension and acts to reduce motor torque. The web will relax between the winding roll and the paste roll, but the tension roll will not see the tension decrease because it is isolated from the slack by the paste roll. As a result, the web will be slack when the knife attempts to complete the transfer cycle. The knife will not sever the slack web and no transfer will be made. To make the transfer, it will be necessary to stop the machine in order to make a manual transfer at the rewind. This problem can be corrected by specifying a current memory circuit which is enabled when a transfer is initiated. The current memory circuit will hold the winding motor current constant until the transfer is complete, thus assuring proper web tension for the transfer operation. The rewinder is returned to its normal tension control mode when the knife severs the web.

Draw Rolls

Draw rolls are added to a coater to allow differential tension or to motivate the web when there are long expanses of web between driven sections. Draw rolls can be nipped, S-wrapped with high friction covers, single rolls with high wrap and high friction covers, vacuum rolls, or vacuum tables. Nip rolls or vacuum rolls are also used to overcome the boundary layer of air that accompanies a moving web. All must be designed to handle the differential tension that the process requires of the section.

Nipped Draw Rolls

A nipped section before the rewind will isolate the preceding sections from any rewind transients. This section should be designed to handle maximum machine tension in the event the web breaks between the rewind and the nip. The motor can be sized to handle full machine tensions using the motor overload capabilities since the machine will be stopped immediately in case of a web break. This section's power supply must have the same capability to handle overloads as the section motor. In other words, the full tension will be handled only for the time that it takes to bring the machine from process speed to a stop.

If a nipped draw roll is selected as the lead section (Fig. 6.4), the operator may elect to run with the nip open or bypass the section if the nip is causing defects, or worse, if it is causing web breaks. With the nip open, the web will slip at very low levels of differential tension and may cause scratches on the coated web.

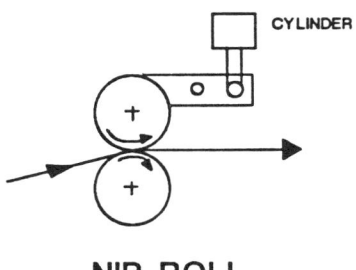

NIP ROLL

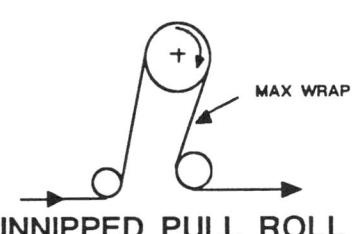

UNNIPPED PULL ROLL

Fig. 6.4. Nip roll and unnipped roll

Elastomer Draw Rolls

Unnipped elastomer or rubber-covered rolls use the friction of their covering to hold the web. These covers can become glazed, causing a reduction in the coefficient of friction which allows the web to slip. Roll slipping against a coated web can cause scratches on the coating. Elastomer roll coefficient of friction will be reduced by factors such as aging, high temperatures, abrasive coating on the web, slippage of any web across the roll, or by a web with a dusty coating which lubricates and clogs the pores of the elastomer roll. All of these factors take months to reduce the section's control to a level where the web breaks away and slips. It is necessary to examine the elastomer rolls in a preventative maintenance program to determine if they are becoming "polished" and ready to slip. Operators renew the surface of these rolls without removing them from the machine by sanding the roll surface with a fine sandpaper while the roll is run at very slow speeds.

S-wrap Draw Rolls

S-wrap rolls and high-wrap single rolls are most effective on slow-speed machines, especially if plastic webs are involved. Paper webs can use these rolls at higher speeds because of the porosity of the paper web.

Vacuum Rolls and Vacuum Tables

In cases where unnipped rolls cannot control the web and where nips will damage the coating, vacuum rolls or vacuum tables are used to motivate the web. These sections can control the web by driving against the uncoated side of the web. A vacuum roll is a hollow roll with a porous shell. The inside of the roll contains a vacuum chamber (or vacuum box) which limits the vacuum area to coincide with the width of the web and the angle of wrap of the web around the roll. So for a wrap angle of 180°, the vacuum chamber's envelope must be a little less than 180°.

Vacuum rolls are too often designed into a machine with too little wrap angle. A vacuum roll with a shallow wrap is incapable of controlling the web and frequently creates web-tracking problems, tension interaction problems with other sections, and web scratches. All drive rolls should be designed with maximum wrap, and the vacuum should be added to enhance the drives' ability to control the web. Large-diameter rolls (12-in. diameter or larger) with a minimum of 180° wrap is a good size. If that much wrap is not possible, work with the vacuum roll manufacturer to determine how much differential tension is possible, given the coefficient of friction of the roll covering, the coefficient of friction of the web, and the process speeds of the machine. Large-diameter vacuum rolls seem to handle larger values of differential ten-

sion at lower levels of vacuum and thus handle the web more gently (Fig. 6.5).

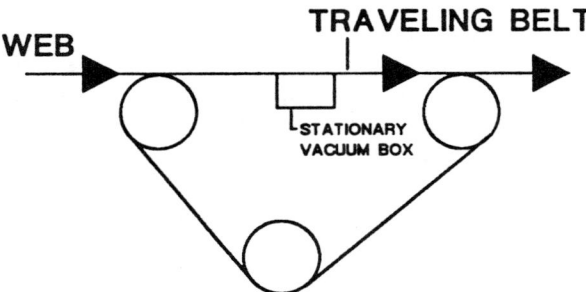

VACUUM TABLE DRIVE

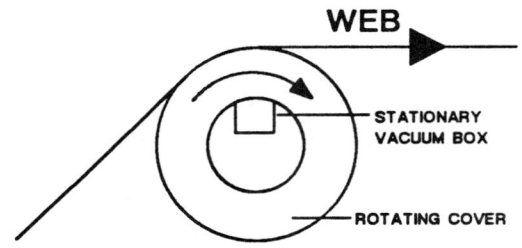

VACUUM ROLL

Fig. 6.5. Vacuum table drive and vacuum roll

One word of caution on the use of vacuum rolls: Even though the vacuum roll or vacuum table is in contact with the bottom of the web, a rough screen surface may cause the web to dimple and to impart a screen pattern to the wet coating. This is especially true if a vacuum roll is located immediately after the coating station. However, fine pattern screens may not cause a dimpling pattern. If dimpling is encountered, the section may be bypassed.

Vacuum tables are frequently too short to control the web during acceleration, deceleration, and splicing. Vacuum tables consist of a long belt which supports the web as it travels over a vacuum box. This table is used to apply tension to the web when the web is running in a vertical or horizontal plane where no wrap on a roll is possible. The vacuum table is usually located between impingement or flotation dryers, at the entrance of a dryer, between air turns, or at a dryer exit as the web descends into another part of the process.

Vacuum tables have been used on coaters up to 1000 ft/min. They are used with a vacuum loop control and a sonic detector to control the tension of wet web without contacting the web. Flat vacuum tables are similar to the old Evans Rotabelts which were used on fourdriniers in the past.

Vacuum tables and vacuum table vacuum pumps must be sized for process speeds of the various products and then used as designed. If the adjustable vacuum end dams are adjusted beyond the edges of the web, the vacuum system is open to atmosphere and the vacuum drive loses control. Machine personnel can tap into vacuum roll vacuum systems for auxiliary functions such as web cleaners. The total vacuum demand of the new load exceeds the capacity of the system and thereby reduces the ability of the vacuum drive system to control the web.

Vacuum rolls and vacuum belts are designed to function at high machine speeds. Vacuum levels must be increased as machine speeds are increased so that the boundary layer of air that is carried with the web can be evacuated from the interface that exists between the roll surface and the web. Vacuum rolls have been used at coating speeds in excess of 3000 ft/min.

Isolation

Draw rolls also separate a tandem coating operation into discrete process zones. This zoning capability is called isolation. Isolation is only valuable when the need for different tensions actually exists at each part of the process. Isolation may be necessary if other processes are included in the coating machine, such as: (1) laminating thinner webs to thicker webs, or webs of different characteristics to the basic substrate, (2) wetting the web to the point where it becomes weak, or (3) heating a plastic web to the point where it becomes pliable.

Isolation may take the form of speed control or tension control. Careful consideration of what the process needs will lead you to select the form of tension control that is needed to get the best results.

Any coater, draw roll, or laminator may need motoring and regenerative capability, especially if the exit tension is higher than the entry tension or if the primary purpose of the section is to provide tension isolation.

Isolation also insulates the effects of poorly wound rolls on the unwind from the coating section. But poorly wound rolls produce a web with edge cracks and with wrinkles which may break down at the isolating draw rolls or coating blade.

Many off-machine coaters run no differential tension from the unwind to the reel (rewinder). They have eliminated the draw rolls which precede the coaters because of the draw rolls' ability to cause web breaks. These machines seem to run better without the extra complexity of isolation sections. For that reason, isolation is grossly overrated on coating machines that simply coat web in a tandem mode with no other process to deal with.

Here are some examples of different tension zones:

1. Unwind tension as opposed to rewinder tensions

2. A loosely wound roll on the unwind may be run at a lower tension than the entry side of the first coater

3. The entry side of the coater versus following the coater where the wet paper web is apt to stretch

4. The entry side of a laminator versus the exit side of the laminator. If two webs of different characteristics are being laminated, one web may require a significantly lower tension than the other. The resulting laminate may or may not require a higher tension at the laminator exit, depending both on the laminating process and the tension required to successfully track the heavy laminated structure.

Machinery builders and drive vendors may apprehensively insist on the safety of isolation for fear that a user does not understand his own process needs. The idea of isolation is a carryover from the early days of coating. Isolation is not always necessary, but that need isn't carefully evaluated, and so the use of isolation carries on when new machines are built. Draw rolls may still be required to motivate the web when the distance between driven sections is very long. It is up to the user to evaluate the process needs and to decide what is appropriate for the coating machine. The funds saved on two draw roll sections complete with tension-controlled drives can provide valuable "extras" on a project which is tight for funds.

Guiding Equipment

Web Guides and Web Spreaders

Web guides are devices which laterally position the web in the machine; web spreaders are devices which remove wrinkles from the web at critical points in the process.

The web frequently approaches the process off-center for a number of reasons. Web guiding systems sense this lateral offset and make the necessary corrections to present the web to the process in the proper lateral position. Rolls, coming from the paper machine, are sometimes wound offset on a reel spool. The web from the offset-wound roll must be positioned properly in order to get the best coated results.

Web guides are total systems consisting of sensors, controllers, and positioning devices. There are a multitude of guiding systems, all of which have been successfully applied. Each web guide vendor can suggest a web guide system that will solve your guiding problem. Vendor advice should be solicited before the machine design is firm and certainly before a machine startup indicates the need for web guiding. Articles listed in the reference section of this chapter give sound basic guidelines in selecting web guides. The following are some areas where web guiding systems are used:

1. After an unwind stand and just before the entry point of the coater

2. On the entry point of a laminator which is laminating two webs

3. On the entry point of a laminator which is laminating web to an extruded web

4. On the entry point of the second coater in a tandem coating operation (following a long dryer tunnel)

5. After a web is flipped over and just before the next process point

6. At the entry point of the rewind.

The spreader roller or curved-axis roller can be used in many applications on a given web process. The roller can cause different effects depending on its placement within the web (Fig. 6.6). These various uses include: removing wrinkles or smoothing, slit separation, web or material expanding, equalize tension for webs to be coated or laminated, wet felt conditioning in paper mills, and wet felt wash cycle.

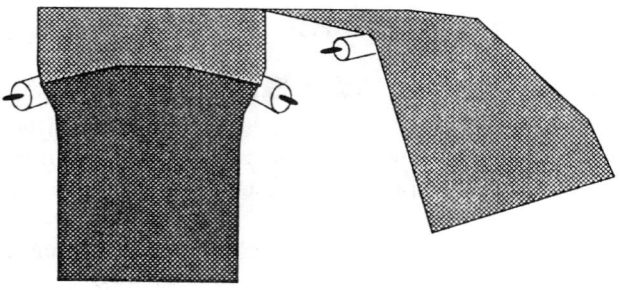

Fig. 6.6. A curved roll removing wrinkles

Web spreaders are required to present a flat and uniform web surface to the process and result in a uniform coating layer and drying. Web spreaders are potential problem solvers if installed after an unwind or before a nip roll, coater, or rewind. Web wrinkling and web spreading are discussed by Bruce Fiertag in papers referenced at the end of this chapter.

All of these applications require specific bits of information that will permit the correct selection of the required roller and method of installation. The information necessary to define a wrinkle-removing application can only be provided by the careful examination of the existing line. The following information must be provided for

a wrinkle-removing installation. How much will the web have to be spread to remove the wrinkles? How much is the web wrinkled? How long of a distance to the spreader roll is required? As the web is stretched, the action to remove wrinkles occurs in a definite length of the web, and the entry span to the spreader roll has to be at least this long (Fig. 6.7).

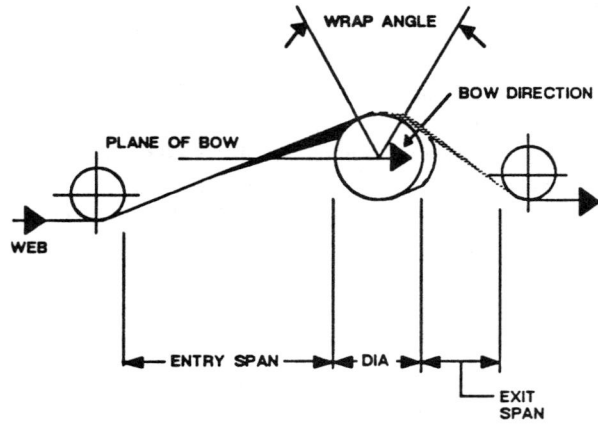

ENTRY SPAN: the distance between the spreader and the idler roller just preceeding the spreader roller

EXIT SPAN: the distance between the spreader roller and the idler roller just following the spreader roller

Fig. 6.7. Web nomenclature

Other questions necessary to answer include: Web speed required? Web tension required? How can the roller be mounted? To specify cover and special materials, specifications of environmental conditions, chemicals present, temperature, wet or dry, etc. must be supplied. If this information is supplied, along with a sample of the material, a realistic selection can be made. On new applications, the web path and experience can be used to establish these parameters. On existing web applications, exact measurements can be made by the customer or supplier.

Unwind Guiding

In unwind applications, web guides assure correct material alignment to a predetermined point. The stand and the first idler move laterally as a complete unit (Fig. 6.8). Some unwinds have hand-operated guides called side-lay to allow manual positioning of the web. Other unwinds are automatically positioned by an automatic web position controller.

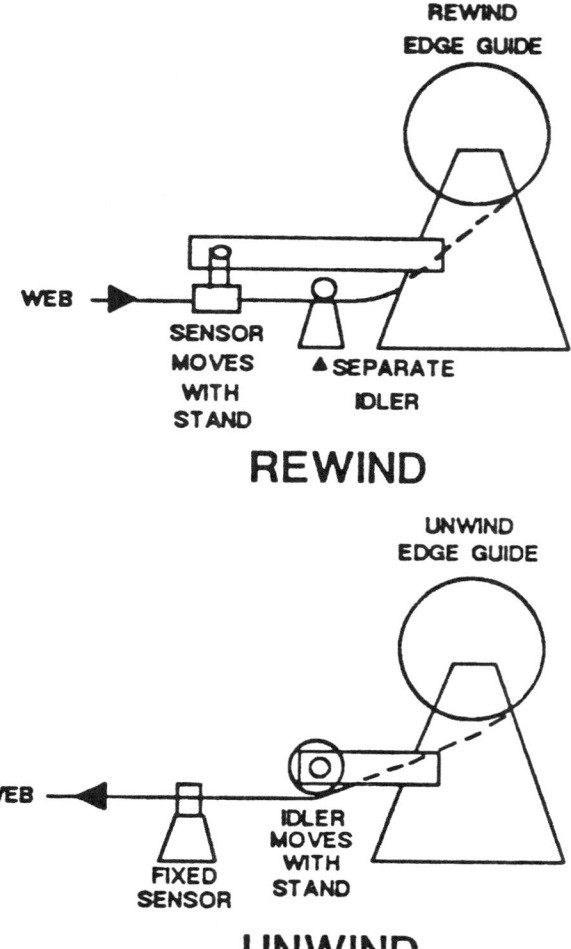

Fig. 6.8. Unwind and rewind

Intermediate Guiding

While there are many factors affecting the selection of an intermediate guide, the Kamberoller or Bowed Roll is generally applied where long, free entering spans are available. An offset pivot guide is recommended when sufficient entering span is unavailable or for materials with a higher modulus of elasticity.

Rewind Guiding

In rewinding applications, the web guide assures evenly wound rolls. A sensor is rigidly attached to the rewind stand so both move laterally in unison. A fixed idler is required between the sensor and rewind stand. This is referred to as "chasing" into rewind (Fig. 6.8).

Consider using an oscillating edge guide on the windup to distribute the coating bead edge stress which comes from various types of coating applicators.

Tension Control

Tension control is necessary when the coater accelerates and decelerates. Each machine has a wide range of inertias which must be accelerated from zero speed to process speed and then decelerated to zero speed at the end of a coating run. The high inertia sections may not follow the acceleration ramp as closely as the sections adjacent to them. When all the sections in a coating machine vary, the result is tension transients between adjacent sections and a potential for web breaks.

Tension control is a valuable tool to compensate for factors which are within the drives control and which may go undetected except through changes in web tension. The older analog speed-regulated drives drifted so much that web tensions changed radically during the coating process, so much so that web breaks would occur without some form of tension control.

Tension control allows the operator to thread the web through the machine and get the machine in production much sooner than with a pure speed controller. During the coating operation, tension control corrects for the normal manufacturing variations which occur in most paper webs. The variations include the paper web's tendency to stretch when wet during the coating process. It corrects for tension transients which occur when the coating blade is loaded against the web and coating roll. It allows a machine to handle webs of varying width, weight, and caliper. This feature increases the flexibility of the coating line since it allows different webs to be spliced and transferred without shutting the line down. This translates into the operator being capable of making quick, knowledgeable, and repeatable adjustments to tensions when grade changes are made.

The added advantage of tension control is that the operator can thread the machine, turn on the tension control system to pretension the web, and then accelerate to process speed much more quickly than other systems that may require the operator to be very precise in the machine setup.

Constant Torque

If torque is kept constant as the unwinding roll diameter decreases and the winding roll diameter increases, the unwind tension will increase on one side of the pull roll while the rewind tension will decrease on the other side of the pull roll. This tension difference across the pull roll is called "differential tension." When the tension differential becomes large enough, the sheet breaks away from the pull roll and slips. The sheet slippage is such that it runs *slower* than the pull roll. As the unwind diameter continues to decrease, the unwind tension will continue to increase while the rewind tension will decrease and the sheet will continue to slip and run at slower speeds. Rolls on the unwind may cinch and telescope if a loosely wound roll is subjected to high unwinding tensions. Cinching is the slippage between layers of web within the roll and can cause scratches on the paper surface within the roll.

Actually, many draw rolls slip at a maximum of 3/4 PLI differential. Much of the ability of the draw roll to control designed differential tension depends on the absolute value of tension they are controlling. It is difficult for a draw roll to control differential tension if the entry side is at 0 PLI and the exit is at 1 PLI. If the basic tension level is at 1 to 1½ PLI, the draw roll may be able to control its designed differential. Figure 6.9 shows a constant torque unwind system.

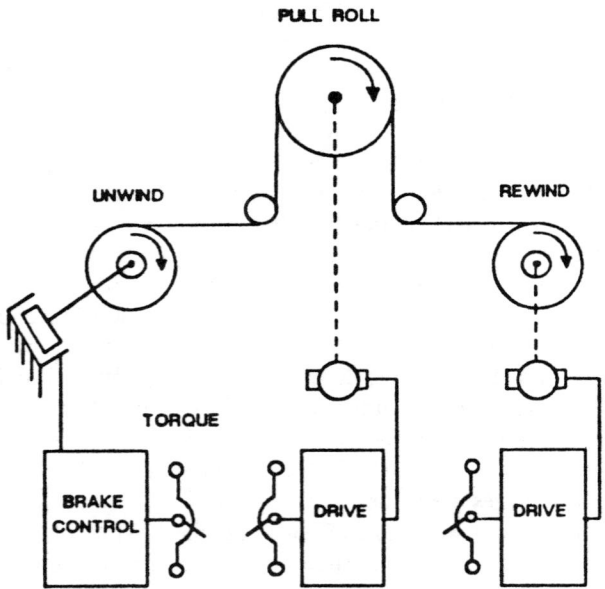

Fig. 6.9. Constant torque unwind

Tension controllers maintain tension by modifying section speed and torque. If the tension being controlled is higher than the yield point of the pliable web, then the tension controller may very well cause the web to stretch more than acceptable even to the point of a web break. Conversely, if the web tends to shrink, then too low a tension may allow the web to shrink. There are three types of tension controls that are applied to drive control systems:

1. Motor current control (Fig. 6.13)
2. Dancer roll or swing roll control (Fig. 6.10)
3. Force transducer or load cell control (Fig. 6.11).

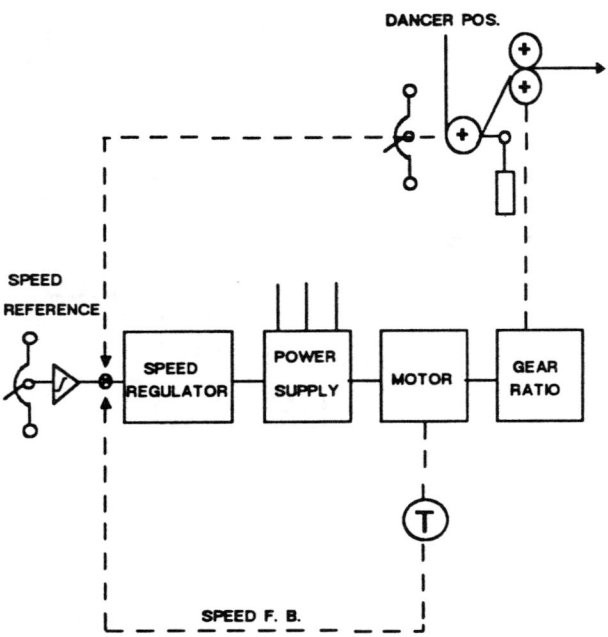

Fig. 6.10. Speed regulator with dancer trim

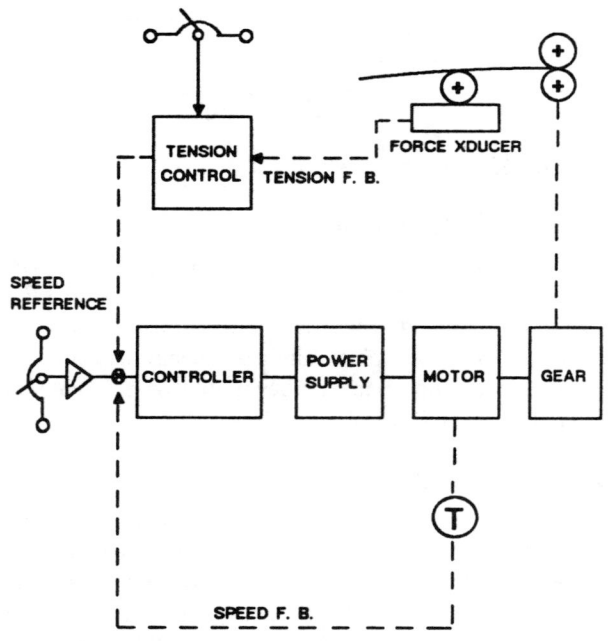

Fig. 6.11. Speed regulator with tension trim

These systems have been used successfully over the years. Best results are obtained when the designer knows the advantages, disadvantages, and some of the pitfalls to avoid with each system. In addition to these systems, there are sections within the coater that can control tension. These sections include: unwind or rewind, nip roll when closed, calender stack, a blade coater as long as its blade is loaded before acceleration to speeds over 500 ft/min, and impregnating coaters.

The following sections become less effective for tension control as web speed increases due to the lubricating effect of the boundary layer of air which is carried with the web:

- Open nip rolls
- Vacuum sections with small wraps or low vacuum
- Controlled gap coaters (impregnating coaters, gate roll coaters)
- Coaters which apply coatings which behave as a lubricant
- An unfelted dryer section.

A dryer section without felts will allow the web to slip at speeds over 500 ft/min. A dryer section with half of the dryers felted (felt side contacts only the uncoated side of the web) may be effective as a tension control at speeds up to 1000 ft/min. Felt tensions on dryers that follow coaters are usually kept low to prevent coating damage. Sometimes these felts are removed when the dryers become plastered with coating after a web break. When the web breaks, the coater deposits a heavy layer of coating on the tail of the web that will dirty the dryers and felts as it is dragged through the machine. The resulting cleanup effort will require so much time that the operators may not put the felts back on the machine. Recent dryer drive systems use load sharing between the dryer drive and a separate dryer felt drive. This approach has allowed tension control at high speeds.

Basic Speed Regulator

Constant diameter sections usually use speed regulators as the primary control. Current control is used as a vernier or trim adjustment of the speed controller. The combination of speed control with current trim is a very stable method of controlling constant diameter tension. Dancer roll control or tension roll control are used as a vernier or trim adjustment for effective and stable tension control.

Speed regulation with current trim is commonly called current compounding or speed droop. This type of tension control works very well on heavyweight products that run over a limited speed range. In other words, a forgiving product or process used with this forgiving

drive can produce excellent results with minimum complexity and minimum cost.

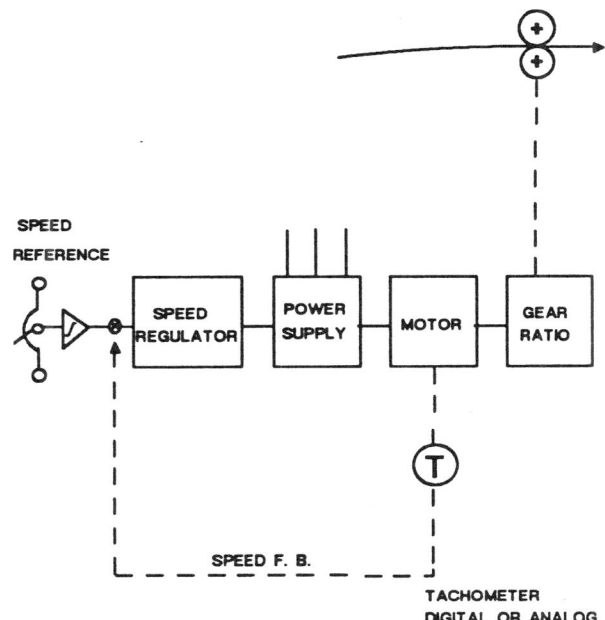

Fig. 6.12. Basic speed regulator

In one case, a multi-section coating drive was purchased with digital speed control. The operator could add tension control using the speed-tension selector switch, but the sections were all speed-regulated. In a speed-controlled mode, the machine was far too cumbersome for the operators, and threading was simply too time-consuming. The operators adjusted the individual sections manually when process speeds were reached. But whenever a splice and transfer was made, or when machine speeds were changed, the operator was required to manually rebalance tensions within the machine. This was time-consuming. To help the situation, tension control was added. With tension control, the coater ran with far less difficulty than when run with pure speed regulators. Since the addition, several web path changes have been added easily because of the web tension control capability.

In the following cases, a speed regulator may be the only tool available for producing good results (Fig. 6.12). When to use pure speed control:

1. When web stretch or shrinkage is intolerable during processing. This is typically a product or process requirement rather than a machine and drive control consideration.

2. When the web is stretched to a specific ratio between adjacent sections

3. When the web is so weak that it can barely support its own weight

4. When controlling the lead section of the machine

5. Where a process creates or treats a pliable web whose characteristics change with each roll.

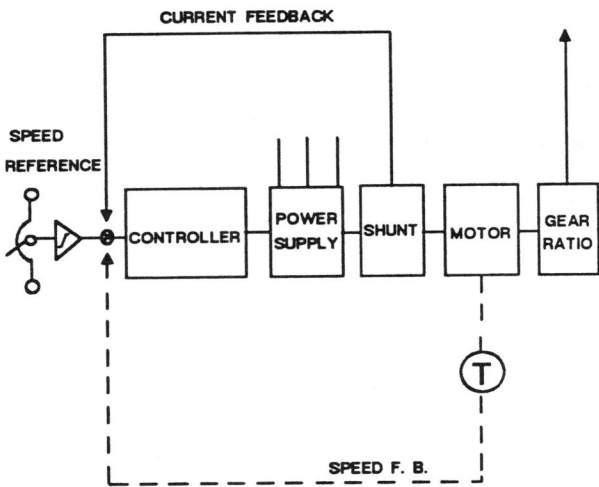

Fig. 6.13 Motor current control

Current regulators are inexpensive yet effective. Unfortunately they have no direct tension readout, no storage, and do not recognize where the load comes from (bearings, tension, rotary joints, or doctors). If a reel cleaning doctor is used only during a reel transfer, it will add load to the section. If a current regulator is used for tension control, then some of the drive torque will be diverted from establishing tension to overcome the losses of the doctor blade. Tension will be lost at the critical point of transfer, and a web break can occur. A better choice for a tension control is the force transducer or storage-type roll. If a current regulator is already in place, then at the very least a current signal correction should be made to compensate for the intermittent addition of load due to doctor loading.

Bad bearings on an undriven paper roll (idler roll) which leads into a section can raise havoc with the web tension between sections. Cold oil in gear reducers and cold grease in bearings are two of the load factors that change as the machine warms up. The changing friction characteristic of steam joints and water joints represent changing load losses on surface windups, steam dryer drums, and chill rolls, all of which result in web tension changes when current regulators are used.

Motor current regulation is a simple concept and the simplest to implement. It may be the only way to control web tension if process conditions do not allow a tension sensor to touch the web. A current regulator controls the current to a fixed value as set by the operator. A current regulator also uses total motor current to infer the amount of tension in the web. However, total motor current is only approximately equal to actual tension since total motor current includes tension load and the losses in the system. A current regulator can't differentiate between tension and the load losses. If load losses in the system are a small part of motor current, then a current regulator will be an effective way of controlling web tension. If load losses are a big part of total motor current, a current regulator should be used only if no other method of tension control is possible.

The increased current needed for acceleration of a section to run speed will cause a decrease in tension unless inertia compensation circuitry is part of the controls strategy. Inertia compensation is especially important on unwinds and rewinds where roll inertia is constantly changing as roll diameter changes.

Motor torque is a function of motor field strength and motor current. Starting with a cold field means the motor torque can vary significantly as the motor field heats up and the motor field excitation decreases. The operator must adjust the tension pot as the motor warms to keep tension at the proper value.

The current controller is frequently used as a tension controller on constant diameter sections when it is impossible to design a dancer roll or tension-sensing roll into the web path. This may occur simply because space is not available or because the coating is still wet at the point where the roll is needed.

Rewind Tension Control by Current Control

A simple current regulator can be used if a center rewind has simple specifications (taper that varies with diameter), the tension isn't critical, and if it has no more than a 2:1 buildup. Any center rewind with a range of 0.5 PLI to 3 PLI and a buildup of more than 5:1 with constant tension needs a more sophisticated control system such as a constant horsepower system. A more sophisticated system senses motor current to infer motor torque and includes a tachometer on the windup motor to determine windup speed. With these values, a calculation is made to control the input power to the motor. In such a system, as the windup roll diameter increases, the speed of the windup decreases and the motor current increases in proportion to the diameter increase. Motor torque increases to keep the tension constant as the roll diameter increases.

Losses in the system are due to friction, windage, and magnetic losses of the motor, as well as the losses of the windup. All these losses are a function of speed. Losses and the speed of the system are at their highest when an empty core is on the rewinder. These losses can be equal to or exceed the magnitude of the tension power requirements. Therefore, it is important to keep external losses to a minimum to maintain constant tension. Loss compensation circuitry is important for this kind of system.

Dancer Roll Control

The dancer roll is a force balance system where the force of the web pulling in one direction is balanced by the loading of the dancer pushing in the other direction. The dancer has the ability to store web to minimize the influence of transients which are generated by acceleration, deceleration, unwind splices, or rewind flying transfers. The dancer control system also compensates for step changes in load losses due to doctor loading, changes in nip loading, changes in coater blade loading, and higher than normal condensate loads in a dryer by sensing a change in tension and adjusting the section speed to reestablish tension to the preset value.

Each motion of the web is sensed by the roll position feedback potentiometer which is connected to the dancer roll mechanism. The feedback signal causes the braking effort to maintain the dancer in a set position, thereby maintaining tension at a constant level. On an unwind, as the unwind roll diameter *decreases,* the web tension at the unwind will tend to *increase.* When the unwind web

tension increases, the dancer roll will raise causing the unwind roll position potentiometer to sense that change. The roll position feedback to the controller will cause the brake controller to reduce the torque to the unwind brake causing the tension to return to the desired value. When the tension decreases to the proper value, the roll will lower to the preset position. The dancer roll is called a position regulator and the dancer control is called a position control system.

In Fig. 6.14, the rewinder has a constant tension system on the unwind and the rewind. This constant tension system uses "dancer rolls" to sense increases or decreases in web tension. Dancer rolls are most commonly used for controlling unwinds and rewinds because these sections can produce splicing and transfer transients. Transients are a result of speed mismatches of the incoming roll during a splice-producing slack; the slack is controlled by the dancer roll.

Other transients can cause the rewinder to temporarily slow down during the steady-state operation of the coating machine.

1. If the operator rotates the turret on a centerwind, the tension will increase and the windup will slow down. The taper tension system will believe that the roll diameter has changed and will send a signal to the tension controller to reduce tension. That reduction in tension can be great enough to cause a soft spot in the roll. The roll will

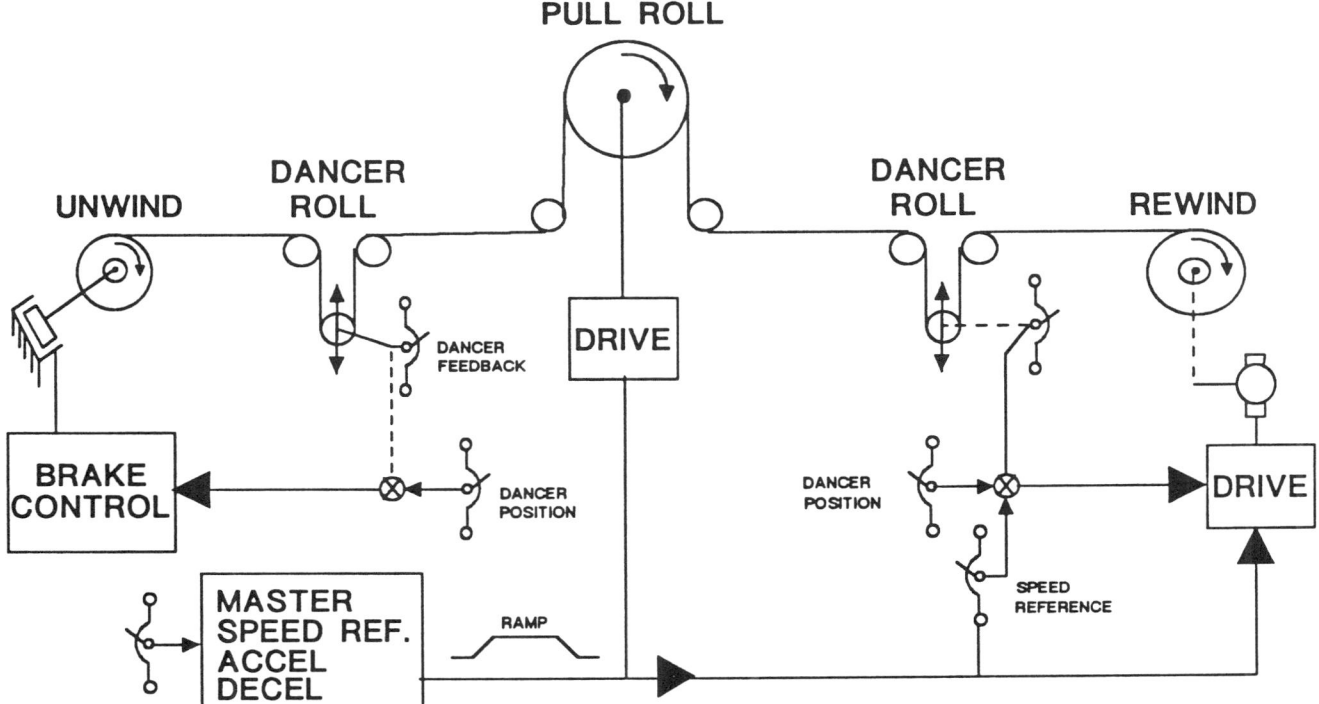

Fig. 6.14. Constant tension unwind

tend to telescope as it continues to wind. Therefore, it is important to inhibit the taper tension controller when the operator indexes the turret and enable the tension controller when the operator-induced tension transient passes. Interlocks from the turret indexing drive are usually inserted in the taper tension circuit to disable and enable it when the turret is indexed.

2. Acceleration or deceleration can produce tension transients which can also fool the taper tension controller. The controller must also be inhibited during acceleration or deceleration periods and enabled when the transient period is over.

The control of constant tension instead of constant torque eliminates most of the disadvantages of the constant torque system.

With increased machine speeds, drive vendors and machinery builders want to ensure that the dancer has adequate storage to respond to transients. The dancer measures web tension directly and therefore is not as affected by system losses as is the current regulator.

The mechanical design plays a big part in the success of a dancer system. The system is fairly simple electrically, requiring only a position-sensing transducer, a tension amplifier, and a summing point into the tension-regulating system. The mechanical system of a dancer is more complex than a force transducer or current regulator.

The dancer roll must be designed so it satisfies the full range of tension specified. Extra rolls are needed to fix the wrap angle on the dancer roll through the full range of storage. The dancer must be mounted on a pair of arms and is generally loaded by the use of a low-friction cylinder (it is sometimes loaded by weights). The loading of the dancer sets the amount of tension which will be applied to the sheet. Sheet tension is varied by increasing or decreasing the loading of the dancer roll.

The loading of the dancer creates some unique problems for the mechanical design people. First, the sheer mass of the dancer can create sheet tension excursions when transients cause the dancer to move in one direction or the other. Similarly, the force from the weight of the dancer and the dancer arms can exceed the force required to control the minimum specified tension value. There have been cases where the friction force in the dancer mechanism exceeded the force required to control the minimum specified tension. Designers frequently forget the weight of the dancer roll and support arms in their tension calculations.

The friction in a pneumatic cylinder, which loads the dancer, sometimes far exceeds the force required to control the minimum tension. There are friction-free cylinders available today. The pneumatic system for these cylinders must always be pressurized. The diaphragm will jam between the cylinder body and the piston if it is operated without air pressure. A jammed diaphragm will give the dancer a varying frictional characteristic which will disturb the force balance of the dancer system. The dancer may move excessively and so violently it will bounce off its stops and force a shutdown until the source of the problem can be identified.

The pneumatic system can add to the system problems. If the air pressure capacity is low, the pressure will change as the dancer moves. The line pressure will increase as the air is compressed and decrease as the air expands. The increase or decrease in line air pressure changes the force on the dancer and thus also changes the web tension. To deal with this capacity problem place a surge tank in the pneumatic line between the pressure regulator and the dancer piston. The problem of excess air cylinder friction can be handled by removing the excess chevron packings, chrome plating the piston barrel, and by using only Teflon® coated packings where packings are necessary.

The use of a pneumatic system for loading a dancer at the rewind provides a taper tension system. A taper tension system reduces the windup tension as the roll builds from the core to the maximum diameter. Taper tension is used to prevent blocking or sticking together of paper layers and to prevent damaging buildup of forces within the roll.

Taper tension systems on older machines used electro-mechanical devices called motor-driven rheostats (MRH) or ratio detector to provide changing signals to the rewind system during the buildup of the roll. Modern drive systems use solid-state circuitry or a computer to provide the same function. The newer systems have more capability and flexibility than the older electro-mechanical devices.

A ratio detector performs a few tasks:

1. It compares the line speed to the winding roll speed and calculates the speed of the winding roll. Roll diameter is obtained inferentially while the roll is building without the need of a rider roll touching the web.

2. The roll diameter signal is used to provide the rewinder with a taper tension signal. This signal can be used to shape the tension winding profile of the roll as a function of the roll diameter. A shaped tension profile reduces the stresses within the wound roll, thereby reducing the tendency to damage the product within the wound roll. If rewinder tension is controlled by a force transducer, the electric taper tension signal is used directly by the tension controller. If the rewind uses a dancer roll, then the taper tension signal is converted to a pneumatic signal before it is delivered to the dancer roll pneumatic loading system.

3. The ratio detector also provides inertia compensation circuitry for the acceleration and deceleration of the

rewinder roll. When the roll is accelerated or decelerated, motor torque must be increased or decreased to cause the roll to go to target speed. The amount of inertia compensation signal will increase as the winding roll diameter increases.

Force Transducer Control

The force transducer system is more expensive electrically but is not as costly mechanically as the dancer roll for tension control. Two force transducers are mounted on a roll which is in the web path. Transducer rolls often can be fitted into an existing web path without difficulty. Their signal is summed and sent to the tension regulator which adjusts motor power output to satisfy the operator's tension setpoint. Unfortunately, some hydraulic transducers get out of calibration easily. They have no storage (can't absorb transients), require responsive drive system (can be a problem on retrofits and rebuilds), and may be a sensitive spring mass system even with damping.

The output of each cell can be displayed to show if an unbalanced load is being applied by a nonperfect web, e.g., a cambered web which may have a "soft" edge and a "hard" edge. This type of web would track through the machine tight on one side of the machine and loose on the other side. Additional rolls may be supplied to fix the wrap angle of the web over the roll to ensure that the force being measured is always directly proportional to web tension. It is important to maximize the force vector that is measured due to web tension and to be sure that the tare-to-signal ratio is compatible with the transducer manufacturer's specifications. It is also important to keep the wrap angle constant on tension sensing rolls which are located on a multi-product machine where multiple threading paths exist. If the wrap angle is different for each web path, then the web tension feedback will be incorrect for all but one web path.

Since the force transducer directly measures web tension, it ignores many of the load losses which bother the current regulating system. However, rolls mounted on transducers must be balanced very precisely. Dynamic roll balance is especially important on rolls that are used as tension feedback devices for tension control systems. Unbalanced tension-sensing rolls will falsely indicate a tension variation each revolution when no tension variation exists. The adverse effects of tension roll unbalance can be reduced by using the viscous damping systems which are part of the force transducer manufacturer's product line.

Force transducers are now applied to all sections of the machine from the unwind to the rewind. The fast response capability of modern drive systems now make force transducers effective in those applications where at one time the storage capability of dancer rolls was absolutely required. The drive vendor, machinery builder, and user must integrate their efforts to ensure that the electromechanical design of the drive system maximizes the response capability of the modern drive.

Any section which is used to control tension has the same characteristics as a good lead section. It can handle large values of differential tension at all machine speeds and can handle transients that are generated by acceleration, deceleration, splicing, and transferring. As in the case of the lead section, the tension-controlled section in a machine must be able to transmit the torque from the motor, through the roll, to the web without allowing the web to slip.

How to Deal with Sections with Limited Capability to Control Tension

To deal with sections with limited capability to control tension first select a practical tension range for the machine. If all webs are run at 1 PLI, a range of ½ to 2 PLI instead of ½ to 3 PLI is more practical. Too large a tension range will have the following effects:

1. It will eliminate current regulators. Current regulators are limited as to how much range they can accommodate.

2. It will eliminate force transducers. These may be marginal; the transducer selection may be difficult if it is to satisfy the highest and lowest tension and still provide adequate transducer tension overload capacity.

3. It may also compromise the ability of the dancer roll to handle the high end or low end of the tension range. Dual dancer loading systems are used to provide a high range and a low range of tension. These systems have met with mixed success and with mixed enthusiasm at the mill level.

Generally, tension control specifications call for a tension control system to provide ±10% tension trim separately adjustable for each section to allow optimization of the tension control for each section. Ten percent tension trim may cause sections to become unstable and lose control of the web. Experience shows the following values of speed trim are adequate for machines where the total speed differential from the unwind of the machine to the windup of the machine is less than 1 ft/min.

Typical speed trim values for constant diameter sections are as follows:

Draw rolls: 3 to 5%, usually

Blade coaters: 3 to 5%

Single dryers which back up
high-velocity dryers: 0.5 to 1%

Dryer nests which are unfelted: 1 to 3%

Dryer nests which are felted: 1 to 3%

Calender: 1 to 5%.

These tension speed trim values must exceed any drift in an analog system. If the drift of the drive system exceeds the tension trim, tension control will be unpredictable. Also, typically there is too much tension trim incorporated in the dryer tension regulator, which causes web control problems during acceleration and deceleration.

Differential Tension

Differential tension is the difference in the web tension between the ingoing side of a section and the outgoing side of the same section (Figs. 6.15 and 6.16). Every section is limited to the differential tension that it can control. When that differential tension limit is exceeded, the web will "break away," the section will slip against the web, and the section will no longer control web speed.

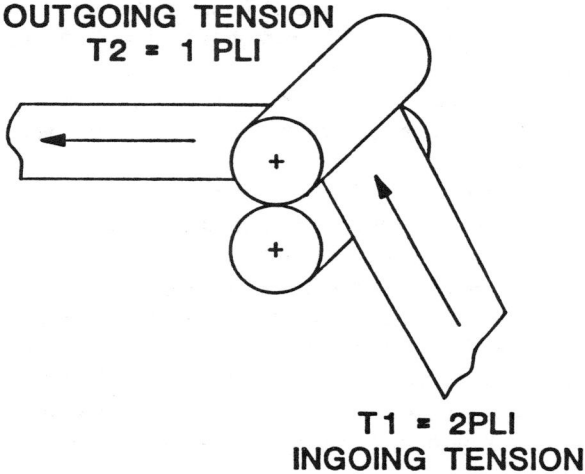

$$T = T1 - T2$$
$$= 2 - 1$$
$$T = 1 \ PLI \ (MOTORING)$$

Fig. 6.15. Differential tension-motoring

Most coating machines tend to run best when all the tension throughout the machine is set at the same level. For instance, if 1 PLI is run at the unwind, then 1 PLI should be used at the coaters, the draw rolls, and the rewinder. Running the same tension throughout a machine is called a "flat tension profile." A flat tension profile reduces the differential tension that a section must handle during steady state and will increase the machine's chances of handling tension transients during splicing, transferring, acceleration, and deceleration of the web. If a section is set at a large enough differential tension, then a splicing transient can cause the section

differential tension to increase beyond its capability to control the web, and the section will slip.

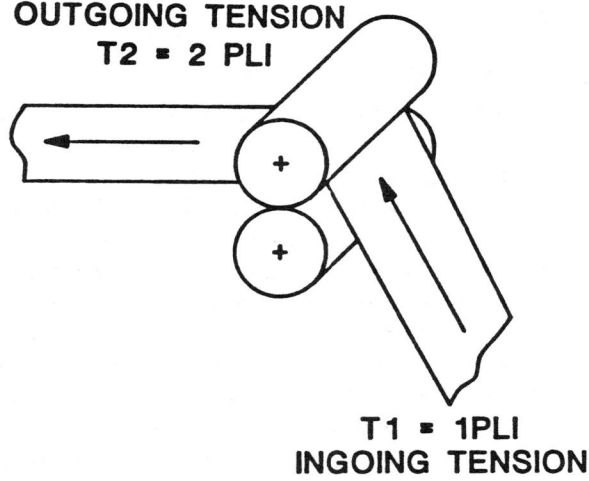

$$\triangle T = T1 - T2$$
$$= 1 - 2$$
$$\triangle T = 1 \ PLI \ (REGENERATIVE)$$

Fig. 6.16. Differential tension-regenerative

Stall Tension

Stall tension is the tension applied to the web when the machine is not running. At rest, the web is kept taut. Stall tension is used on a coating machine to establish an acceptable level of web tension before the machine is accelerated to process speeds.

To obtain the stall tension, the unwind and rewinder are energized to pull the web taut throughout the machine. Stall tension is usually a fixed percent of operating tension. Some stall tension systems are proportional controllers for the benefit of the operators since these controllers are a gentler control than integral controllers. Proportional control does not suffer from the "reset windup" or saturation that are characteristic of integral controllers. Therefore, proportional control is the control of choice when the stall tension system removes the slack normally left after threading the web through the machine. A stall tension system using integral control may tend to break the web if too much slack is left in the machine when stall tension is applied.

After the proportionally controlled stall tension applies the desired tension, the system is converted into the more accurate integrating control as a function of the operator's run command. Sometimes, unwind/rewinder stall tension systems cannot influence tension on the interior of long coating lines. Some companies have found it helpful to place the entire coating line into stall tension.

The commutators of drive motors can burn if the machine is left in stall tension too long (15 minutes or more). Drive vendors can limit the length of time that a machine can stay in the stall tension mode by de-energizing all drives through a time-delay relay.

Sources of Coating Variations

Drive System

It takes a great deal of patience and a familiarity with the coater to identify if the source of coating defects, such as coating chatter, is caused by a mechanical or process problem.

One defect, coating chatter, appears as periodic light and dark coating bands on the web. Coating chatter is unacceptable for any type of coater. If the coating stand is not isolated from surrounding machinery, then the vibrations caused by other machinery will vibrate the coating stand and cause the chatter to be transmitted to the web. Coating chatter can be caused by the roll, the motor, the control equipment, the motor power supply, and power transmission equipment such as couplings, gearboxes, chains, belts, and universal joints. Unstable feed valves, piping, positive displacement pumps, and unstable pump drives can also cause coating variations.

If the coating roll is connected to the drive motor by gears, particularly if the drive motor is lightly loaded, the profile of the gears may show up as chatter on the coating surface. With light loads, the torque requirement may shift between motoring and braking and when the load shifts, the gear teeth will first drive on the forward side of the tooth and then hold back on the back side of the tooth. The changing of the tooth contact must pass through the backlash region of the "slop" or clearance. That clearance is responsible for causing a nonlinear velocity which results in coating chatter. Some operators compensate for slop by increasing the tension at the ingoing side of the coater. The increase in tension increase results in a motor load and keeps the gear in contact with one side of the tooth. If that does not solve the problem, then the gears must be replaced.

Chatter which comes from gears, gear couplings, timing belts, or chains can be eliminated by replacing the gear devices with zero backlash devices and direct-drive motors. All couplings must be aligned properly or the misalignment can cause velocity perturbations which translate into coating variations. Gear reducers can be eliminated by connecting the coating roll directly to the drive motor through a zero backlash coupling.

Chatter caused by the coating roll may be attributed to conditions such as defective or worn bearings, roll unbalance, or a nonconcentric shape. Roll unbalance may

be due to buildup of dirt or coating solids on the coating roll. This problem is difficult to diagnose. Bent roll journals or a roll kept stationary too long can sag and result in a nonconcentric roll. To avoid the sag, rolls are frequently supplied with a "Sunday" drive or "inching" drive to keep the roll rotating during long periods of inactivity like a weekend shutdown or plant shutdown.

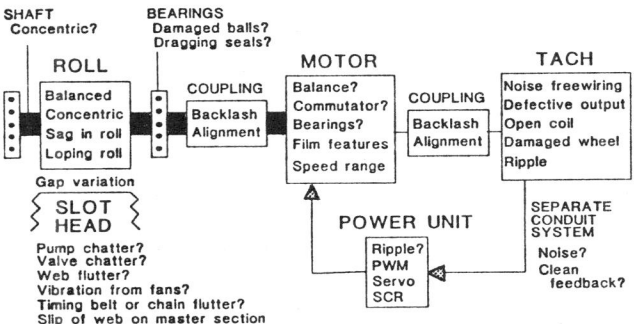

Fig. 6.17. Drive-related defects

It is important to be aware of any changes in the process that might have occurred just before a defect is noticed. The following are some of the questions to ask:

- Was a coating roll changed?
- Were process air filters changed?
- Was new equipment added to the machine?
- Was maintenance performed on the machine?
- Was the instrumentation recalibrated during a plant shutdown?

Additional chatter can be caused by poorly installed bearings, damaged bearings, work power transmission equipment, and worn steam and water joints on rolls near the coater.

The drive system can also provide a source of variations. To avoid these variations, the drive motor may require precise balancing, special bearings, or skewed armature slots to satisfy extremely precise or sensitive coating needs. Such a motor is said to have "film industry features" for smooth rotation. A servo quality motor may be required. (See Table 6.7 for drive comparison.)

The output of the power supply to the motor may contain enough ripple to affect coating quality. Single-phase SCR-type power supplies produce a very high ripple at low coating speeds. Power supply ripple can be reduced by going to three phase SCR, PWM (pulse width modulated) power supplies, or servo quality power units. Many servo quality drives are designed to operate at extremely low RPM without "cogging" or creating velocity perturbation.

Coating quality can also be affected by the couplings or belts that drive velocity feedback devices or by the "ripple" of the velocity feedback device itself. If digital devices are used as a velocity feedback, then its signal converter may contain enough ripple to influence drive system velocity. If a digital reference is converted to an analog reference for the drive, the resulting analog output may contain an unacceptable amount of "ripple," thus affecting coating quality.

Roll drift is another source of variation. To adjust for roll drift conduct, a simple zero speed test for coaters with analog power supply-type drives. For this test, put the line speed reference voltage to zero and set all power unit outputs to zero volts. Next put the line in the run mode at zero speed, then at zero speed adjust the individual drives so no section moves more than 15° in 15 min. This process ensures no section will leap when the motor contactors are closed, ensuring the acceleration start will be very smooth. Roll drift is still a problem for all static power supplies with analog regulators because of the inherent drift of these systems. Tension control can compensate for this drift condition. Still, even with static power supplies, drift should be checked periodically and not ignored.

Drying Equipment

Air Drying

The dryer air-handling system can also affect coating quality if it causes the web to flutter at the coating head. The effects of flutter may be corrected by adding an isolating roll or by increasing the wrap angle on the coating roll. Coating chatter can be caused by vibrating dryer fans if they are mounted in the ductwork of the machine and the ductwork is connected to the coating stand. This practice is common on lab or test coaters where space is limited.

Flotation dryers are designed to transport the paper web using a cushion of air instead of idler rollers which are used by air-impingement dryers. At one time, it was believed flotation dryers required less tension to move the web through the dryer zone than the more traditional air dryers. Tension required to transport web through all air dryers is consistent with the normal running tensions of a web as defined by the TAPPI horsepower constants. If the process requires very low tensions or if air drying consists of a series of long sections, the web may need help getting through the drying zones without mistracking. In an impingement dryer, the web-carrying rolls may have to be driven. Impingement dryer idler rolls are driven by catenaries which form between the idler rolls. These catenaries are a combination of the weight of the web, the web tension, and impingement air velocity. If the weight of the web and the force of the impingement air exceeds the web tension, the web will form deep catenaries between the web-carrying rolls, and the rolls will stall without driving the web-carrying rolls. It is possible to have a machine sending web into the dryers with no web coming out of the dryers. In this case, the web tension is not enough to overcome the losses of the dryer.

Finally, the air system design plays a big part in whether the web will track through the air dryers properly. If the process air sweeps across the web to the exhaust system, the web may slide off the machine at low levels of tension. This can happen with air flotation or impingement dryers. If the process air balance lifts the web off the idler rolls in impingement dryers, web tracking will be unstable and web breaks will occur during transient periods. It is very important for machine designers and process air designers to understand that successful web transport depends on the efforts of all the disciplines of coating system design.

Cylinder Drying

Older cylinder dryer drives are connected only to the dryers. Cylinder dryers which run at low tension values (1 to ½ PLI) are more apt to slip than dryers which run at higher levels of tension, but all dryers seem to slip at higher machine speeds. Felted and unfelted dryers may intermittently affect web tension at higher speeds. An inspection of the dryer ammeter will show section load increasing and decreasing as the dryer grabs and releases the web. Keep the tension trim on these dryers at a very low value (1 to 3%) to keep the dryer from generating tension transients during the coating process.

Dryer tension is controlled to about 500 ft/min. At that point, the coating blades are loaded, and the machine is accelerated to process speed. The problems with the dryer occur when the web breaks away from the dryers during the acceleration process and continues slipping during steady-state operation. The tension controller will integrate the prevailing error which exists between setpoint and actual tension until the tension controller output is saturated. The controller will stay in saturation and the dryer will slip until the stop button is depressed and the dryer decelerates to about 500 ft/min. If the tension trim is set above 1%, the dryer speed will be badly out of speed match with the web. At about 500 ft/min, the dryer will stop slipping and a transient will occur. With a large speed mismatch, it is doubtful that the dryer drive can correct the resulting transient before the web breaks.

Today's computer-controlled drive can be programmed to sense when the dryer tension control system is saturated. It can be programmed to switch to speed match when the line is at a process speed and can be programmed to return to tension control when the stop button is depressed.

Modern drying section drives consist of a drive for the dryers and a separate drive for the felt. The total load of the section is split between the dryer drive and the felt. The addition of the felt drive seems to have eliminated the slippage that is seen on older felted dryer sections. A drive load-sharing system prevents damage to the web.

The condensate level of the steam dryers may vary. Higher condensate levels represent a higher water load, i.e., more load losses. This varying load will result in tension changes. Similar changes in water removal from chill rolls or surface windups represent similar changes in tension. In the case of the surface windup, changes in water load will also affect the quality of the wound roll.

Fixed or rotary siphons which are not set close enough to the inner surface of the dryer drum cannot evacuate the steam condensate. The first evidence of condensate buildup is the increase in the torque which the motor must provide to accelerate the section. In some cases, the drive system may go into current limit and be unable to accelerate the section to process speed. This type of problem is most likely to surface during the initial startup of the machine. Broken siphon pipes or leaking rotary joints may prevent effective condensate removal with the same results as described above. Some rotary joints may stick and slip during operation which represents a variable torque load to the drive system. If the torque disturbance is severe enough and is at a frequency to which the drive can respond, the drive may become unstable.

Rewind Telescoping

The finished coated paper can become loose as it winds and traps air. The roll may cinch and telescope because the tension is too low to wind a quality roll. Telescoping occurs when the layers of web move laterally (sideways) on the core. If the web does not telescope on the rewind, it may telescope while it is moved or on the unwind of the next machine. The telescoping can be so bad it may make it impossible to remove the coated paper in roll form from the machine.

At one time the only tool for eliminating air entrapment within the wound roll was the rewind tension system. Operators increased tension until the wound roll would not telescope. But many processes could not, and cannot, tolerate high tension levels without damaging the product. The use of the controlled pack roll, together with lower tension requirements for the winding roll, is producing more high-quality wound rolls than it was possible to produce in the past.

Some machinery manufacturers are using vacuum roll as winder drums on high-speed winders. Vacuum rolls may find their way to high-speed coating machines. Vacuum rolls help eliminate air entrapment.

Wrinkling

Wrinkles in the basestock can be caused by nonuniform basis weight, web chamber, uneven cross-machine web caliper profile, an uneven moisture profile, and internal web stresses which occur during papermaking process. These uneven profiles can often be felt in the jumbo roll as "corrugations" or hard spots. The defect can appear as wrinkles when the sheet is wet at the coater. The wrinkling can lead operators to believe the coater is out of alignment (which also can cause wrinkles), when in reality the web is out of alignment with the machine. If the wrinkles form before reaching a nip point or the blade, they may crease and cause a web break. Devices such as bowed rolls, smooth rolls, rolls with spiral wound masking tape, rolls with a spiral wound ground-in pattern, crowned rolls, and concave rolls are used to help eliminate wrinkles.

Cylinder drum dryers can induce wrinkling if the dryer is running at speeds lower than the web speed. This can happen if the tension trim system reduces the section speed when the web breaks away. This is typical for unfelted dryers and cylinders which back up high-velocity dryers. High-velocity dryers are usually run with a negative air pressure, which has an additional lifting effect on the web, causing it to float and break away from the dryer drum. Wrinkling is blamed on the turbulence which is caused when the boundary air layer is disturbed by the slow running dryer. If reducing the tension trim does not eliminate the wrinkling, then sometimes switching the dryer from tension control to speed match will sometimes eliminate the wrinkling. If this is done, the operators must remember to switch back when they intend to stop the line.

Nips on off-machine coaters can cause web breaks when web wrinkling occurs during a flying splice or during normal operation from a nonuniform web. These nips may be opened to avoid these web breaks. The nip roll before the rewinder is not opened because it prevents a web break at the rewinder from causing a web break within the coating machine. It is much simpler to thread the web a short distance from the last pull roll to the rewinder than to rethread the entire machine.

Sometimes the difficulty of identifying the cause of wrinkling comes from the preconception that the basestock is not a problem. The web is almost never identified as a possible cause of wrinkling because the assumption is the specifications for the basestock are met. The real causes for wrinkling may be:

1. Basestock specifications not complete or realistic
2. Basestock specifications are not policed

3. Basestock is not in compliance with specifications
4. Some machine rolls are misaligned
5. Unsupported and undriven draws are far apart
6. Both web and machine are at fault.

Impression Marks and Scratches

One source for impression marks or scratches on the web or coating is lumps of dirt or dried coating built up on the drive rolls. Depressions in rolls can also be a source. Depressions from an operator cleaning a lump of coating off a roll surface with a razor blade or utility knife may gouge the cover; the resulting gash may mark the web with an impression mark. If the spacing between the repeated marks can be determined, it is a simple matter of calculating the roll diameter and therefore which roll is the source of the defect. If the drive section has two rolls, then be aware of both diameters.

Another source for impression marks is a web break in the dryer. After such a break, the operator may try to splice the two ends together on a drive roll. The cover of the drive roll or vacuum screen can be damaged if the butt ends are cut against the roll. In the case of the vacuum screen, the damage may be severe enough to cause the damaged screen to puncture the web. To avoid this possibility, the drive roll can be supplied with a splice table.

Roll Shape

1. Paper rolls which are stored directly on the floor or on skids develop flat spots. It will be impossible to set an accurate gap between paste roll and jumbo roll if the jumbo roll is not concentric and a splice attempt will be unsuccessful. Poor quality rolls that are soft and have flat spots may telescope and cause web breaks during acceleration.

2. Paper rolls which are stored over a shutdown period may become soft and difficult to accelerate without telescoping or cinching.

3. Paper rolls with many and poor machine splices may be nonconcentric and may contain enough defects to cause web breaks.

Gear Reducers and Power Supply Selection

Some gear reducers have high losses and will give a braking effort to a drive section. This is especially true for worm reducers or for gear reducers which have very high reductions. The losses may be high enough to provide the necessary braking effort that would normally be provided by the section's power supply. If this is the case, this braking effort must be replaced when new gear reducers are supplied during a rebuild. If, as part of the rebuild, unidirectional power supplies are provided in-

stead of regenerative power supplies, then the master section may not be able to maintain control of web velocity, and tension control will be unpredictable. If an existing drive system has unidirectional power supply and high loss gear reducers, the reducers may provide the necessary braking to prevent a section from having to be overhauled. If after a drive speedup, the new reducers do not supply the necessary braking, regenerative power units will be required.

Many engineers are grateful for the gross overload capability of motor generator (MG) sets. Consider the following case history. A new 200-in., 4000-ft/min coater was equipped with two Yankee dryers. The dryers were driven by a 75-hp motor which reached its current limit during acceleration due to dryer inertia and high condensate levels in the dryer. These motors were powered from MG sets. The problem of acceleration was solved by increasing the current limit to 400%. The drive successfully accelerated the drum to 4000 ft/min in 100 seconds. At the end of the acceleration period, the normal running load of the drive motor dropped to 30% of the full load rating of the motor. With today's constraints on power units, the same problem would not have been solved so easily. A 75-hp motor may still be used, but the power unit must be specified to handle an overload.

On a second occasion, the speed of a coater was increased by 50%. This speedup occurred after the normal running loads of the individual sections were documented and evaluated. It was determined that all existing motors and power units could be used. The gear ratios of the existing gear reducers and the belt ratios to the unwind and rewind were also changed. The unwind drive and rewind drive could be overloaded to 200% of full motor nameplate rating when they were called upon to accelerate large jumbo rolls. The existing MG sets accepted the work without incident. Since the existing power units were not replaced, this saved considerable cost for new equipment, downtime, and construction for removing the old equipment and the installation of the new equipment.

Drive Application Data

Integral vs. Proportional Operation Control

There are times when simply reducing the range of tension trim is not enough to allow a tension-controlled section with marginal tension-control capabilities to regain control of the web.

All modern controllers use operational amplifiers as the heart of the control system. These amplifiers are more precise than the old proportional amplifiers, but the integrating attribute of the amplifiers may never allow these marginal sections to perform properly. There has been success when tension controllers are converted from integral to proportional control.

The softer characteristic of the proportional controller together with the tension trim limit has allowed sections with marginal tension control capabilities to control tension. For high-speed coating lines, the combination of proportional control and tension trim limit may be particularly effective for dryer drums. On any line, where unnipped rolls such as coaters and pull rolls tend to breakaway easily, the combination of proportional control and tension trim limit may provide stable performance to a machine that cannot otherwise support the demands that an integrating controller places on it.

Proportional control and tension trim limit may be the control of choice when operating personnel do not have an active system for inspecting the covers of elastomer rolls. It may also be the control of choice for correcting for the reduction in coefficient of friction which normally occurs with time, temperature, or abrasion. The goal is to have a machine that runs consistently, and if a machine performs better with the less precise proportional system than it does with integral control, then use the proportional control.

Speed-monitoring System

Modern speed-monitoring systems consist of a microprocessor-based controller which accepts a speed-related signal from each driven section and some undriven sections. These systems are available with a wide range of alarm and report generation capabilities. Speed-monitoring systems are ideal for older coating machines which do not have process computers providing this function. They are also be ideal for modern coaters where the operator wants a continuous display of speed, which is especially important during machine startup and troubleshooting.

A web speed-monitoring system should include a speed signal from a web driven roll, often called a paper roll or idler roll. The web must drive the roll without slipping, or the speed signal will be inaccurate. To assure accuracy, select a roll with a high degree of web wrap.

Roll slip can be detected quickly by monitoring the speed differences between the master section and the web support roll. Some monitoring systems will provide an alarm or computer output to indicate slippage beyond acceptable limits. Monitoring an undriven idler roll is acceptable on slower speed machines running under 500 ft/min. On machines above 500 ft/min, measure the speed of tension-controlled sections. It is important to document the acceptable performance parameters for a given coating machine to know when unacceptable roll speed may be contributing to product defects.

Selecting a Tension-indicating System

All tension-measuring systems measure the total tension that is applied to the web (Fig. 6.18). As long as the width of the web is the same, the tension can be displayed as unit tension or as total tension. The operators and engineers can see at a glance the machine tensions to use for roll balancing and troubleshooting. It is common to see tension meters on a coating line calibrated to read 0 to 3 PLI. Since unit tension is used directly in many calculations, a meter that reads in unit tension is preferred as long as the web width is the same.

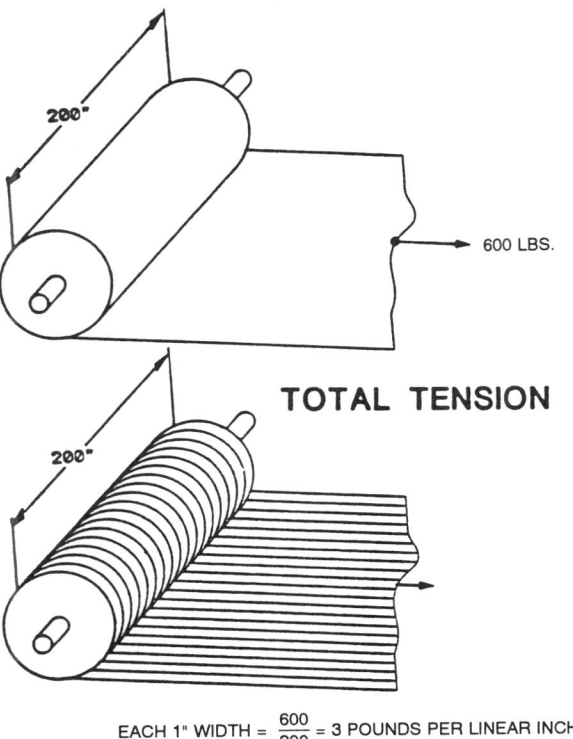

EACH 1" WIDTH = $\frac{600}{200}$ = 3 POUNDS PER LINEAR INCH

Fig. 6.18. Tension indicator

Many machines run multiple web widths and still require a tension-indicating system that is accurate for all widths. For machines that run multiple widths, a tension-indicating system which is calibrated to read total tension is more meaningful than a PLI indicator. A chart which tells what value of PLI to run for different web widths is helpful. However, the meters will only be correct when the machine is running a 200-in. web. Therefore, a total tension system will be much more helpful to the operator. Fortunately, today's computer-controlled drives can accommodate varying web widths in their menus and display the exact tension in whatever units operating personnel prefer.

A crude approach to displaying unit tension on older drives is to use a selector switch to select varying web widths, e.g., 80-100 in., 90-110 in., and 100-120 in. of web width. This gives the operator an input to recalibrate the meter based upon web width.

Wiring Design

Today's high-speed machines, which run products with well-defined specifications, need high-performance drives, fast-responding regulators, and minimum downtime to produce a quality coated product at least cost. This combination demands a carefully installed and noise-free drive control system.

Some drive system users disregard the drive vendor's recommendation to supply a separate conduit system for digital and analog tachometers, potentiometers, and other low-level signal devices. Instead they combine the signal wiring with control and power wiring. The resulting signal distortion and noise causes poor drive performance when individual sections drift and make step changes in speed whenever a process motor is started or stopped. It is considered too expensive and too much trouble to run separate conduit. The practice was common in the old multi-section drives that were slow to respond to noise and because the process was forgiving. However, combining signal and power wiring produces a machine-tracking characteristic which results in each section tracking a machine acceleration ramp at different rates and to a different end point (Fig. 6.19).

It is strongly recommended that signal, control, and power wiring is separated as old drives are replaced on older machines. For reference, refer to the TAPPI recommended guidelines in TIS 0406-12 "Recommended Electric Wiring Practices for Automatic Control and Regulator Systems."

Operation Requirements

Stopping

Stop modes also influence the horsepower selection of drive system motors. Emergency stopping rates are extremely fast and may be dictated by regulatory bodies such as OSHA or by the plant safety department. Rapid stopping of large inertias like the inertia on large-diameter rolls on the unwind and rewind sections may exceed the power needed simply to control tension. Horsepower for these sections are based on the worst case needs.

Stopping time can be anywhere from 60 to 90 seconds. If the section drive system can handle stopping time as a temporary overload, then it is not necessary to size the drive for the full-machine tension requirements.

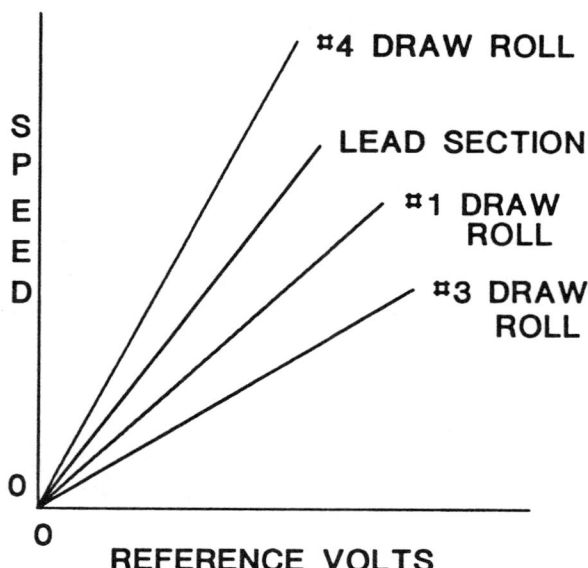

Fig. 6.19. Tracking curve no.1 (actual machine performance data)

On transfer unwinds, deceleration requirements are especially severe during an emergency stop. Additional requirements occur after a splice. At this time, the drive motor is at maximum speed and at minimum field strength. The braking effort is severely reduced under weak field conditions. Additional braking can be obtained by the addition of mechanical braking and by increasing the drive motor field strength to maximum as rapidly as possible. Deceleration rates of 300 ft/min/s/s are recommended for unwinds.

Machine Speeds

The machine speeds will affect the selection of drive motor horsepower and the power transmission equipment which connects the motor to a driven roll. Operating speeds will affect the capacity of the drying zones and the coating system and the coating fluid batching and holding systems. Speeds are expressed in feet per minute (ft/min), meters per minute, or inches per minute.

Acceleration

The acceleration rate for the unwind must be high enough to allow the drive to quickly accelerate good quality rolls to line speed, but also low enough not to cause conditions such as telescoping, dishing, starring, or cinching of the accelerating roll. All of these conditions can cause web breaks, poor control of tension, and edge cracks. They also can cause coating scratches of pre-coated webs as the layers of web slide past each other within the wound roll. Acceleration rates of 100 ft/min/s/s are recommended for unwind stands. The motor is also necessary to bring the expired core to a rapid stop after a splice is made. If not stopped quickly, the web left on the core can interfere with the newly spliced web and cause a web break.

Motoring and Regeneration

Motoring is the ability of the drive system to add torque to the web. Regeneration is the ability of a drive system to hold back or provide a braking effort to the web. A motor can motor or regenerate, and motorgenerator sets can motor or regenerate, but today's SCR power supplies will be supplied for regeneration only if the need for regeneration is specified.

The user should examine the current operating techniques in order to establish a desired tension profile for the machine. This information will be used by the drive vendor to select section horsepower and the type of power supply for each section. This information is important whether the task is a drive rebuild or a new drive purchase.

Regeneration is necessary for a number of applications. All high-inertia sections such as Yankee dryers, steam dryers, and calender stacks may need regeneration for deceleration. If tension on the exit side of these sections is higher than the entry tensions, then regeneration will be needed for effective tension control.

An unwind needs to motor and brake when controlling tension. Some unwinds, which use armature control, provide a motoring effort when the unwinding roll is reduced to its smallest diameter at low-tension values. At that point, the losses of the electro-mechanical system are at their greatest and may exceed the value of braking effort required to maintain tension. The drive motor will supply a motoring effort to maintain tension control at the proper level. Motoring effort is also needed whenever the jumbo roll is accelerated to line speed prior to a splice attempt.

Center rewinds need to supply motoring effort when they are winding web, and need to regenerate when the roll is braked to a rapid stop. The rate of deceleration of large rolls of paper on rewinds, and unwinds may be limited to a rate which prevents the roll from cinching or telescoping.

Characteristics of Power Units

The selection of a proper power unit is not a trivial process. If an existing motor generator set or SCR drive with regenerating capability is replaced by a non-regenerating power unit, then machine performance and existing production output will be adversely affected until the necessary corrections are made. Production loss is not well tolerated at any plant site.

Motorgenerator (MG) sets are the oldest form of power supply for multi-section drive systems. They have the ability to provide positive or negative current to the motor allowing the drive system to motor or regenerate.

SCR power units can be ordered with unidirectional or with added regenerative capability. If regenerative capability is required, it must be specified. If replacing MG sets, regeneration may be required but may be overlooked. Errors have been made during drive rebuilds when the user and vendor neglected to establish the tension profile for a machine and ended up supplying unidirectional SCR power units in place of MG sets for ALL sections on the machine. This includes the master section which needs motoring and regenerative capability for effective web velocity control.

Without regenerative capability, isolation will not be possible, and it will be impossible for the lead section to maintain velocity control of the web. There are documented cases where regeneration was added to dryer section and calender stack power units after it was discovered that they were being overhauled by high rewind tensions.

Servo type power units (Pulse Width Modulated-PWM) are supplied with regenerative capability in sizes that are common to the converting industry.

Specifications for a drive system should list which sections require regeneration and which do not. It should not be assumed that regeneration is inherent. Frequently, vendors redesign their drive systems for financial reasons that may include a redesign that changes a drive supplied with regeneration into a unidirectional drive.

Do not assume that a positive motor current during process conditions means the section is not being overhauled. Instead, compare motor current without web on the machine to the motor current under various process conditions. If the motor current under process conditions is less than no web motor current, then the section is in a braking or regenerative mode. A regenerative power supply is the power supply of choice.

Sizing Power Supplies

Power supplies should be sized to the motor capabilities and to the needs of the load. Rapid emergency stops will place further quick-stop demands on the section power supply. Since all machine sections have low friction and high inertia, it is necessary to provide regenerative power supplies to ensure rapid emergency stops and controlled normal stops. DC-drive motors can sustain a 150% overload for one minute. Power supplies for low-inertia sections should have the same overload capacity.

High-inertia sections such as unwinds, rewinds, calenders, and dryers may need drive systems (motor and power supply) which can accelerate and decelerate the section quickly to run speed and then to rest.

High-inertia section drive system capability should be based on the need to accelerate and decelerate the inertia load in a reasonable time. Above 2500-ft/min machine speeds, the inertia may require additional drive capacity for acceleration (suggested rate, 40 ft/min/s). This power capability is a continuous rating, since a 150% rating for one minute isn't enough for acceleration or deceleration of high-inertia loads. Therefore, a 200% continuous overload capability for motors and power supply may be more appropriate for these sections.

Today's power units are most likely to be SCR with an isolation transformer. Drive vendors do not design SCR power units for each motor size, but instead design them to handle a range of horsepower loads. If the design section hp is less than the maximum rating of the power unit, then the power unit has extra capacity.

A user may wish to use some of the extra capacity when a helper drive is added to a section or when machine speed is increased. However, a vendor may limit the user's access to this extra capacity if the armature components are supplied and the isolation transformers are based on the section design horsepower instead of the maximum rating of the power unit.

Most machines in a paper mill are eventually rebuilt, speeded up, or reconfigured. If the user does not invest funds in the beginning to guarantee maximum power capacity and specify that all armature components and isolation transformers match the maximum rating of the power unit, then future machine changes may require complex drive changes, machine downtime and construction time. It is also advantageous for the user to establish the maximum KW rating of each power unit and then establish the cost of increasing the transformer size and armature loop components in each power unit. This ensures that the maximum power capacity of the unit is available. The cost of increasing the size of components today is far less than replacing the same components tomorrow.

Master Section Determination

The master section is the lead section of the machine (Fig. 6.20). It establishes the speed of the web in the coating line. If the grip is broken, the section loses its ability to control web velocity. Therefore, a good lead section must maintain this grip at all speeds up to the maximum design speed even when subjected to transients during acceleration, deceleration, or at any stable speed. The lead section must be capable of adding torque (motoring) or holding back torque (regenerating) to the web in order to control web speed.

An example of a good master section on a coating line is the coating drum. A typical coating drum is 38 in. in diameter and will have a circumference of about 10 ft. That means for every revolution of the coating roll, approximately ten feet of web will be drawn off the unwind stand, through the pull rolls, to the coater, dryers, and passed along the machine sections to the rewind. When the lead section effectively controls and maintains the web speed, then we know the proper coating thickness is applied, the coating is dried properly and wound rolls of good quality are produced.

It is common to see one coater running at a considerably higher load than the other. On some coaters, this may be due to differences in blade loading since one coater coats the felt side and the other coats the wire side. Also, the position of a coater in the drive train may cause one coater to contribute more tension load than the other.

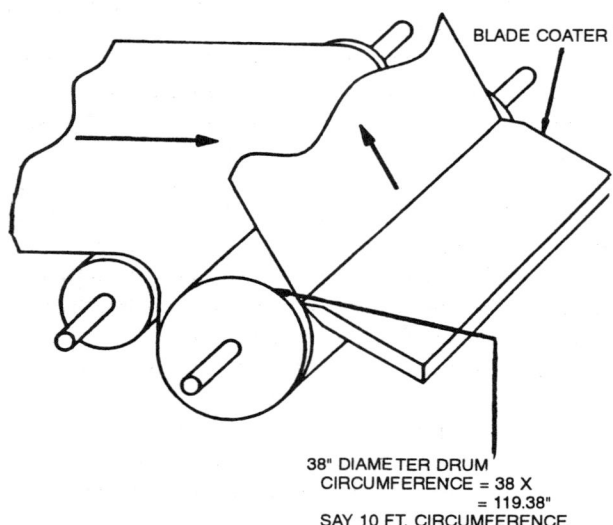

BLADE COATER

38" DIAMETER DRUM
CIRCUMFERENCE = 38 X
= 119.38"
SAY 10 FT. CIRCUMFERENCE

Fig. 6.20. What is a master or lead section?

If the lead section is unable to control web velocity, and the web slips, then defects such as thick or thin coating, overdrying or underdrying, web scratches, coating scratches, and poorly wound rolls can be expected. Web breaks are common when web slip causes tension control systems to become unstable. Worse yet, it may be impossible to track the web as it slips through the machine longitudinally and laterally. When slip exists, the machine may operate at lower speeds, but transients will cause the web to slide from side to side and eventually break. If slip is bad enough, it may not be possible to perform even the simplest task such as threading a machine.

In summary, the lead roll in the machine must be able to transmit the torque from the motor, through the roll and to the web without letting the web slip.

Small dryer sections, whether felted or unfelted, have been poor lead sections for off-machine coaters especially if they are subjected to even moderate values of differential tension. Dryer felts are frequently removed after the first web break when coating contaminates the felts. The dryer will certainly slip without the felts.

As mentioned under cylinder drying, modern dryer section drives are split into two motors per dryer section, one for the dryers and one for the felt. The total dryer drive load is measured and is distributed between the felt drive and dryer section drive, increasing dryer felt drive load. This causes the felts to "pin" the drive against the dryer, thus exerting good control of the web. This new technique makes a felted dryer a good master section.

Horsepower Requirements

The following is taken from a TAPPI Technical Information Sheet. The horsepower calculation process is well documented in the TAPPI TIS sheets (see bibliography at end of chapter). The following excerpts are from those sheets and are included to clarify the following examples.

Power constants for paper machine drives are commonly expressed in horsepower per linear inch of machine width per 100 ft/min of speed and abbreviated as hp/in./100 ft/min. This is actually a torque figure. Width is taken as wire width on fourdrinier machines. For coaters, width is taken as roll width. Two power constants have been established for each machine section. One is Normal Running Load (NRL) and the other the Recommended Drive Capacity (RDC).

Normal Running Load (NRL)

NRL is the expected running load on each section of the machine when it is operated under normal conditions. This includes such items as:

- Normal vacuum conditions of the vacuum rolls
- Normal nip pressures on nipped sections
- Normal tension in the web or felt entering or leaving a driven section
- Adequate bearing alignment
- Properly lubricated bearings
- A full roll calender stack
- And properly evacuated dryers.

So, the estimate of the total normal running load of any machine becomes the sum of the NRL constants of the various machine sections.

Use NRL for selecting: (a) lineshaft turbines or lineshaft drive motors adding 15% for lineshaft losses,

(b) power supply for a single-power supply-type sectional drive, with proper allowance for losses in section motors, (c) prime movers for generator sets, (d) feeder transformers on sectional drives with proper allowance for losses in the DC motors and power conversion equipment, and (e) section power supplies where multiple indrives are used in a single section.

See drive requirements for paper machines for further guidelines.

Recommended Drive Capacity (RDC)

RDC represents the estimated power requirements for any particular section when operated at maximum anticipated load. These maximum loads may results from:

- High vacuum on the vacuum rolls
- Physical condition of vacuum boxes
- High loading or nip pressures
- Excessive amounts of water in the dryers
- Tight dryer steam joints
- Misalignment or other mechanical difficulties
- High coating blade pressure
- High sheet or felt tension entering a section
- Or any other nonstandard conditions.

Use RDC for selecting: (a) section drive motors for sectional drives or (b) section power supplies for single indrive sections.

The difference between RDC and NRL represents the operating safety margin above NRL, but does not provide for the high starting characteristics that may be peculiar to certain sections. Nor does it provide for unusual operating conditions created by peculiar machine characteristics.

Unwind Drives

Unwind horsepower is generally based on tension requirements and the drive to accelerate the jumbo roll to run speed before a splice attempt. Remember, no sheet tension is available to help accelerate the jumbo roll before a splice attempt, so the unwind drive must supply all the torque to accelerate the jumbo roll. TAPPI Technical Information Sheets (TIS) that recommend drive constants for winders and paper machines also list the standard methods and hp constants for selecting unwind drives. Unwind losses account for about 10% of the torque that is needed for controlling tension. Therefore, unwind motor hp can be reduced slightly over the calculated need. For instance, if the calculated tension horsepower is 80 hp, a 75-hp motor will be acceptable.

Refer to Tables 6.2 to 6.6 at the end of this chapter for formulas and example calculations.

Draw Rolls

Draw rolls or pull rolls split the machine into tension zones and help threading. Draw rolls are also designed to help motivate the web when there is a long span between two driven points. Draw roll motor hp is sized based on the differential tension requirements and section losses. Nip losses should not be forgotten if the draw roll is a nipped section. Differential tensions of 1 to 1½ PLI are normal for machines which are designed for a maximum of 3 PLI; differentials of 2½ PLI are used on machines designed up to a maximum of 5 PLI. This represents approximately 0.003 hp/in./100 ft/min for each PLI of web tension.

Nipped draw rolls are used in all paper coaters and are limited to low-speed machines. A nipped section before the rewind isolates the preceding sections from any rewind transients. This section should be designed to handle the maximum machine tension in the event the web breaks between the rewind and the draw roll nip. The motor is sized to handle full machine tensions using the motor overload capabilities since the machine will stop immediately in the event of a web break. This section's power supply must have the same capability to handle overloads as the section motor. In other words, there will be full tension only for the time it takes to bring the machine to a stop.

Initial running loads on blade coaters are quite high during startup when various blade pressures are tested for coat weight control. Higher than normal tensions are run until the machine operators develop what levels of tension and blade pressure are required for each product. Typically, blade pressures run at 3-6 PLI or 10-12 PLI, depending on the type of blade coater used.

Dryer

Selection of the drive motor and dryer running loads hp is based on the recommended TAPPI drive constants for the various types of coaters and paper machines. Cleaning doctors will increase the section load predictably, as defined by TAPPI constants, as long as the dryer surface is clean and the blade pressure is maintained at 1 PLI (TAPPI constants are based on 1-PLI doctor loading). On dryer surfaces which are contaminated with dry coating, dryer doctor loading can triple the section running load. Other factors which increase dryer section load include poor condensate removal and the load from felted dryers when they are used to control tension on a coating line.

The recommendation is to double the TAPPI drive constants for doctors on dryers that follow coaters and to increase hp by including the tension load of the dryer section (0.003 hp/in., 100 ft/min, per PLI). Note: TAPPI recommended drive capacity (RDC) for dryers contains some doctor load.

Doctors on dryers are frequently run at blade pressures higher than the recommended 1 PLI (e.g., 3 PLI). This additional loading is used to remove dried coatings that accumulate on the dryer surface. It is questionable if there is a way to compensate for extreme conditions of doctor loading, but adding hp to a section is one way to help reduce the amount of trouble. Dryer RDC values are based on assumed loading of doctors on one third of the dryer drums at any one time.

Add 0.0015 hp/in./100 ft/min for each metal doctor and 0.002 hp/in./100 ft.min for each laminated plastic doctor for each doctor used over the assumed 30% in use in order to obtain the section RDC. Add the values for each doctor to obtain the proper drive RDC factor. Some drive designers never reduce actual motor horsepower below the calculated RDC. These factors are slight increases over the factors used for clean drums. Dryer normal running load (NRL) does not include doctor loading in its calculation.

Felted dryers or unfelted dryers in groups of three or more can affect tension, especially at speeds under 1000 ft/min. Therefore, add 0.003 hp/PLI/100 ft/min for each PLI of differential tension that a dryer may be required to produce in the web. Felted dryers may break away at speeds between 1000 ft/min and 1500 ft/min. One PLI is a normal differential on coaters designed for 3-PLI maximum tension, and 1½ to 2 PLI is normal on coaters designed to 5 PLI.

Selection of dryer horsepower should be based on the possibility the dryer will contribute to sheet tension control at lower speeds. As stated before, coater dryer felt tensions are kept low to prevent damage to the coating, but low felt tensions result in slip.

Modern dryers may contribute to web tensions at all speeds.

Calender Stacks

Calender stack drives should be sized to power the maximum number of rolls which can be accommodated in its frame. The drive should also consider the possible addition of a swimmer or expander roll. If an expander roll is already part of the calender, some thought must be given to driving the roll when the oil in the roll is cold. This represents additional torque requirements which again emphasizes that motor HP should not be reduced below the calculated RDC values.

In some cases, a reduction in horsepower may be possible if a separate inching drive using a high torque ac motor is used to inch the calender while the oil is being heated. Inching drives can also be helpful when maintenance is performed on the stack.

Reel

Reel section horsepower can be selected using the latest TAPPI horsepower constants for a paper machine. Section horsepower can be reduced slightly if the actual tensions at the reel are less than 3 PLI. Add 10-25% horsepower for the characteristic water joint friction of water cooled reel drums.

Centerwind windups share many of the same characteristics of the unwind. Horsepower selection is based on the same criteria. However, unlike the unwind, rewind losses of 10% must be added to the tension hp needs. So if the tension hp is calculated to be 80 hp, the actual requirement at full tension may be 88 hp. At this point, a drive designer must evaluate a number of questions:

1. Do I need a 75-hp motor or a 100-hp motor? A 100-hp motor would be appropriate for this application.

2. Is the maximum tension specified in a realistic value? If not, what is a practical value? Does the user have data on actual running tensions?

3. The rewind alternates between two drives which allow one drive motor to cool when the other is winding. Will this reduced duty cycle allow the drive motor to cool sufficiently?

4. If taper tension is used, will the amount of taper that is used reduce the torque requirements enough to select a 75-hp motor? If not, then a 100-hp motor may be required.

Example Drive Specification

Let's look at a simple machine that unwinds rolls of paper and rewinds it. We'll call it No. 24 winder that serves the No. 24 coater. Keep in mind this discussion that can also be applied to our No. 24 coater.

- General Specifications:
  - Name of Machine No. 24 Solvent Coater
  - Code Classification Class 1, Division 1, Group D

- Speeds
  - Machine Run Speed: 20-300 ft/min, current operating speed is 190 ft/min
  - Thread Speed: 0-20 ft/min (adjustable)
  - Auxiliary Run Speed: 20-300 ft/min
  - Section Jog Speed: 15 ft/min (adjustable)
  - Acceleration Rate: 20-40 ft/min (adjustable)

- Machine Stop Modes:
  - Emergency Stop: dynamic braking at 100 ft/min/s/s
  - Normal Stop: 20-40 ft/min/s/s (adjustable)
  - Coast Stop: variable

- Web Specifications:
  - Web Width: 50 in. minimum to 68 in. maximum
  - Web Materials: 1 mil. to 5 mil. paper and poly; 0.5 mil poly for future products
  - Web Tensions: 0.25-2 PLI or a total tension of 12.5-136 lb total

- Core Dimensions: Unwind—3 in. and 6 in. ID
  - Rewind—6 in. ID

- Roll Diameters: Unwind—30 in. maximum
  - Rewind—30 in. maximum

- Roll Weights: Unwind—3800 lb
  - Rewind—3800 lb

- Utilities:
  - Electrical: 460 volt, 3 ph., 60 Hz power, 120-volt control
  - Compressed Air: 100 PSI at the compressor, 60 PSI maximum design pressure at full production operation

Safety note: All electrical equipment is rated at class 1, division 1, group D per NEC and NFPA. If equipment is not manufactured with these ratings, then purging is acceptable so long as they satisfy the ISA (Instrument Society of America) and NFPA (National Fire Protection Agency) recommendations for purging. Equipment operating outside the solvent environment may be rated as ordinary wiring.

This machine is typical of equipment that can perform a number of functions: it can rewind poorly wound rolls, it can be used to remove defects from the wound roll, or it can be used to slit the rolls to customer-specified widths before shipping.

The No. 24 Coater is designed for a top speed of 300 ft.min, will handle rolls that are 30 in. in diameter wound on 6-in. ID or 3-in. ID cores, at ¼ to 2 PLI tension. The coater will run continuously, while our winder will stop at the end of each roll to remove the rewound roll and to load a new roll onto the unwind.

Case 1: It will rewind poorly wound rolls and be used as a quality control device, then it will start and stop frequently. What speed should it be designed to run? Frequent starting and stopping means that the machine will never reach elevated speeds and to design for elevated speeds would require more costly drives. Therefore, we will design the machine for 300 ft/min. We may need two winders to service the continuous coater. However, if the coater runs for only one shift, we can run our winder for three shifts to service one 8-hour coater shift.

Case 2: If, however, the machine will be used to slit web for shipment, it will be shut down infrequently. Therefore, if our machine is to keep up with the steady stream of finished rolls from our coater, it must be designed to process rolls at higher than coater speeds. A good rule of thumb is to design our slitter/winder for 2-3 times the speed of our No. 24 Coater.

Let's design the winder/slitter for 600 ft/min. In either case, the winder will be designed to handle maximum roll diameters of 30 in. on cores that are 3 in. ID or 6 in. ID with web tensions of ¼ to 2 PLI.

Case 3: The economy is suffering, capital money may be limited, and project approval may be hard to get. The production department needs a winder to improve output and has other items of high priority which it would like to include in this year's budget. The total cost of all items exceeds the department allocation.

Now, let's look at Fig. 6.9 for our No. 24 winder using a constant torque unwind and constant torque rewind. Our winder is arranged as follows:

1. The unwind has a manually operated torque control for a brake.

2. The pull roll is a driven roll that "isolates" the unwind from the rewind and has a manually operated speed control on its drive. This section is the master section for the winder and establishes the velocity of the web through the machine. The master controls web velocity by allowing one foot of web to pass from the unwind to the rewind for every foot the circumference of the master moves.

3. The rewind has a manually operated torque control for its drive.

The operator places the roll in the unwind and "threads" the web through the machine, around the pull roll to the rewind where the "tail" of the web is attached to the rewind core. We now have a 30-in. roll in the unwind which is threaded through the machine and attached to a 6-in. core on the rewind.

The operator then sets the unwind torque control to yield the proper unwind tension and sets the rewind torque control to yield the proper rewind tension. To yield the same tension at the unwind as we have on the rewind, unwind torque is *five times* greater than the rewind torque because we need five times more torque on a 30-in. roll than we need on the 6-in. core.

This is a very simple machine, is relatively inexpensive, but has a few serious drawbacks. First, the operator must *constantly* reduce the torque setting of the unwind brake as the unwinding roll diameter *decreases* from the maximum diameter to the core. At the same time, the operator must *increase* the torque setting of the rewind drive as the roll diameter of the winding roll increases.

When the unwinding roll is reduced to 6 in. and the rewind roll increases to 30 in., the torque on the rewind will now be five times more than the torque on the unwind.

Calculation of the torque requirements for the unwind brake and the rewind drive is as follows:

1. Calculation of maximum torque due to web tension of 2 PLI maximum. Calculation is based on a 30-in. diameter maximum roll size. Roll radius is 15 in. 2 PLI max. tension x 68 in. max. web width = 136 lb total tension. 136 lb x the roll radius of 15 in. = 2040 in.lb max. torque required of the unwind brake and rewind drive. This value establishes the maximum torque required of the brake and drive. The brake and motor must be capable of dissipating this amount of torque without overheating.

Checking the individual brake and motor loss will determine which brake or motor can be used. Torque required for a 6-in. roll is one fifth of 2040 or 136 lb x 3-in. radius = 408 in.lb.

2. Calculation of torque due to web tension of 1/4-PLI tension at 30-in. diameter roll. Torque at 64-in. web is: 2040 x 0.25/2 = 255 in.lb, torque at 50-in. web is: 255 x 50/64 = 199 in.lb.

Calculation of minimum torque on a 50-in. wide web at 1/4 PLI on a 3-in. ID (3 1/2-in. OD) core = 50 in. x 0.25 PLI x 1.75 in. or 21.9 in.lb. This value represents the least amount of torque required to obtain 1/4 PLI on a 50-in. web at a 3-in. ID core. If the drag torque (residual torque) of the brake exceeds that value, then it may be necessary to substitute a motor in place of the brake. The motor can provide braking effort or motoring effort to overcome residual drag. Total torque range for the unwind is 60/50 x 30/3.5 x 2/0.25 = 82.3 to 1.

Summary

Good machine performance can be achieved only when the sheet coming to the coater is reasonably uniform and consistent. A uniform web will result in fewer web breaks and more uniform coating and drying. Each mill must establish specifications for caliper profile, basis weight, and moisture profile which give the best results for the off-machine coater and the supercalender which follows. In addition, close attention must be paid to holes in the web, edge cracks, and machine splices. Experience has shown that machine efficiency is poor where basestock specifications do not exist or are not policed. Conversely, rigid control of practical specifications improves machine efficiency.

This chapter discusses the fundamentals of web transport drive control and some mill operating conditions. Web transport engineers must thoroughly understand the needs of each mill site so they can supply a drive which will deliver maximum flexibility at least installed cost. Understanding begins by observing actual operating conditions and techniques in the mill and then applying established guidelines to select the proper web transport systems. Engineers should use the operator's log and instruments such as ammeters, voltmeters, speed indicators, and recording oscillographs to characterize existing machine performance before specifying the parameters of a new machine. Characterization of an existing machine for the purpose of determining actual process needs is one way of avoiding the mistakes which come from those who specify machines from guidelines only.

Table 6.1. Tension data for typical converting materials

Material	Tension (Lbs./inch/mil.)	Paper & Laminations	Tension PLI
Aluminum foils	0.5 to 1.5 (1.0 average)	20#/R-32.54 gm/m^2	0.50 to 1.0
Cellophanes	0.5 to 1.0	40#/R-65.08 gm/m^2	1.0 to 2.0
Acetate	0.5	60#/R-97.62 gm/m^2	1.5 to 3.0
Polyester	0.5 to 1.0 (.75 average)	80#/R-130.1 gm/m^2	2.0 to 4.0
Polyethylene	0.25 to 0.30		
Polypropylene	0.25 to 0.309		
Polystyrene	1.0		
Saran	0.05 to 0.20 (0.10 average)		
Vinyl	0.05 to 0.20 (0.10 average)		

Substrate	Approx. Tension PLI	Substrate	Approx. Tension PLI
<u>Polyester</u>		<u>Cellophane</u>	
0.0005"	0.25	0.00075"	0.5
0.001"	0.5	0.001"	0.75
0.002"	1.0	0.002"	1.0
Nylon/Cast Propylene <u>(Non-Oriented)</u>		<u>Paper</u>	
0.00075"	0.15	15 lbs./ream (3,000 sq.ft.)	0.5
0.001"	0.25	20 lbs./ream	0.75
0.002"	0.5	30 lbs./ream	1.0
		40 lbs./ream	1.5
<u>Paperboard</u>		60 lbs./ream	2.0
8 pt.	3	80 lbs./ream	2.5
12 pt.	4		
15 pt.	5		
20 pt.	7		
25 pt.	9		
30 pt.	11		

Table 6.2. Helpful formulas for unwind and rewind drives

1   Roll RPM = $\dfrac{\text{Line Speed (FPM)}}{\text{Roll Circumference (feet)}}$ FPM = feet per minute

2   Motor RPM =   Roll RPM x Gear Ratio (GR)   RPM = revolutions per minute

3   Motor RPM = $\dfrac{\text{FPM x GR}}{D}$

4   HP = $\dfrac{\text{T (ft.-lbs.) x RPM}}{5{,}250}$

5   Winding HP = $\dfrac{\text{Tension (lbs.) x Maximum FPM}}{33{,}000 \text{ lbs.-ft./min./HP}}$ (See Note 2)

6   Tension (lbs.) =   PLI x Width (inches)   PLI = pounds per lineal inch

7   Acceleration Torque = $\dfrac{WR^2 \text{ x RPM}}{308t}$ t = acceleration time in seconds

                                                 WR^2 = inertia in lb.-ft.^{2}

8   Acceleration HP = $\dfrac{\text{Torque x RPM}}{5{,}250}$

9   Acceleration HP = $\dfrac{WR^2 \text{ x (RPM)}^2}{(1.615 \text{ x } 10^6) \text{ x } t}$

Gear Reduction is a torque multiplier.

Gear Reduction is a WR^2 "squarer".

NOTE 1: Losses must be added to values calculated with the above formula.

NOTE 2: Centerwind Motor HP =

$$\frac{\text{Tension (Total lbs.) x Max FPM}}{33{,}000 \text{ lb.-ft./min./HP}} \quad \text{x} \quad \frac{\text{Max. Wound Roll Diameter}}{\text{Core outside diameter}} \quad \text{x} \quad \%\text{Taper}$$

Table 6.3. Calculation of horsepower for an unwind

Example: Accelerate an 86" diameter roll to 6000 FPM in 50 seconds.

Roll diameter = 86" = 7.16 feet
Roll Width = 200"
Tension = 4 PLI
Assume losses of = 30 HP
WR^2 of Roll = 237,000 lb = ft.^{2} max.
Acceleration time = t = 50 seconds

Tension/winding HP $= \dfrac{\text{Tension (lbs.) x Max. FPM}}{33,000 \text{ lbs.-ft./min./HP}}$ (Formula 5, Table 6.3)

$= \dfrac{4 \text{ PLI x 200" x 6000 FPM}}{33,000} = 146 \text{ HP}$

ACCELERATION HP:

1. Roll RPM $= \dfrac{\text{Line Speed (FPM)}}{\text{Circumference (FT)}} = \dfrac{6000}{7.16 \text{ ft. dia.}} = 267 \text{ RPM}$ (Formula 1, Table 6.3)

2. Acceleration HP $= \dfrac{WR^2 \text{ x } (RPM)^2}{(1.615 \text{ x } 10^6) \text{ x } t} = \dfrac{237,000 \text{ x } (267)^2}{(1.615 \text{ x } 10^6) \text{ x } 50} = 209 \text{ HP}$ (Formula 9, Table 6.3)

HP REQUIRED TO ACCELERATE A JUMBO ROLL TO SPLICE SPEED:

HP = Acceleration HP + losses = 209 + 30 HP = 239 HP

Motor HP if the drive system can sustain a 200% overload for 60 seconds:

HP $= \dfrac{239 \text{ HP}}{2}$ = 120 HP, i.e., a 125 HP motor

Check for deceleration: full roll deceleration HP, assuming deceleration HP = acceleration HP

Unwind HP for deceleration: (Deceleration is regenerating)

HP = -209 + 30 -146 (tension) = 325 HP

If using a drive with a 200% overload capacity, then

HP $= \dfrac{324}{2}$ = 160, i.e., a 150 HP motor

Check TAPPI HP RDC $= 0.0125 \text{ x } \dfrac{6000}{100} \text{ x } 200 = 150 \text{ HP Motor}$

Check for acceleration of a full roll after threadup

Acceleration HP = Acceleration HP + Losses - Tension HP =

HP = 209 + 30 - 146 = 93 HP, use a 150 HP motor due to the deceleration required.

Calculation of Machine HP for a Coating Machine

EXAMPLE COATING LINE GENERAL SPECIFICATION

Type of Machine:	30-40 lb Publishing
Line Speed:	500 FPM to 3000 FPM
Thread Speed:	50-125 FPM adjustable
Jog Speed:	25 FPM
Web Width:	220 inches
Machine Tension:	1/2 to 3 PLI
Acceleration Rate:	20 FPM/sec to 40 FPM/separately adjusted
Deceleration Rate:	20 FPM/sec to 40 FPM/separately adjusted
Draw Range:	± 10% of Line Speed Reference (Adj)
Tension Control:	± 10% of Line Speed Reference (Adj)

Table 6.4. Recommended horsepower constants

	NRL	RDC	REMARKS
Draw Roll	-	0.005	
Blade coater puddle	0.006	0.012	Can be reduced slightly when information
Blade coater applicator	0.008	0.012	indicates that less horsepower is adequate.
Dryer, 120-144 in.	0.0024 w/o doctor	0.035 w/cleaning doctor	
Dryers (inch)			
72	0.0012	0.0035	
60	0.0012	0.0027	
48	0.0012	0.0021	Add 0.003 per in./100 ft/min per PLI.
42	0.0012	0.0020	
36	0.0012	0.0017	
Dryer doctor	0.0010	0.0015	
Last draw roll	-	-	Size based on tension range.
Reel	0.008	0.0138	With a water joint, add 10-25% for friction
	0.008	0.0230	and water load.
Roll coater			
Gate roll	-	0.0041	
Distributor roll	-	0.005	
Coater roll	-	0.010	

Power supplies: Should be capable of 200% regeneration to meet the deceleration rate requirements for emergency stops and normal stops of high inertia sections.

Table 6.5. Example of calculations of HP for a coating machine

---

Unwind based on 3 PLI maximum: Tension HP $= \dfrac{3 \times 220 \times 3000}{33,000} = 60$ HP

Unwind acceleration HP based on 40 FPM/sec/sec and $WK^2 = 275,000$ lb-ft^2

Accelerate HP $= \dfrac{275,000 \times (134)^2}{(1.615 \times 10^6) \times 100} + 10\%$ losses $= 31$ HP $+ 3 = 33$ HP

Decelerate HP $= -31 + 3 - 60 = 88$ HP. If a 60 HP motor can handle 150% for 100 seconds, then use a 60 HP motor.

Check with TAPPI constants: $= 0.0125 \times \dfrac{3000}{100} \times 200 = 82.5$ HP
(TAPPI HP RDC)

Draw Rolls from TAPPI constants: HP $= 0.005 \times \dfrac{3000}{100} \times 220 = 33$ HP - SAY 30 HP

Check for differential tension of 1 1/2 PLI: HP $= \dfrac{1.5 \times 220 \times 3000}{33,000} = 30$ HP - Check

---

Coater RDC: HP $= 0.012 \times \dfrac{3000}{100} \times 220 = 79$ HP, i.e., 75 HP.

Air Dryer Backup Drum : HP $= 0.035 \times \dfrac{3000}{100} \times 220 = 232$ HP, i.e., 250 HP.
(RDC HP - 144")

Dryer Section: six 60 inch dryers with six doctors (minimum HP is 150; recommended HP is 200 HP)

RDC constant $= $ (NRL/dryer x 6) $+$ tension load$+$ (doctor load x 6)
$= (0.0027 \times 6) \quad + 0.003 \quad\quad + (0.001 \times 6)$
$= 0.0162 \quad\quad\quad + 0.003 \quad\quad + 0.006 \quad\quad\quad = 0.0252$

RDC HP $= 0.0252$ (RDC constant) $\times \dfrac{3000}{100} \times 220 = 166$ HP

Reel is water cooled with one doctor:

RDC $= 0.0138 + 0.0015 \times 1.25 = 0.019$
RDC HP $= 0.019 \times 6600 = 126$ HP

Table 6.6. Example of how to use power constants

30-40 lb Publication
3000 FPM
220 inches wide
3 PLI tension

$$K = 220 \times \frac{3000}{100} \text{ or } 6600$$

	NRL	NRL RDC	RDC HP	Min. HP	Base HP	Dryer Tension RDC	Doctor RDC	Total RDC	TDC
Unwind (3 PLI)	-	-	-	-	-	60	-	-	-
Draw roll	-	0.005	-	33	30	-	-	-	-
Coater	0.008	0.012	53	79	75	-	-	-	-
Air dryer (144")	0.0024	0.035	16	232	250	-	-	-	-
Draw roll	-	0.005	-	33	30	-	-	-	-
Coater	0.008	0.012	53	79	75	-	-	-	-
Dryer section[1]	0.0162	0.0252	107	166	150	0.0162	0.003	0.006	0.0252
Draw roll[2]	-	0.003/PLI	-	60	60	-	-	-	-
Reel	0.008	0.0138	52.8	91	100	-	-	-	-
Reel[3]	0.019	126	125	-	-	-	-	-	-

$$\text{Tension HP/PLI} = \frac{(1)\,(220)\,(3000)}{33,000} = 20 \text{ HP/PLI}$$

[1] six doctors (6-60")

[2] one doctor, must support 3 PLI

[3] with doctor and joint

Table 6.7. Comparison of drive features

	1 Phase SCR DC	3 Phase SCR DC	DC Servo	Brushless Servo	Vector AC
MOTORS	DC motor XP motor ok	DC motor XP motor ok	DC motor XP motor ok DC servo motor XP motor or purge	Brushless Servo motor Purge motor Area	AC induction Servo motor XP motor ok or Purge motor area
MICRO CONTROLLER	- Least complex - Needs lowest level of diagnostics	- Mid-range controller complexity - More diagnostics available	- Mid-range controller complexity - More diagnostics available	- Higher complexity - More diagnostics	- Highest complexity - Highest level diagnostics
HORSEPOWER	1/4-7.5 HP	3-600 HP	1/4-30 HP	1/4-16 HP	1-50 HP
APPLICATIONS	- Use in low HP general applications - Medium performance level	- Use in mid-high HP general applications - Medium performance level	- Use in low-mid range general & high perform-applications when response or smoothness is critical - High performance level	- Same use as DC servo & high speed position applications - Highest performance level	- Same as brushless in higher HPs - Very high performance
SPEED RANGE	20:1, Cogging at low end	30:1, cogging at low end possible	1000:1	2000:1	1000:1
GEAR REDUCTION	Necessary	Necessary needed if motor torque sufficient	Little or none needed if motor torque sufficient	Little or none needed if motor torque sufficient	Little or none
REGENERATIVE	To AC line	To AC line	To snubber or DC bus	To snubber	To snubber or DC bus
POWER FACTOR	- Low 0.68 avg. - Higher at higher speeds	- Low 0.68 avg. - Higher at higher speeds	0.95 constant	0.95 constant	0.95 constant
EFFICIENCY	- 72-78% at full power output - Efficiency decreases with dec. speed	- 72-78% at full power output - Efficiency decreases with decreased speed	- Up to 80% depending upon bus regeneration - Fairly constant over speed range	- Highest efficiency up to 83% - Constant over speed range	- Up to 80% efficient depending upon motor
SPEED FEEDBACK	Resolver, pulse tach or analog tach	Resolver, pulse tach or analog tach	Resolver, pulse tach or analog tach	Pulse encoder	Pulse encoder
PRICE COMPARISON (Non-XP Motor)	Least expensive	- 3 phase SCR DC & DC Servo roughly compare in price - Mid to upper range expensive - Application economies available in DC servo system	- 3 phase SCR DC & DC Servo roughly compare in price - Mid to upper range expensive - Application economies available in DC servo system	- More expensive than 1 phase SCR - Less than DC servo & 3 phase SCR	Most expensive

Bibliography

Resources

1. Bentley, J.M., "Design and Application of the Mechanical Drive Train," 1986, 1991 Paper Machine Drives Seminar Notes, TAPPI PRESS, Atlanta.

2. Borden, D.L., "Thermal Considerations in Geared Drives," 1975 Engineering Conference Proceedings, TAPPI PRESS, Atlanta.

3. Brewer, C.L., "The Paper Machinery Builders View of Direct Tension Regulation Systems," IEEE, May 1966, Niagara Falls, Ontario, Canada.

4. Davies, I., "Paper Machine Dry End Operations," 1986, 1991 Paper Machine Drives Seminar, TAPPI PRESS, Atlanta.

5. Feirtag, B.A., "Selection Report of Bow for Curved Axis Spreader Rolls," Oklahoma State University.

6. Feirtag, B.A., "Summary Report on the Use of Curved Axis rolls for Wrinkle Prevention on Paper and Plastic," Oklahoma State University.

7. Feirtag, B.A., "Selecting a Web Guiding System," Oklahoma State University.

8. Grenfell, K.P., "Tension Control on Paper Machines and Converting Equipment," 1963 IEEE Engineering Conference, June, Boston.

9. Jamison, R.A., "Tension Control for Film Winding," Reliance Electric.

10. Kheboian, G.I., "Off Machine Coater Drive Requirements," 1971 Engineering Conference Proceedings, TAPPI PRESS, Atlanta.

11. Olshansky, A., "Winder Applications," 1986, 1991 Paper Machine Drives Seminar, TAPPI PRESS, Atlanta.

12. Shelton, J.J., "Lateral Dynamics of a Moving Web," Doctoral Thesis Submitted to Oklahoma State University, July 1968.

13. Timmerman, D.N., "Gear Drives—Selection and Application," 1975 Engineering Conference Proceedings, TAPPI PRESS, Atlanta.

14. Trueb, T. and Derrick, R.P., "Application of Drive Systems," 1986, 1991 Paper Machine Drives Seminar, TAPPI PRESS, Atlanta.

15. Veres, R.P., "Web Tension Control for Continuous Processing Lines," Reliance Electric.

16. Veres, R.P., "Winder Drive Applications," 1986, 1991 Paper Machine Drives Seminar, TAPPI PRESS, Atlanta.

17. Basic reference information pertaining to ratings, practices, standards, design, testing and manufacture may be obtained from: American Gear Manufacturers Associations (AGMA), One Thomas Circle, N.W. Washington, D.C.

18. 1992 Paper Machine Drives Short Course Notes, TAPPI PRESS, Atlanta.

TAPPI Technical Information Sheets (TIS), TAPPI PRESS, Atlanta:

0406-12: "Recommended Electrical Wiring Practice for Auto Control and Regulating Systems."

0146-13: "Winder Power Requirements."

0406-14: "Regulating SystemsDefinition of Terms."

0406-15: "Environmental Factors Effecting the Performance of Regulating Systems."

0206-01: "Drive Requirements for Laminators."

The following TI Sheets are being discontinued but can be used for reference until an active new TIS is issued. For more information, call the TAPPI Library at 404-446-1400.

0406-05: "Power Requirements for Fourdrinier Machines."

0406-06: "Power Requirements for Tissue Machines."

Chapter 7, Section I

Brush Polishing of
Mineral Coated Surfaces

Jack A. Perry, Q-Jet Systems Inc.

Introduction

Brushes have been associated with the production of pigmented coated paper and board substrates since their beginnings at Champion Papers near the end of the 19th century. Over the decades, the reasons for brush polishing have changed from smoothing wet coating during application to polishing dried coatings to obtain gloss and smoothness without base sheet densification. The following sections define the process of brush polishing and its operating variables. Like all process systems, "brushing" has both positive and negative application parameters. As basis weights continue to decrease, and recycled fiber furnish contents increase, brushing may play an increasing role in coated surface finishing.

Brush Polishing Equipment

Figure 7.1 illustrates two established brush polisher configurations for printable coated paper and paperboard operations. Of the two, the backing drum assembly is most frequently applied to paper grades, while the single brush and backing roll combination is favored for paperboard weights. Web support is necessary in all instances for paper grades. Multi-nip single brush and backing roll installations are also practical and may provide a valuable finishing component depending upon the coated product desired.

The majority of installations, regardless of basis weight, are multi-brush units consisting of 3-6 brushes, depending on the results desired. Both the backing drum and brushes are driven. Brushes are composed from a variety of materials ranging from synthetic fibers to natural horsehair. Bristle stiffness is varied between the brushes with the stiffest brush being the first to contact the incoming coated surface, and the softest brush the last. The speed of each brush is varied according to its stiffness, with the stiffest brush running at the slowest speed, and

the softest at the highest speed. Each brush is individually driven through a directly coupled gear case and electric motor with speed control. Dust extraction equipment is also part of any brushing operation. This may include the final brush on a multi-brush unit acting solely as a duster.

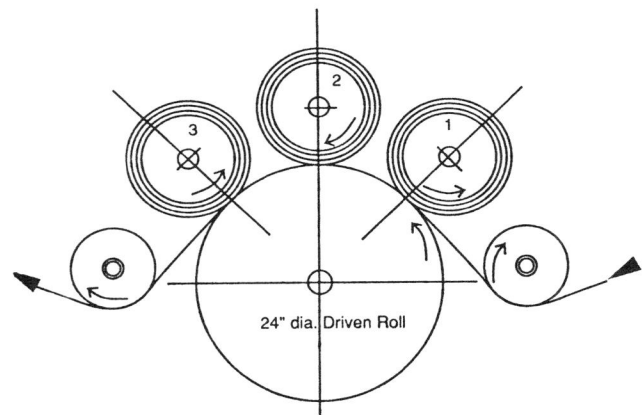

Fig. 7.1. a. Brush polishing machine

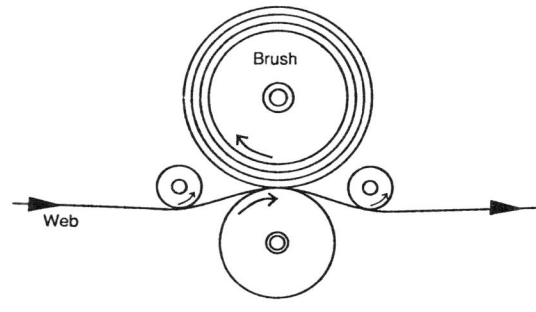

Fig. 7.1. b. Brush polishing machine

Table 7.1 summarizes the operating variables of brush polishers as a group whether single or multi-nip units.

Table 7.1. Brush polisher operating variables

- Number of brush nips
- Web speed
- Brush tip speed
- Brush stiffness
- Brush pressure
- Brushing nip length
- Sheet moisture content
- Sheet surface temperature
- Coating formulation
- Base sheet formation
- Base sheet smoothness

Coated paper brushing applications have been cyclical. They started out on paper applications; gravitated to paperboard grades; then expanded into such specialty grades as paper printing plates. During the last three decades, brushing in combination with both soft-nip calendering (SNC) and supercalendering has been well documented. In some instances, such processes have been patented. Regardless of its negative characteristics, such as (a) the addition of another operation and (b) brush wear, brushing provides an action which other surface finishing methods do not. Table 7.2 outlines the process results, in terms of coated substrate properties, obtained through brush polishing.

Printable coated paper and paperboard substrates finished solely and/or partially through brush polishing operations, exhibit the following characteristics: (1) reduced ink and varnish consumption, (2) quicker ink set times, and (3) improved bending and folding performance.

These results should be expected, given the gloss and smoothness provided without sheet densification. Surface smoothness alone should predicate better ink mileage which dictates quicker set times as well as lower varnish consumption. Reduced densification of folding boxboard grades generally leads to improved bending at the score characteristics. This ultimately has an impact on both top and underliner fiber furnish formulations in multi-ply grades of boxboard.

Table 7.2. Brush polishing results

- Surface gloss
- Surface smoothness
- No bulk reduction
- Improved conversion characteristics

How are brush polishable coatings formulated and applied? Or, "Is it all in the coating?" The real answer is a combination of all the variables interacting to make a better product. Table 7.3 lists the components of a coating applied to produce a 60 lb/3300 ft^2 coated one side (C1S) label paper and the results achieved by brush polishing.

The data presented in Table 7.3 are from a 60-lb C1S label paper. The brushing operation consisted of six brushes acting on the paper running at 1000 ft/min. The brushes were run at 5000-ft/min surface speed. Gloss increased from 18 to 23 points, and smoothness improved by 23-30 points.

Summary

Brushing polishing has a place in finishing coated paper and paperboard. It adds another step and another maintenance expense. In some cases, it could be the best

Table 7.3 Air doctor coated and brush polished label paper

Coated Paper Tests	Coating Formula	Brushed Paper Tests
Basis Weight 60 lb/3300 ft^2	Dispersant 0.35	Caliper 3.85
Caliper 3.9	Pigment System[1] 100.00	Brightness 80.0
Brightness 80.0	Binder System[2] 16.00	Gloss 48-52
Gloss 27-29	Lubricant[3] 0.40	Smoothness 35-39
Smoothness 58-69	Coloring 0.15	
	Percent Solids 50%	

(1) Number 1 coating clay standard brightness.
(2) An 8°C Tg Vinyl Acrylic fine particle size latex.
(3) An emulsion consisting of 10% Carnauba Wax, 3% anionic dispersant, 87% water.

finishing method for the grade, particularly on secondary fiber-containing products. Some indications are that brushing is best applied to boxboard and specialty paper production; however, it may also add a significant contribution to the production of lightweight coated publication papers in conjunction with soft nip calendering.

Chapter 7, Section II

Steel-to-steel Calendering

Ronald L. Fox, Converters Paperboard Co.

The calendering process is a mechanical method used to transform the properties of paper and paperboard from one characteristic state to another. Calendering is accomplished by passing a web between all or part of a complement of rolls stacked vertically, with loads applied on each journal. This term applies to the use of cast-iron rolls, with chilled, hardened surfaces or inert, nonresilient materials. Calendering reduces the caliper of the sheet and increases the density of the sheet by compaction occurring in the nip. Increases in gloss and smoothness or other property changes are the result of the paper or paperboard physical characteristics, both internal and external, particularly for coated sheets.

Calendering a sheet in-line with the paper or paperboard making process, or after coating in-line to the paper or paperboard making process is referred to as machine calendering. Machine calendering coating basestock improves holdout of the wet coating film.

To give higher finish and greater smoothness to machine calendered sheets, shallow boxes filled with water at a constant level are placed at the uptake side of a nip to transfer to the sheet at the pressure zone. These are called calender boxes. With coated paperboard the term wet finished is used when this is done to a coated side, and to further improve this effect a solution of zinc sulfate is also applied. In addition to this, other calender sizes applied in this fashion are emulsified wax, solutions of starch, alginate, polyvinyl alcohol (PVA), and other synthetic products. These are used to improve properties such as sizing, smoothness, porosity, fiber pick, and strength.

Chapter 7, Section III

Soft-nip Calendering of Coated Paper and Paperboard

Jack A. Perry, Q-Jet Systems Inc.

Introduction

Today we call it "Soft-nip Calendering" (SNC). In the '60s and '70s, it was known as gloss-calendering and sometimes as thermo planishing. Without question, it is a spinoff from older process disciplines: cast coating, thermoplastic molding; and it is a supercalender replacement.

Soft-nip calendering is a method of finishing coated paper or paperboard in line with the coating and drying processes. In sequence, the web is coated; dried to optimum moisture; and finished by hot thermal calendering in a one- or two-nip thermofinisher historically called a "gloss calender." Various degrees of gloss may be obtained by employing this process without materially densifying the web. Ink holdout and ink set time may be controlled by adjusting both the coating formulation and the thermal finishing intensity.

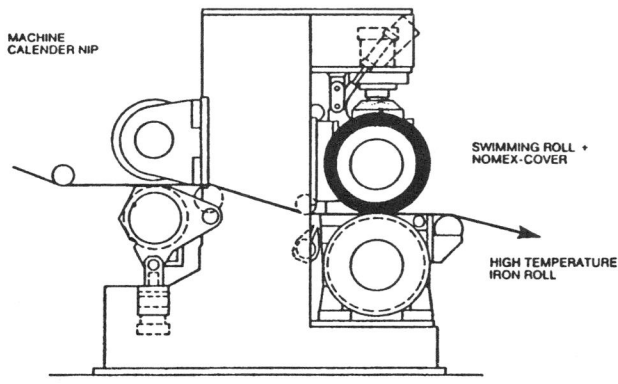

Fig. 7.2. Single-side soft-nip calender

Figure 7.2 illustrates a single-side SNC for paper machine coated substrates. Such units are normally installed on the paper machine following coating and drying prior

to the reel. Increasing levels of gloss and smoothness are obtained without materially densifying the web.

An SNC in its simplest form consists of a smooth, highly polished, hot roll or molding drum and a resilient pressure roll which form a nip loaded by a suitable system. The hot roll may be chrome plated or just ground to a very smooth surface finish and is heated by steam, oil, or other controllable means to temperatures ranging from 200 to 600°F. The pressure roll may be either a plastic- or rubber-covered internally cooled roll, or a conventional filled roll. Pressure roll surface hardness ranges from 5 to 30 P&J. The pneumatic loading system provides operation nip pressures from 500 to 4000 pounds per lineal inch. The web is processed in such a way that it is pulled out of the finishing nip directly away from the hot roll surface so there is less opportunity for the coating to adhere to the molding surface.

To compare the SNC with its calendering competitors, Vreeland, Jewett, and Ellis developed the graph in Fig. 7.3.

Figure 7.4 illustrates the results obtained from one nip of a machine calender versus one nip of an SNC. This clearly defines one of the major assets of soft-nip calendering: Sheet density is not materially increased. Also, it ensures that appreciable percentages of valuable sheet strength properties remain.

Regardless of its basis weight and composition, each coated web has a basic formation characteristic which is definable as density variation, manifesting itself as high and low areas across the web. A conventional steel-to-steel nip machine calender affects only the high or denser areas and leaves the low, less dense portions untouched. Subsequent printing of the coated surface results in mottled ink vehicle absorption. Conversely, finishing in the resilient nip of an SNC results in finishing of all areas of the coated surface. Such papers print with cleaner halftone fidelity and uniformity of solids.

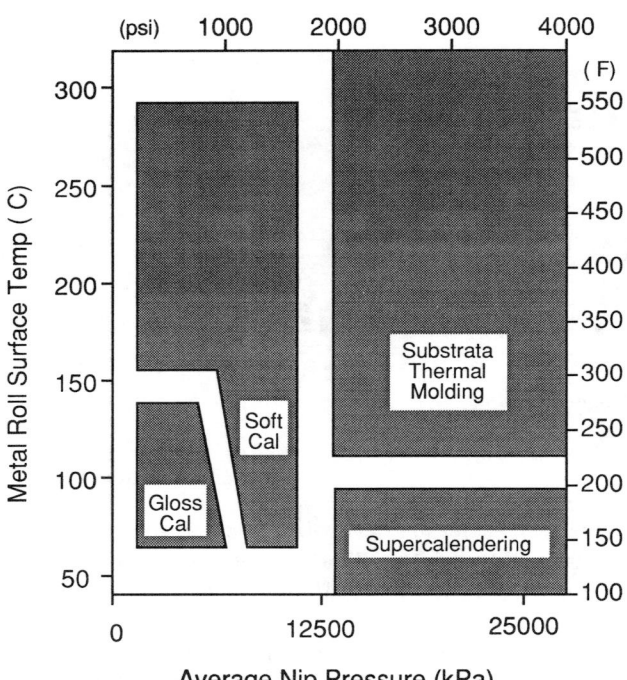

Fig. 7.3 Regions of resilient roll calendering

Caution! The previous is by no means intended to imply that the SNC is the answer to the papermaker's formation variation challenges. On the contrary, it again emphasizes the important contribution of fine papermaking to ultimate coated paper quality.

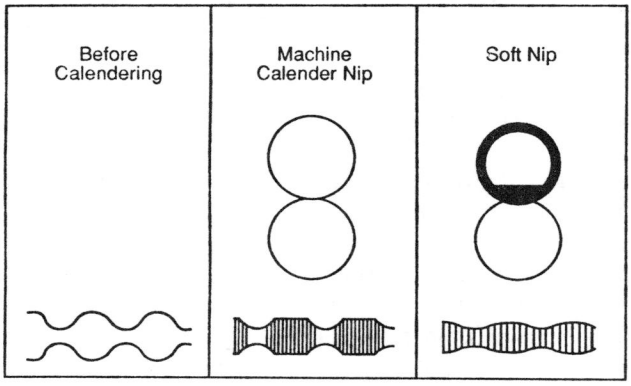

Fig. 7.4. Comparison of cross-direction (CD) caliper of web after being run through a machine-calender nip and a soft-nip calender

The construction of an SNC is of great importance to its successful operation. Of particular importance are the cover and core of the pressure roll as illustrated in Fig. 7.5, which compares a cotton- or wool-filled supercalender roll to a resilient covered pressure roll of an SNC.

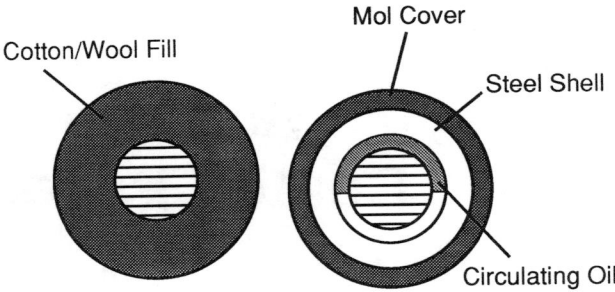

Fig. 7.5. Cross sections of a conventional cotton roll and a MOL covered roll

As each of the rolls in Fig. 7.5 is stressed during operation (calendering), both are deformed as illustrated by Fig. 7.6. The work done generates heat inside the resilient cover at a point about 3-4 mm below the outer surface. Here, supercalender filled rolls degrade, and SNC pressure roll covering compounds are differentiated. The molding roll of the SNC is pumping Btus through the sheet into the resilient cover. This heat must be removed to prevent degradation of the cover. The chemical (polymer) composition of the resilient roll is consequently very important and one of the most proprietary components of SNC technology. Employing a heat exchange liquid, such as oil, to extract Btus from the resilient roll shell was adopted to contribute to longer pressure roll cover life.

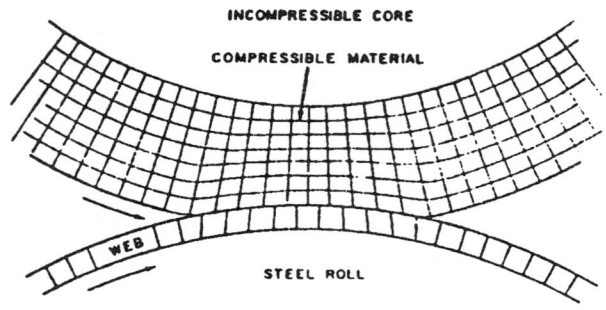

Fig. 7.6. Rolling friction action in resilient rolls of calenders

Figure 7.3 provided a summary of the overall temperatures and pressures employed in the SNC process. Figure 7.7 defines the possible temperature gradients between the heated smooth roll and the oil inside the shell of the resilient covered pressure roll.

TEMPERATURE ALONG AB

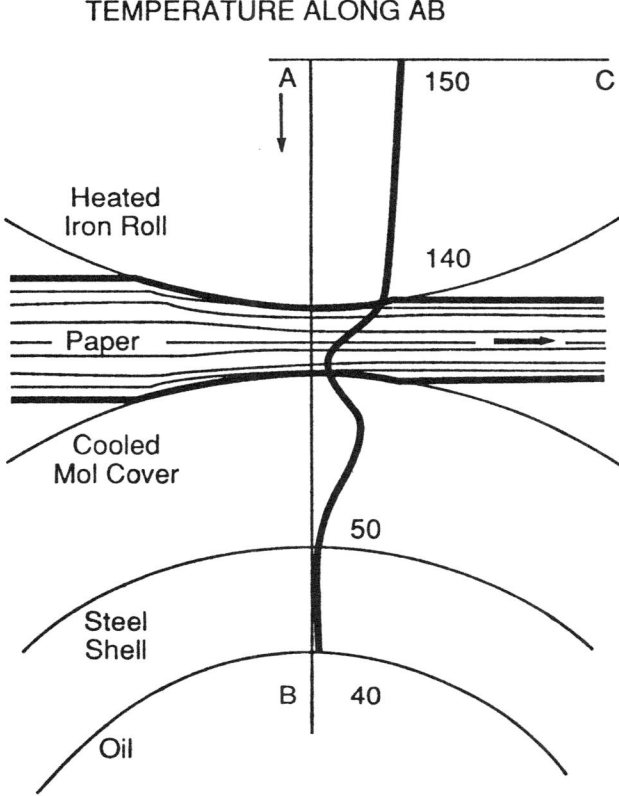

Fig. 7.7. The insulating effect of paper in a calender nip

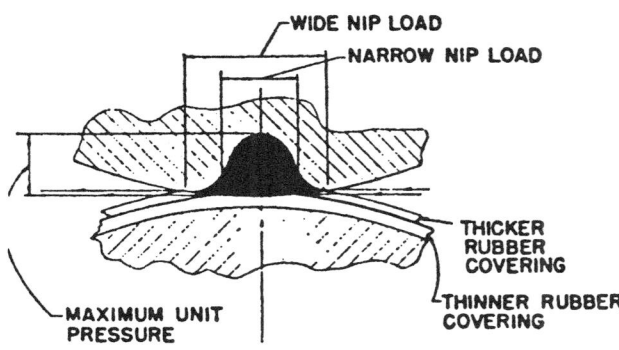

Fig. 7.8. Schematic nip loading chart

A visualization of the effect of nip loading pressure is given in Fig. 7.8. This illustrates the forces at work on the resilient roll cover as the SNC calender operates.

The moisture content of the coated paper entering the SNC nip, may be more or less critical depending upon the type of coating materials used. The moisture content should be controlled by regulation of the drying process which follows coating application, rather than by an additional moisturizing treatment. The coating formulation may be of any composition which will provide the end product desired. Glossy finished surfaces are generally obtained by using high percentages of platy pigments, such as delaminated clay, and thermoplastic synthetic polymers in the coating formula. The total coating formulation is important when gloss is desired.

The relationship between paper gloss development, hot roll temperature, and sheet moisture were illustrated by Vreeland et al. as reproduced in Fig. 7.9. These curves for uncoated bleached hardwood kraft papers are important to any application of SNC technology, more particularly to thermo molding as it relates to plastics technology.

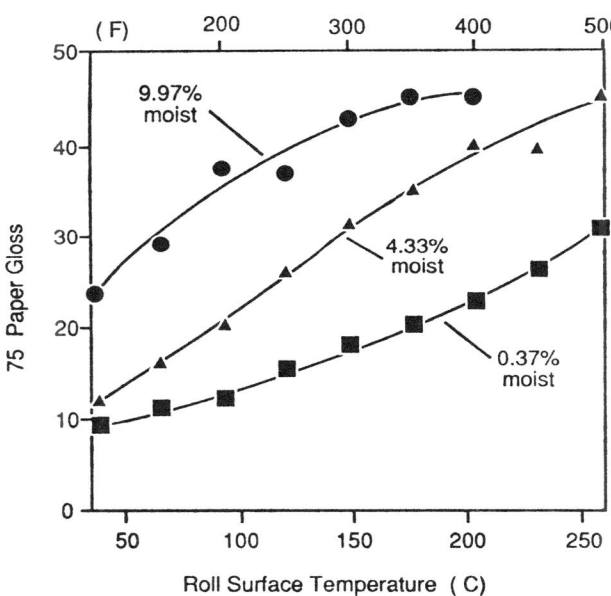

Fig. 7.9. Gloss as a function of roll surface temperature and sheet moisture for an uncoated bleached hardwood kraft (BHWK) sheet

The surface smoothness and temperature of the molding roll are just as important to the operation of the SNC as the resilient roll cover composition, sheet moisture, and coating formulation. Figure 7.10 from Vreeland et al. shows the dynamic effect which varying the surface temperature of the molding roll has upon the resulting paper gloss. Conversely, the side against the resilient roll essentially exhibits no gloss gain.

In order to obtain satisfactory action, the coating must adhere to the hot roll during molding. It is essential that the coating film cleanly release from the molding surface. Simple coating mixtures containing only pigment and binder do not always release satisfactorily. Other techniques are required to obtain release properties. One successful method is continuous treatment of the hot roll to give it release properties. This is accomplished by the addition of oleaginous materials to the coating mixture.

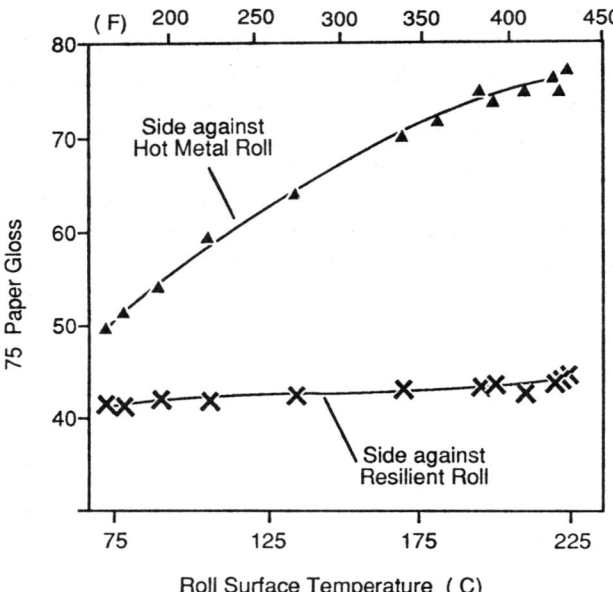

Fig. 7.10. Independence of gloss development between the two sides of the sheet when resilient roll hot calendering

The molding surface may be preconditioned prior to startup by rubbing it with oleaginous material. Thereafter, the roll is kept in suitable condition as the oleaginous material is desorbed from the coating onto the molding surface, which maintains a film to affect release properties. Among the suitable oleaginous materials are vegetable oils, fats, fatty acids, partially saponified oils or fats, fatty acid amines, and dimers of aliphatic ketenes. Sulfonated oils and organophosphates also provide satisfactory release. Table 7.4 lists some common chemicals and compounds which may be used as release agents.

Table 7.4. Gloss calender release agents

- Calcium and Ammonium Stearate and Oleate Dispersions
- Microcrystalline Wax Emulsions
- Polyethylene Glycol Dispersions
- Paraffin Wax Emulsions
- Polyvinyl Stearate Emulsions
- Carnauba Wax Emulsions
- Aquapel Emulsions
- Corn Oil Emulsions
- Sulphonated Castor Oil
- Emulsified Long Chain Fatty Acids such as Stearic and Oleaic

Soft-nip calendering has a number of advantages over both machine and supercalendering. The more important advantages are: (1) installation can occur both on and off the paper machine; (2) less capital investment; (3) less operational cost than a supercalender; (4) fewer operating problems; (5) less bulk reduction; (6) performs on both coated and uncoated papers; (7) higher strength retention, brightness, and opacity; (8) equal gloss and smoothness; and (9) better subsequent web converting characteristics.

At first blush, everything seems simple and easy. While the process is well proven, it represents a major commitment on the part of any manufacturer who puts everything together in one continuous process system. It is no substitute for inadequate coating basestock formation, uneven or irregular coating applications, or insufficient process control. The following operations checklist provides some guidelines for SNC operations.

SNC Operations Checklist

The following variables may be manipulated to obtain better finish in terms of smoothness, gloss, and runnability as to freedom from sticking or picking on the SNC.

1. Binder—All synthetic latex binder systems enable higher gloss and smoothness development for the following reasons:

   A. Polymer hardness affects the operating conditions of the SNC. The addition of crosslinking agents also has a substantial influence on SNC results.

   B. The binder system must be thermoplastic to be moldable

   C. Incorporation of some natural binder accomplishes hot roll release, but detracts from the resulting gloss and smoothness. It is nowhere near as effective as the use of organic release agents.

2. Pigment Type—Easy-to-gloss pigments are best for this process. Anything that doesn't finish well under moderate pressure is a detriment. Platy pigments are best.

3. Operating Variables of the coating and finishing line which affect product quality are:

   A. Base sheet and coating layer moisture content

   B. Base sheet sizing

   C. Coat weight applied

   D. Web smoothness prior to SNC calendering

   E. Raw stock formation and smoothness prior to coating

   F. The type of coating defined by 1. and 2. above

   G. The mechanical operating variables of an SNC which are:

1. Hot roll temperature
2. Hot roll finish. This must be at least 1-3 micro inch for highest gloss and smoothness.
3. Pressure roll durometer
4. Nip pressure
5. Nip length
6. Web speed
7. Coated layer moisture and temperature prior to entering the finishing nip
8. Drive mechanics of the SNC
9. Number of nips used

4. A most important factor which influences the SNC's finish of a coating is the introduction of release agents into the coating formula at levels ranging from 0.1% to 2.0% based on the dry pigment.

5. Every finishing operation has operating requirements which dictate finished coated surface characteristics. It is necessary to adjust the process by trial and error to achieve optimum results. This adjustment period begins during initial process startup, and includes all factors which may be variable.

Plastic Molding Under Pressure With Heat

Polymer chemists classify the polymer systems employed to produce molded objects by Glass Transition Temperature, commonly referred to as Tg. A simple definition of such a characterization is "the temperature at which a polymer begins to flow." Our interest as SNC finishing engineers is to cause the outer surfaces of the coated paper to flow under pressure with temperature. Not only is it necessary for the coating to move, but it must STOP moving quickly upon exiting the SNC nip. Consequently, coatings designed to be SNC finished include much higher synthetic binder contents. When double coating is employed to produce such products, the top coating is frequently bound with an all-latex system selected not only for its binding strength but for its Tg!

Another coating material which contributes to the successful operation of SNCs is plastic pigment. Here, both particle size and Tg make a significant contribution to successful SNC performance in addition to providing good color and opacity. Coupled with platy high brightness clays, a top coating of 8-10 pounds per ream may be finished at coater operating speeds ranging from 1500 to 3200 ft/min for 75° gloss values suitable for number 1 quality. The cost of coating material is insignificant compared to the overall operational cost reduction obtained from a smooth performing coating station followed by an SNC section prior to the reel.

Tables 7.5 and 7.6 define an actual SNC-finished run of a coated one side (C1S) packaging paper grade which was successfully printed by four-color sheet-fed offset, varnished and converted into a bag-type product. It should provide a reasonable starting place for further development.

Table 7.5. Coating formula and operational data

Organic Dispersant	0.4590
Brightness Platy Clay	92.0
Fine Plastic Pigment	8.0
34°C Tg co-polymer latex	18.0
Finishing Aid/Plasticizer	0.35
Coating Solids	62.5
Coat Weight	10 lbs/3300 ft²
Coater	Blade at 1800 fpm
Finished Basis Weight	60 lbs/3300 ft²
Pre-SNC Moisture Content	6.5%
SNC Nip Pressure	1200 PLI
SNC Molding Roll Temperature	275°F
SNC Nip Length	5/8 inches

Table 7.6. Finished coated paper characteristics

Brightness	80 - 81
75° Gloss	82 - 85
Ink Gloss	92 - 92
P.P.S. Roughness	1.0 - 1.1

The relationship between SNC speed at operating pressure and temperature which produces a salable product is dependent on all the components of the process from papermaking through soft-nip calendering. Figure 7.11 provides a graphic example of the results obtained on an 80 lb/3300 ft² sheet of coated flour bag stock. This example relates SNC nip pressure and smoothness, printed gloss, and finished gloss. Observing the data establishes that maximum gloss and smoothness were obtained between 800 and 1000 pounds per lineal inch of nip pressure, all other variables being constant. In the real world, it is necessary to determine what operating variable levels will provide the desired end product values of gloss, smoothness, etc.

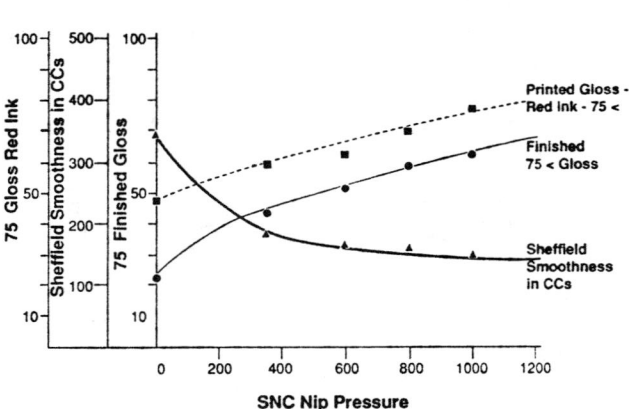

Fig. 7.11. Soft-nip calendering of C1S bag stock

Figure 7.12 describes two generic 80-lb/3300 ft^2 flour bag stocks, one of which was finished by supercalendering and the other by soft-nip calendering. The same equipment was used for both products. The raw stocks used for each were slightly different. Regardless of the number of supercalender nips, or SNC nip pressure, respectable gloss values were obtained. Bulk reduction as indicated by caliper change clearly demonstrates the advantage of SNC finishing in this area.

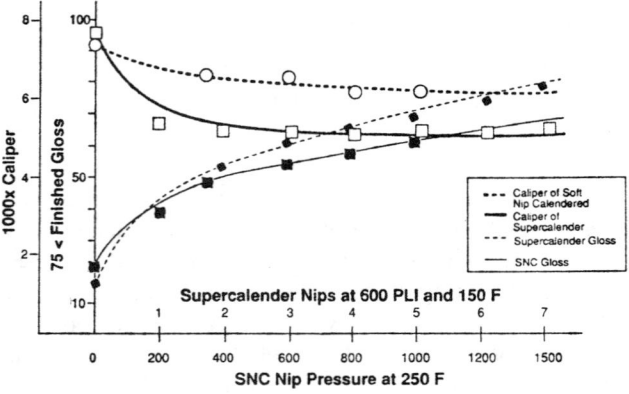

Fig. 7.12. Comparison of gloss development vs. caliper reduction by soft-nip calendering and supercalendering

The paper produced for Fig. 7.12 was made using a 36-in. wide, 80-lb/3300 ft^2 surface-sized basestock on a pilot plant air-knife coater at 800 ft/min to simulate mill conditions. The coating formulation used differed from that defined in Table 7.5 by pigment composition only. In this instance, no plastic pigment or 90 brightness clay was used.

Summary

Soft-nip calendering of coated substrates in line with coating and drying, either on the paper machine or as part of an off-machine conversion process, yields a process improvement alternative to conventional calendering and supercalendering operations. Equipment costs and operational expenses are favorable to SNC installations. Instrumentation and automation devises and techniques are available which enable control of the SNC process during high-speed operations. In recent times, the application of SNCs to in-line coated paper finishing has achieved considerable acceptance, particularly in Europe and more recently, in North America.

The expected expansion of secondary fiber usage, and high alkaline pigment base sheet loading in North America provides an excellent opportunity for the expansion of SNC applications and their attendant gentler action on the paper substrate. Additionally, as the basis weight of coated two sides (C2S) magazine stock continues to be driven downward by increasing postage fees, the SNC will continue to expand its market share because of its ability to provide finished gloss and surface smoothness without significant sheet strength degradation.

Chapter 7, Section IV

Supercalender

Raymond O. Wiener, Simons Eastern Consultants Inc.

The supercalender is an off-machine method of imparting greater gloss and smoothness to the coated sheet surface than is achievable with on-machine, all steel roll calenders. It can either be a single or a double finishing unit with as few as five and as many as sixteen rolls in the stack. The majority of the modern supercalenders used for coated paper are nine to twelve rolls. Specialty coated grades may use fewer rolls.

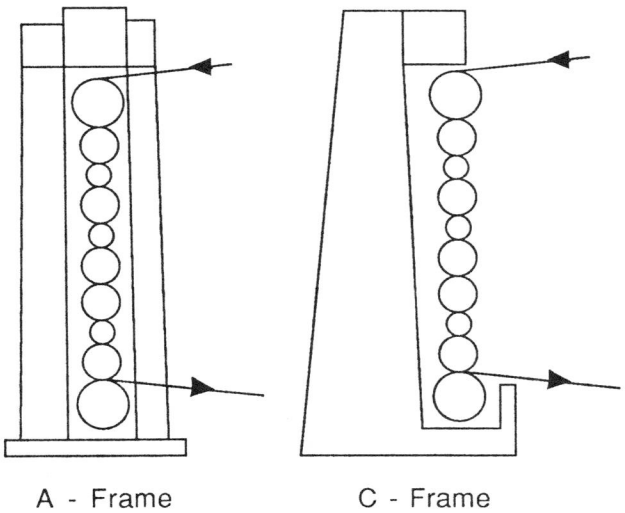

A - Frame   C - Frame

Fig. 7.13 Supercalenders

For paper to be calendered on both sides in one operation, a reversing nip is used. The nip which is formed by two filled rolls in contact with each other is located in a position that results in equal calendering to both sides of the sheet.

The speed range of the supercalender is about 1000-3000 ft/min. Therefore, it is very common to have two or three support the production of one paper machine or off-machine coater.

The two most prevalent styles of supercalenders in use are the A-Frame and C-Frame arrangements. The A-Frame consists of two massive cast-iron or fabricated steel columns on each side of the rolls and joined at the top to support the pressure cylinders. In the slot between the columns, the bearings of all the vertical rolls fit. All the rolls except the "king" roll can move up and down. Roll removal is from the end.

The C-frame or open frame arrangement has one massive, upright frame on each side of the stack. The bearing housings of all rolls except the king roll fit on the front vertical faces of the frames by means of grooves. The rolls are held in place with suspension arms that are swiveled or secured by pins that allow the rolls to be raised or lowered. This type of construction results in easier front-side roll removal and better visibility of the draws and nips.

The basic supercalender consists of a top "queen" roll, alternate filled and metal intermediate rolls, and a bottom king roll. Hydraulic pressure is usually applied to the top queen roll to give operating nip pressures generally in the 1000-2000 pounds per lineal inch range for coated paper.

Some of the newer supercalenders apply hydraulic pressure to the bottom king roll. The advantage of this arrangement is that nip relief can take place in less than one second because the dead weight of the rolls is sufficient to allow quick separation by dropping. The top roll hydraulic application requires longer time for nip relief because the rolls have to be raised.

The drive roll most often is the bottom king roll or the bottom intermediate metal roll.

The generally accepted reason for gloss development has been that with the application of pressure, the metal roll causes a depression in the resilient filled roll. When the rolls are rotating, the filled roll material will start to creep in attempting to return to its normal shape. The creeping motion results in a difference in surface speed between the metal roll surface and filled roll surface.

This speed difference causes a friction which produces a polishing action thus producing a gloss. This phenomenon has been called the "Law of Rolling Friction" or the burnishing effect (1, 2).

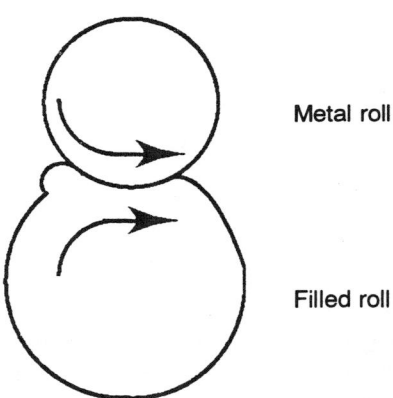

Fig. 7.14 Supercalender nip action

The degree of gloss developed is dependent on speed, number of nips, roll diameters, nip pressure, temperature, moisture, and coating formulation.

Another school of thought for the reason for gloss development is replication of the smooth metal roll's finish onto the paper surface. This phenomenon has been identified by stopping the supercalender with the paper web nipped, waiting a few minutes and then jogging it forward. The surface of the paper that was held in the nip is smoother and glossier than the adjoining surface.

The king and queen rolls are generally of the same construction. The earlier solid chilled iron rolls are being replaced with hydraulically loaded zone control-type variable crown rolls. This roll has internal hydraulic cylinders spaced across the roll face to vary the crown of the roll. The purpose of this action is to compensate for roll deflections due to temperature, roll weight and length, and overhanging loads. This type of roll also will aid in leveling out caliper variations.

The filled intermediate roll construction is generally steel shaft and ends with a compressed natural fiber cover. Cotton is the predominant fiber used. Smaller amounts of other natural fibers, such as wool, are sometimes added to the cotton to vary the hardness, elasticity, heat resistance, durability, and sheet marking properties. Synthetic fibers are being investigated as a filler, but so far have not replaced the natural fibers to any great extent.

Speed, pressure, and temperature all affect the filled roll performance and durability. Each of the various fillers has certain advantages and disadvantages. The type of filler used is usually dependent on the quality of the

coated sheet to be calendered and the properties of the calendered sheet desired.

The intermediate metal rolls are generally chilled iron construction with internal provisions for heating or cooling. The older installations utilized rolls with bored shafts. These are being replaced with rolls that have the capability to internally heat or cool the outer shell directly. Heating is used to initially warm up the stack rolls to proper operating temperature on startup. Cooling is used to maintain operating temperature by removing the heat added by the friction generated in the nips.

Because the supercalender is an off-machine unit, it requires its own unwind and wind stands. Generally the wind stands and unwind stands are at different elevations due to the overall height of the stack. The unwind is usually at the lower elevation with the sheet being fed to the bottom king roll nip.

The modern supercalender sometimes uses the automatic splicing arrangement so that the calender does not have to be stopped to start a new roll of coated paper. This provides the advantages of less downtime and offgrade product. The use of the fast nip relieving characteristics of the king roll-applied hydraulic pressure lends itself well to the continuous system.

The coated sheet increases in width and decreases in caliper when it travels through the nips. To prevent wrinkle formation, rolls are installed that spread the sheet before it enters each nip. These rolls have been called fly rolls, blow rolls, or guide rolls. In addition to spreading, these rolls prevent what is known as calender cuts. By taking the sheet away from the nip so that wrapping of the filled or metal roll does not occur, air does not become trapped between the paper and the roll and cause the calender cuts.

Another function of these rolls is to aid in reducing the intensity of depth of sheet marks due to marks on the filled roll. They do this by not allowing the mark to form twice in the same place on the sheet surface.

Depressions or marks on the filled roll are often caused by wads of paper passing through the nips during a paper web break. The only satisfactory way to remove these is to remove the damaged filled roll and grind it or turn it down to remove the impression. The rapid opening (less than one second) feature made possible by the hydraulically operated king roll is an excellent preventive measure for the avoidance of marked rolls.

At the nips where the major reduction in calender and greatest increase in width occurs, spreader rolls are very often installed. These rolls are used in conjunction with the fly rolls.

Steam showers are often installed, especially at the top one or two nips of the stack. It is thought that the steam softens the coating so that it is more receptive to the development of gloss. The coating formulation itself has a bearing on how much steam is required to develop the degree of gloss desired. Therefore, it is not unusual for a mill to have different coating formulations for a supercalendered sheet and a nonsupercalendered sheet.

Other accessories which include spare roll storage rack, overhead crane for intermediate roll changes, sheet break detectors and web cutoff device, movable platforms, nip guards, oscillating doctors, and on-machine temperature control make up the supercalender system. They are intended to improve the operation, safety, serviceability, maintenance, and control functions of the modern supercalender.

Bibliography

Literature Cited

1. Norman, E.B., *Pulp and Paper Science and Technology*, McGraw-Hill Brook Company, 1962.

2. Schiller, F.E., *Manual of Supercalender Operations*, Miller Freeman Publications, 1976.

Chapter 8

Statistical Process Control for the Coated Paper Process

David J. Damato, RUST International Corp.

Introduction

What is quality? This is a question operations and design people ask themselves on a daily basis. Several quality managers will respond, "Conformance to specifications." Operations managers will respond, "Quality is variation management." Financial managers might respond, "Quality is what it takes to stay in business." Whatever your perspective is, quality can be summed up as the level of reliability necessary to satisfy the customer's expectations at the minimum costs to the supplier. Remember, we are all suppliers to someone.

Some basic fundamentals of managing quality are discussed in quite a large quantity of books and manuals. Most all authors will agree that any quality program requires a sincere commitment of the company's top management in order for the program to work.

The function or purpose of a statistical quality control (SQC) program is to provide a timely commentary on the performance of the paper machine coater system or process. This means that continual feedback of the coated processes operation is imperative at all times during operation. SQC also provides a common language between design, technical, production, and quality personnel to discuss product conformance issues. Statistical process control (SPC) also provides a means to correlate product defect data to the coated process. This system will tell us where the problem is and what is not performing as intended.

An Introduction to SQC Principles and Practices

W. Edwards Deming discussed a list of 14 points for top management in his book titled *Quality and Productivity*. These 14 points describe the commitment this type of program requires between the engineering, production suppliers, and customers alike. All must have an overall understanding of the coated process to better visualize the final product (1).

How does this relate to the process?

The production of quality has to begin at the earliest point in the process in order to fully realize the maximum benefits achievable. This means incoming chemical and material quality must be checked to ensure that the product specification and quality has been maintained by the supplier. The coating formulation must have several procedures for checks and balances to assure constant formulation quality demands are met.

Every manufacturer of coated paper products has, or should have, quality specifications for the "finished" coating recipe such as: coating viscosity and coating solids, etc. These are two examples of coating color quality checks that must be maintained in order to produce a repeatable, high-quality coating for paper. Another example of a coating process check is measuring coating pigment flow to a mixer which can be balanced against pumping time, mixer liquid level, total weight in the mixer, and possibly weigh-bin weight change.

A coater operation must also have its own system of checks and balances to result in a quality operation. Some examples of coater checks are: blade change times, coater head position, applicator roll speeds, air-knife position, coat weight, and coating pickup. The coater dryer system also has a set of quality checks including: air temperature, dryer speed, dryer temperature, dryer steam pressures, and others specific to the method of drying used.

Applications of SPC in the Coated Paper Process

The following is a sample list of coated process parameters, raw or incoming material parameters, and finished quality parameters that can be tracked in an SPC system.

Sample coated process parameters:

1. Coating solids percentage, by weight
2. Coating pH
3. Coating viscosity
4. Coating temperature
5. Coater speeds
6. Coat weight
7. Coat weight moisture
8. After-coater drying temperature
9. After-coater drying air velocity
10. Coater section draws

Sample raw or incoming material parameters (2):

1. Pigments, binders, and additives
   a. Percent solids
   b. Viscosity
   c. Particle size distribution (sedigraphs)
   d. Brightness
   e. pH
   f. Residue solids, percent

2. Bulk materials (starch, $CaCO_3$, protein, dry clay, etc.)
   a. Particle size distribution (sedigraphs)
   b. Brightness

Sample finished quality parameters

1. Basis weight
2. Sheet moisture
3. Paper gloss
4. Heat set ink gloss
5. Color
6. Coating spread

Remember, in all coating process quality programs, each operation must have an established system of quality checks to ensure each aspect of the coating process is operating as it was designed and intended. It is much easier to manage a statistically predictable system than a completely uncontrolled one.

What are the benefits of an SQC program?

The benefits of an SQC program, first and foremost, includes improved and predictable product quality and reduced rework and unit downtime. If you spend time concentrating on increasing the quality of your first production, the amount of time spent on rework will be reduced. This also relates to unscheduled downtime, which any production employee can tell you is the one category that hurts productivity more than any other. Constant process monitoring will help make unscheduled downtimes infrequent as defects are detected and corrective action is planned for.

A natural reaction of people attempting to achieve product quality targets is to alter or "fix" the product. Statistical process control methods and principles tell us to fix the process, not the product. One of the key objectives and purposes of employing SPC methods is to control the process as early in the production sequence as possible. This is accomplished by monitoring the process at each strategic point which are defined by production and process experience. Using these points, finished product quality problems are identified as to their earliest origin in the process and once identified the problems are eliminated by one or more of the following methods: process parameter setpoint adjustment, equipment maintenance or repair, change in process addition point or sequence, or improved operator training and direction. The problem ends prior to resulting in accumulated sub-standard quality production tonnage.

Variability, in statistical process control circles can be a "dirty word." Six simple and straight forward steps should be taken to bring your process under control (3):

1. Identify it
2. Reduce it
3. Eliminate it
4. Improve process
5. Measure improvement
6. Continue to improve... forever.

A good, active SQC program will also result in reduced quality inspection costs, because the time spent on incoming material quality control and process quality control, will reduce the requirement for "every reel testing."

In addition to everyday quality measurement, an active SQC program will help to compare new machine coater production designs and methods. SQC provides a timely pulse of the process as well as defining the overall capability of that process, which provides subtle guidance for strategic process design. This results in reduced operating costs with improved decision making (4).

A typical introduction or explanation of statistical process control might be a kit of tools to help influence decisions related to the coated process by disclosing the natural capabilities of the process (5). As explained earlier, SQC and/or SPC results in the process talking to us. It is a system of process tracking or information management that depicts the natural capability of the process under study.

What Does Statistical process control Mean?

Simply stated, SPC is the practical application of statistics combined with experienced interpretation of continuous and tracked process data on automated systems or manual control charts. This information identifies process limits and provides a tool to identify assignable causes for a process disturbance (6).

The basic statistics calculations utilized are arithmetic mean and standard deviation. This provides the average parameter value to chart or trend, as well as the amount of variation within that average value. Control charts are the graphical tool used to display the chronological order of a process variable. As a process variable is charted, the normal variation occurs statistically and tells us various information about how that process is controlled, and the controllability limits of the process.

There are three basic sources of product variation, which can be directly related to the coated process. They are:

1. Test methods
2. Process variation
3. Raw material variation.

One papermaker has identified that one third of product variability comes from raw materials, one third from test methods, and one third comes from the papermaker's own process (7).

It is necessary to have the ability and tools to control the coating process and to provide the top product quality. To optimize the overall effectiveness, the process is defined by certain quality standards requirements such as coating viscosity, pH and solids, and performance requirements such as color, ink receptivity, heat-set ink gloss, paper gloss, and appearance. Standards are achieved and maintained in areas within the process. Each sub-area should be controlled to a degree that provides the desired end result by controlling a specific parameter special to that area that affects the finished paper. Process people can help direct the process control effort to the most effective area of the process. When looking for process areas to control, ask the question: What will controlling this area give to the stability of this part of the process or the finished product? If the process parameter is measurable objectively and it contributes to the overall process stability and finished coated product, then that process parameter should be included in a constantly maintained statistical control charting program.

All processes can be represented by a relatively simple block diagram. Like most logical block diagrams, there is various input information and one or more outputs. In a statistical process control process block diagram module, there are five basic input categories of information that exist. They are:

1. **Material:** The raw materials of a given process or supply product to a given step in the process. For example, the uncoated sheet for a coater station.

2. **Mentality:** The general attitude of the specific team on a given coater system. For example: If a given crew has had better instruction and feedback to a program, generally they will have a better attitude, or "mentality" toward their group.

3. **Equipment:** The specific process equipment involved in a given process, the coater, or the coating mixer, for example.

4. **Environment:** The general air around the operating system and the enclosure surrounding the specific process area... the coating preparation area...or the coater salvage system.

5. **Operation:** The process and how it functions in its surroundings or in unison with other operating areas (8).

Figure 8.1 depicts a sample process, including input items and the output representing the final product. If the process is determined to be out of statistical control, one of the five input items must be changed to correct the problem. Samples of corrective action are:

Input Item	Possible Corrective Action
Material	Evaluate a different supplier or change material feed rate.
Mentality	Train operators or change work assignments.
Equipment	Repair defective equipment or replace process equipment.
Environment	Improve equipment, process or both, or improve operator accessibility.
Operation	Update process standards or change sub-process setpoints.

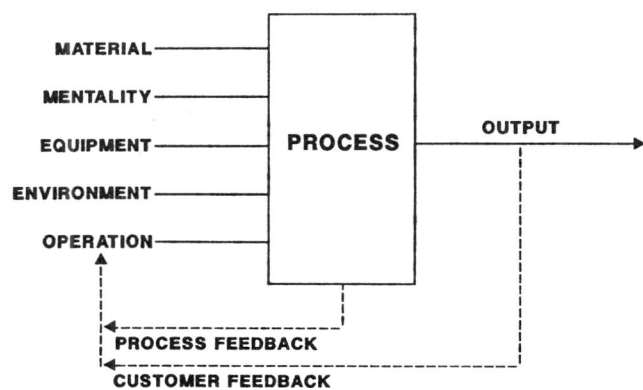

Fig. 8.1. Block diagram

The outputs from a given block diagram usually fall into one of two categories:

1. Process output, the product itself, all good SPC programs should contain references to the finished product. A process variable that is reflected toward a finished product without being aware of the condition of that final product is useless.

2. Feedback, the information on the condition of both the process itself as well as customer feedback. If the customer responds with positive feedback (usually in the form of repeat business), you probably can be assured that the operation is proceeding in a satisfactory manor. Bad news usually travels very fast.

Development of Statistical Control Charts

Normal statistical distribution is referred to as a bell-shaped curve. The peak of the curve is at the arithmetic mean of the data sampled. At 3 standard deviations (3σ), 99.7% of the data sampled is within this area. At two standard deviations (2σ), 95% of the data sampled is within this total area, and so on. A graph of normal distribution is shown in Fig. 8.2. Normal statistical distribution is the basis that statistical control charts use (9).

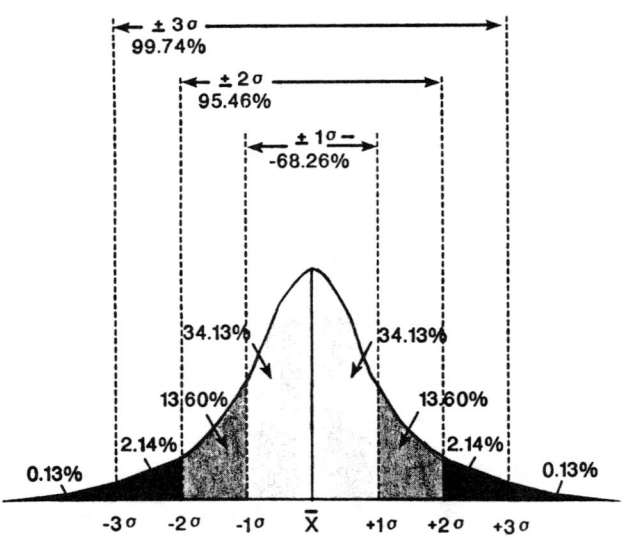

Fig. 8.2. Normal distribution

The following list identifies some major types of charting. Some or all of these methods may be applicable to your specific needs.

1. Process flow charts or flow sheets are schematics of the process and depicts its logical sequence. This type of diagram is imperative in order to accurately depict the logical sequence of the process. Aids in identification of variability in a given process or sub-process. It is from this document that all other charts and diagrams are developed.

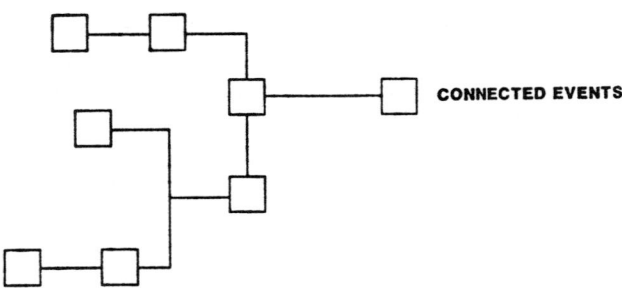

Fig. 8.3. Statistical charting techniques—flow chart

2. Fishbone diagrams is a cause-and-effect type of diagram, which logically depicts major processes and sub-processes or tasks in a given process stream. The fishbone diagrams are also used in process troubleshooting techniques and to depict major, minor, and subordinate processes. This chart is also helpful if target probable causes for unnatural variation (10).

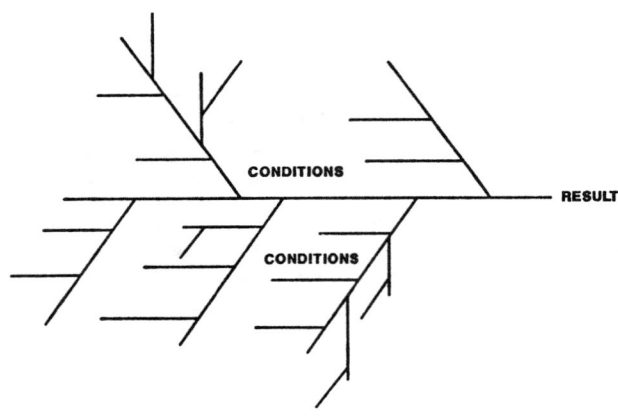

Fig. 8.4. Fishbone chart

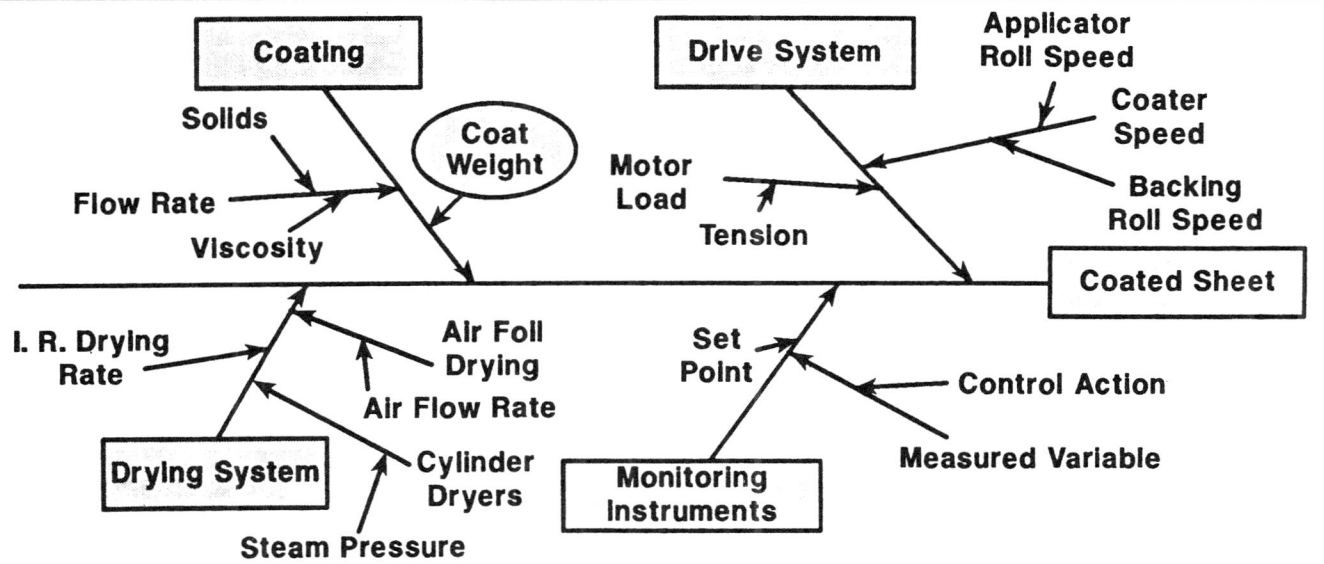

Fig. 8.5. Fishbone diagram with cause and effect

3. Pareto charts are bar graphs which graphically depict the major cause of upset in your process. Tell what, where, how much, and when, about a given set of conditions around a process. The process variation causes are listed and plotted against relative impact and magnitude, frequency, or both. The most probable cause will then be easily identified. These charts also depict the order of magnitude of a process variable variation compared to other tracked variables.

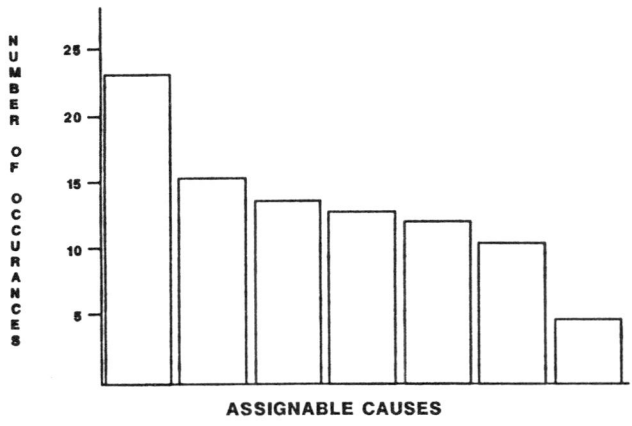

Fig. 8.6. Pareto graph

4. Histograms are simply an expanded Pareto chart plotted against time. This can further identify a continuing process variation or conclude that the problem was attributable to an assignable or chance cause, and will not be repeated.

● Process average and dispersion compared to specs

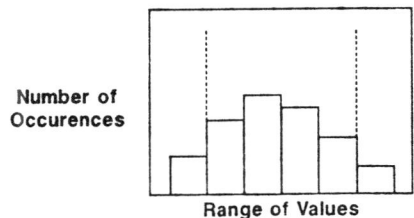

Accurate and Informative Snapshot of Process

Fig. 8.7. Histograms

5. Run charts tells the pulse of the process variable without statistical limits. This chart is the last step before generating a detailed control chart. They also provide one last chance to tune your control chart layout or plans.

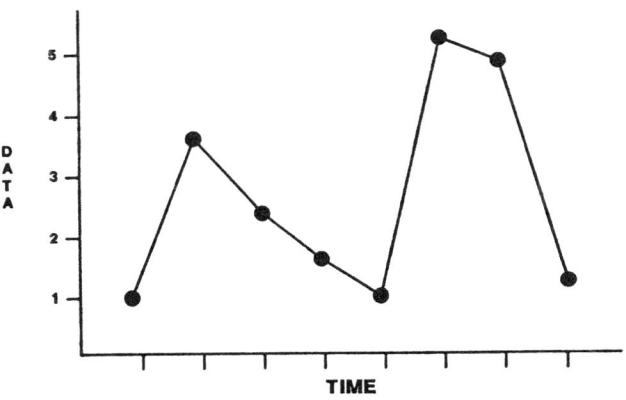

Fig. 8.8. Histogram run chart

6. Scattergrams depict process variable correlation, usually between two process variables. These charts are sometimes referred to as correlation diagrams, i.e., if the compared variables form a somewhat linear relationship, they can be identified as having a nominal correlation and can be further investigated. These charts are also quite helpful to determine process variables that are dependent or react upon another process variables variation.

● **Determine possible correlation between cause and effect**

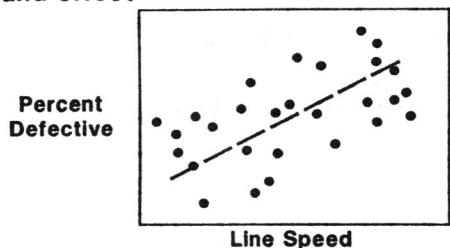

Improved Capabilities for Problem Identification and Elimination

Fig. 8.9. Scattergram

Fig. 8.10. a., b., c., d., and e. Scattergram plottings:

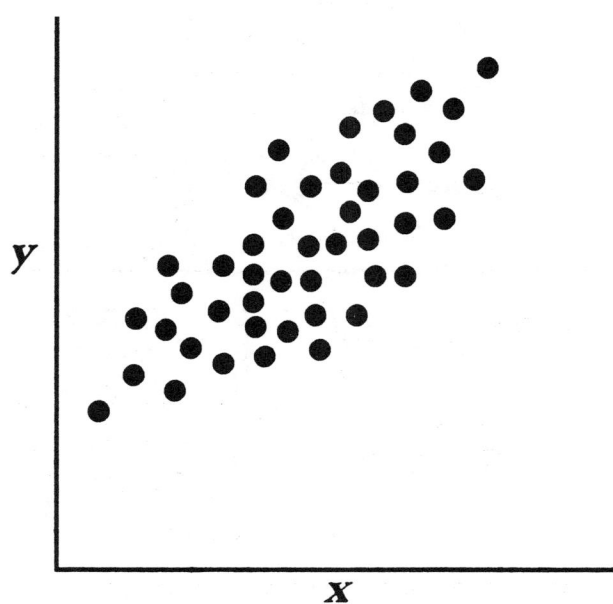

Fig. 8.10. b. Positive correlation may be present

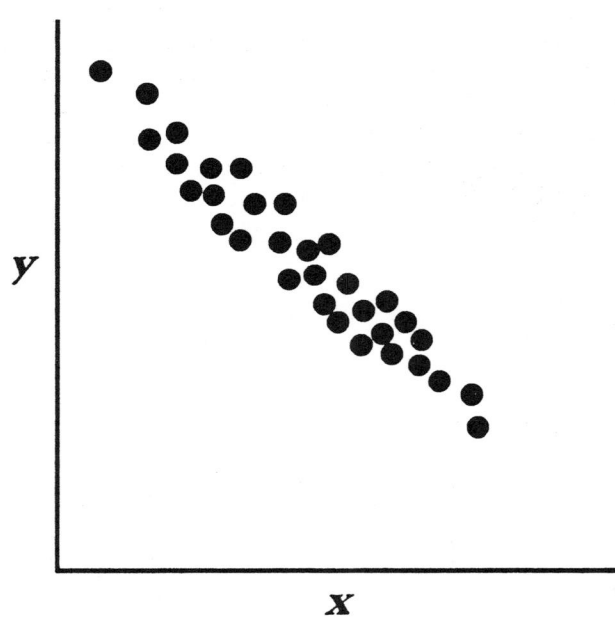

Fig. 8.10. c. Negative correlation

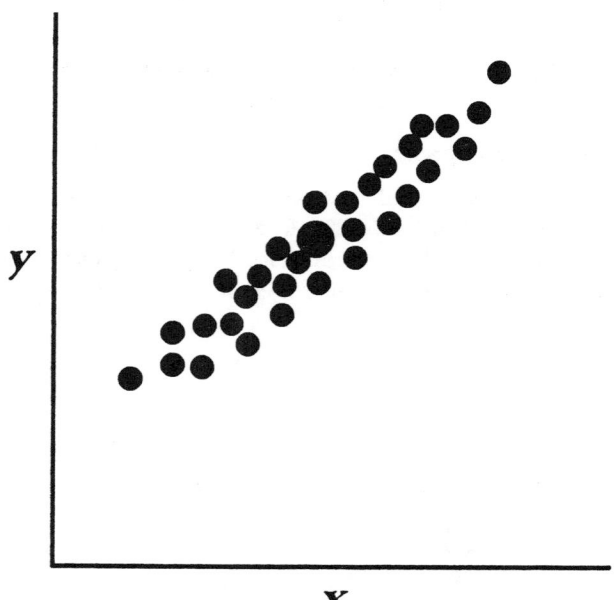

Fig. 8.10. a. Positive correlation

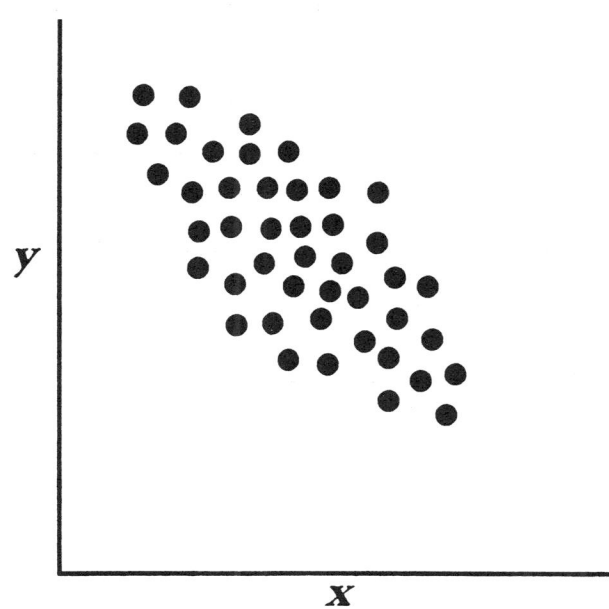

Fig. 8.10. d. Negative correlation may be present

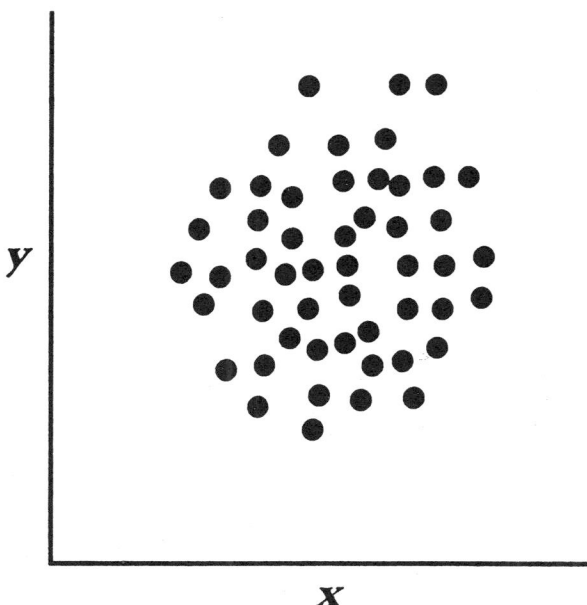

Fig. 8.10. e. No correlation

7. Control charts are the heart of the program and display the capability of a given process. The two basic types of control charts are the X-bar (process variable) and the R-chart (variation range).

- **Identify assignable causes of process fluctuations**

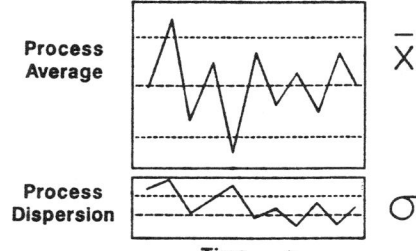

Improved Visibility of Trends Over Time

Fig. 8.11. Control charts

The source of a variation must be identified in any process upset. The process flow diagram, along with the fishbone chart will help eliminate the cause of the variation. Another chart or diagram of another type can then be developed, in the event that you are still not convinced that you have found the cause of the variation.

With this newfound process data in hand, you can now actuate the designated process changes and corrections. This data can also help you to eliminate or at least greatly minimize future assignable causes affecting your process.

What Type of Chart Do I Use?

This is a question that most all developers of a new SPC program ask themselves. First of all, you must design and develop your charts to fit your needs. Details—such as size, shape, and whether or not you want to amplify the impact of your charts by using color backgrounds or not—are the types of decisions that must be made by the designated SPC committee. Remember: Do not design the charting system in such a way that precludes further development or enhancement of the charts (11).

The next step in the SPC program development is to define which process variables are to be charted. The SPC committee in your mill is a good place to start, but remember your Pareto chart data along with your histograms and fishbone charts will tell you what items should be considered. If the data cannot be numerically evaluated and measured, it cannot be statistically controlled. Don't waste your time on items that you can't objectively measure. Another important point is to spend your time on a process variable that will allow you to experience the most return from your effort. Consistency dilution water flow, for example, is of little use or meaning in your SPC program as compared to cross-machine moisture profile percent variation. Spend your efforts where it will provide your team with the most valuable information.

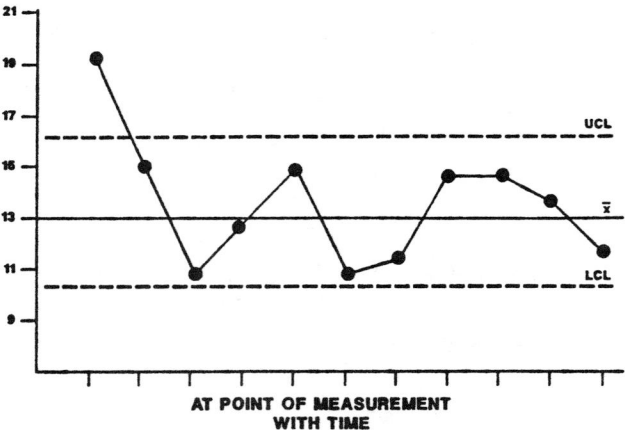

Fig. 8.12. Statistical charting techniques—control chart. (courtesy of A.S.T.M., publication STP 15D)

Now that the preliminary charting decisions are made, you are ready to begin to develop your control charts. The first process variable chosen should have a relative importance in your final product quality to provide a good beginning point in your SPC program. Begin to simply collect data, at the same time of the day for a complete month. As shown in Fig. 8.12, plot the data points on a graph, and calculate the average value of the data points. A straight line, representing the average value, should be overlaid to depict the arithmetic mean on the graph. You can now calculate statistical control limits, X-bar and range, for your graph by using the simple formula listed in Table 8.1 (12).

You will note that on the top of the table embedded in Table 8.1, the quantity of sub-group sample points are listed. This is the quantity of values that were used to obtain the average value. You then follow that value down along the table to select the applicable factors for your control limits and plug these values in the formula. You now have derived your control limits for that period. Just like showing the average value for the month on the chart, you also plot the upper and lower control limits on the graph. You now can easily determine which data points and what time of day the out-of-control data points (Fig. 8.13) occur and special or assignable causes can be determined, refer to the following list.

Common Causes Associated with Things Which are:

Normal	Not Changing
Natural	Steady
Stable	Predictable
Undisturbed	Consistent
Alike	Uniform

Special Causes Associated with Things Which are:

Unnatural	Shifting
Disturbed	Unpredictable
Unstable	Inconsistent
Non-alike	Out-of-the-ordinary
Mixed	Different
Erratic	Important
Abnormal	

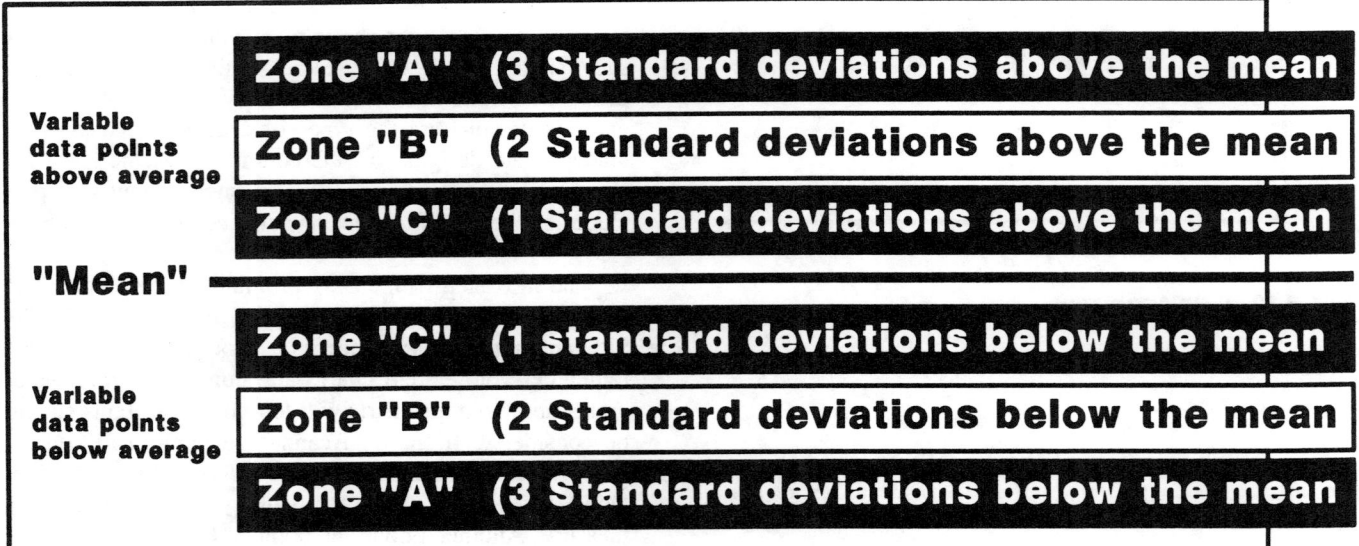

Fig. 8.13. Eight rules for statistical control chart use

Table 8.1. Control limit formula

$$UCL_R = D_4\overline{R}$$
$$LCL_R = D_3\overline{R}$$
$$UCL_{\tilde{X}} = \overline{\overline{\tilde{X}}} + \tilde{A}_2\overline{R}$$
$$LCL_{\tilde{X}} = \overline{\overline{\tilde{X}}} - \tilde{A}_2\overline{R}$$

where D_4, D_3, and A_2 are constants varying by sample size, with values for sample sizes 2 to 10 shown in the following table:

n	2	3	4	5	6	7	8	9	10
D_4	3.27	2.57	2.28	2.11	2.00	1.92	1.86	1.82	1.78
D_3	*	*	*	*	*	.08	.14	.18	.22
$\tilde{A}_2$	1.88	1.19	.80	.69	.55	.51	.43	.41	.36

$\overline{x}$ control chart CL $= \overline{\overline{x}} = 60.0$

$\phantom{\overline{x} \text{ control chart }}$ UCL $= \overline{\overline{x}} + A_2\overline{R}$

$\phantom{\overline{x} \text{ control chart UCL}}= 60 + 1.88 \times 0.21\ 0.07$

$\phantom{\overline{x} \text{ control chart UCL}}= 60 + 0.4$

$\phantom{\overline{x} \text{ control chart UCL}}= 60.4$

$\phantom{\overline{x} \text{ control chart }}$ LCL $= \overline{\overline{x}} - A_2\overline{R}$

$\phantom{\overline{x} \text{ control chart LCL}}= 12.940 - 1.88 \times 0.21$

$\phantom{\overline{x} \text{ control chart LCL}}= 59.6$

R control chart CL $= \overline{\overline{R}} = 0.21$

$\phantom{R \text{ control chart }}$ UCL $= D_4\overline{R}$

$\phantom{R \text{ control chart UCL}}= 3.26 \times 0.21$

$\phantom{R \text{ control chart UCL}}= 0.68$

$\phantom{R \text{ control chart }}$ LCL $= D_3\overline{R}$ (none)

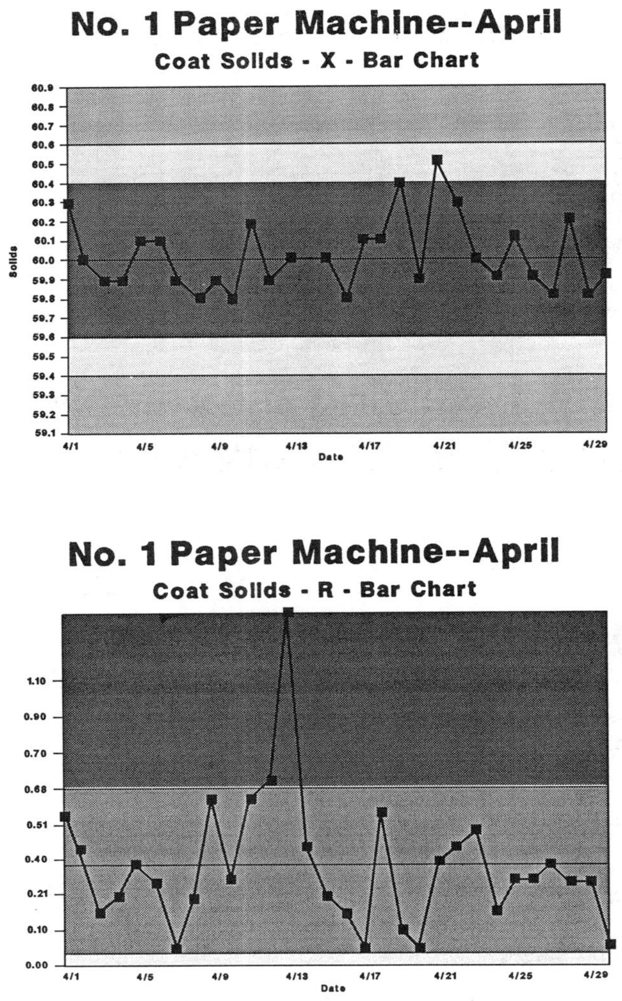

No. 1 Paper Machine--April
Coat Solids - X - Bar Chart

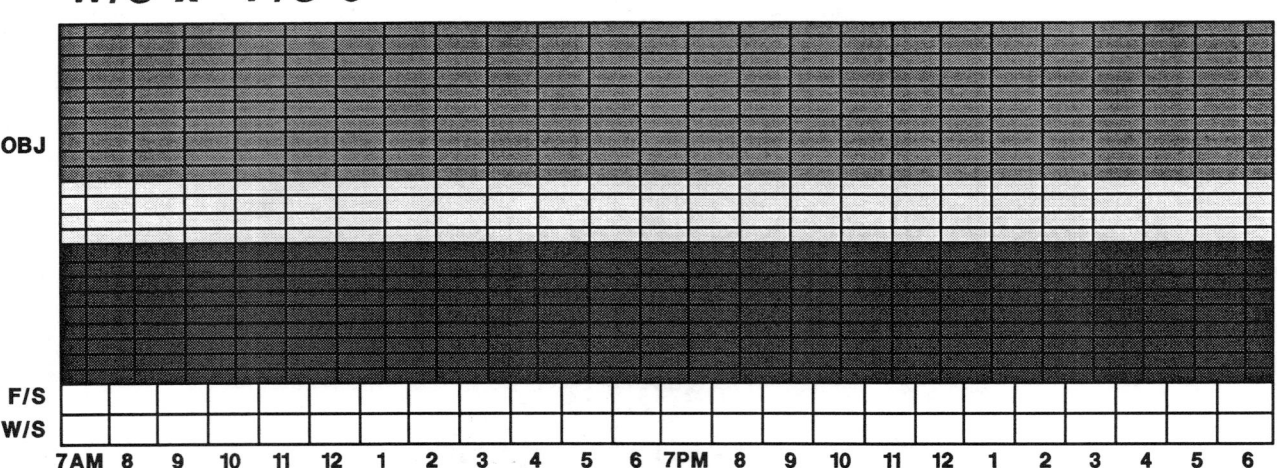

No. 1 Paper Machine--April
Coat Solids - R - Bar Chart

Fig. 8.14. Sample control chart showing X-bar and range for the same process parameter

W/S-x F/S-o

Fig. 8.15. Paper gloss control chart

The required companion for the X-bar chart is the range chart. This simply is a chart using new control limits and plotting the amount of variation or range of the individual data point. The range chart is developed and control limits calculated in the same manner as the X-bar chart.

Figures 8.14 through 8.18 show sample control charts. The best choice is the one that best fits your needs and process data sampling methods (13).

After the Charting

After all of these managerial decisions are made, there is one thing remaining to be done and that is to provide operator feedback of the SPC systems' results. Constantly communicate to your operating crews the status of the program and the results that affect their production unit. The crews sometimes are asked to perform new tasks and, by providing prompt and up-to-date feedback on their efforts, they receive visual evidence (control charts, for example) of a SPC program results. It is also important that SPC managerial teams are receptive to constructive comments by operating crews for different process variables that if tracked can provide even more valuable process information. The users know the details of their system.

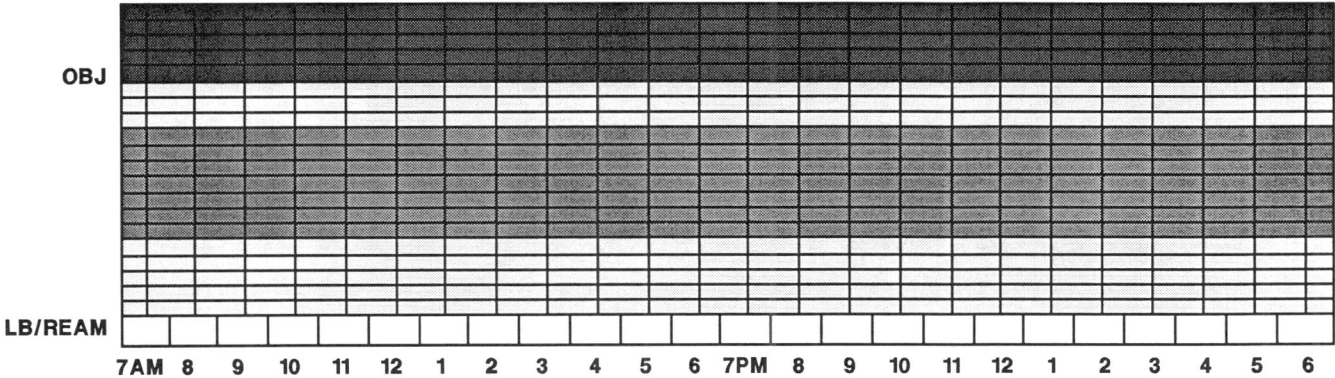

Fig. 8.16. Coat weight control chart

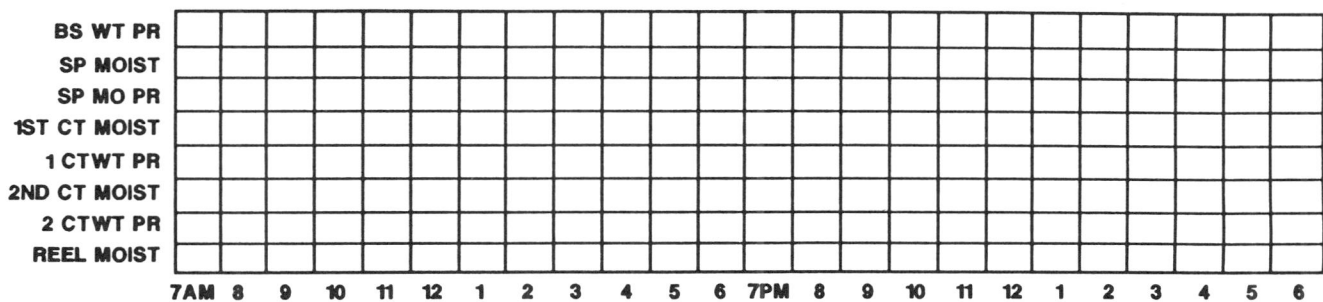

Fig. 8.17. General control chart

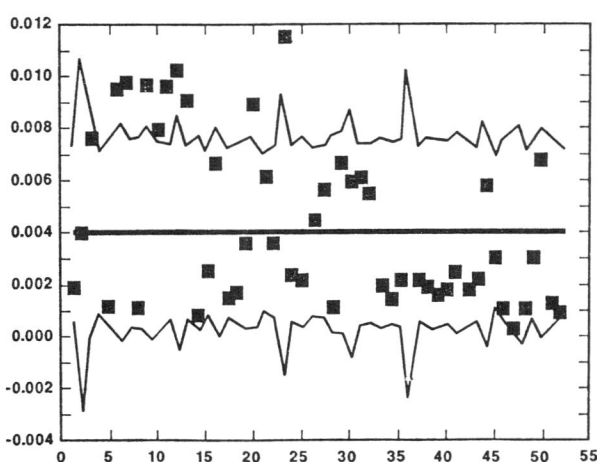

Fig. 8.18. Shewhart control chart

Bibliography

Literature Cited

1. Deming, W.E., *Quality, Production & Competitive Position,* Massachusetts Institute of Technology, Center for Advanced Engineering Study, Cambridge, MA.

2. Dow Chemical Co., Midland, MI.

3. A.S.T.M. Statistical Methods #STP 15D, #04-015040-34, Philadelphia, 1963.

4. A.S.T.M. Statistical Methods #STP 15D, #04-015040-34, Philadelphia, 1963.

5. Juran Institute.

6. Partrick, K.L. and Joscey, M., "Special Report," Pulp & Paper, April 1987.

7. RUST International Corp., Birmingham, AL.

8. Grant, E. and Leavenworth, R.J., *Statistical Quality Control,* McGraw-Hill, New York, 1974.

9. *CRC Handbook of Probability & Statistics,* 2nd Edn., CRC Press, Boca Raton, FL.

10. Ishikawa, K., *Guide to Quality Control, Asian Productivity Organization,* Tokyo, Japan, pp. 91; 30, 1987.

11. Continuing Process Control, Ford Motor Co., Statistical Methods Office, Operation Supports Staff, Dearborn, MI.

12. Smith, K.E., Pulp & Paper, January 1985, pp. 132-135.

13. Continuing Process Control, Ford Motor Co., Statistical Methods Office, Operation Supports Staff, Dearborn, MI.

Resources

"Profile," *Tappi J.* 69(12): 126 (1986).

Crosby, P., *Quality is Free,* McGraw Hill, Book Co., New York, January 1980.

Dybeck, M., "A.I. and SPC Team up in Process Control," *INTEC,* December 1988, pp. 3236.

Everar, R.S. and Blake, J.W, "Statistical Aids for Improving Process Control," *Southern Pulp & Paper,* April 1985, reprint.

Ishikawa, K., *Guide to Quality Control,* Asian Productivity Organization Quality Resources, White Plains, NY, Sixth Printing, 1989.

Moudy, R., "Mead Kingsport Finds SPC is a Way of Life," *American Papermaker,* January 1989, pp. 2023.

Rau, F.C., "The Total Quality Process," *Chemical Processing,* November 1988, pp. 8886.

Shorter, W.W., *AT&T Statistical Quality Control Handbook, 11th printing,* Western Electric Co., Inc. AT&T Technologies, Inc., Indianapolis, IN, May 1985, Publication Number 700-444.

Smith, K. E., "Statistical Controls Improve Quality, Boost Productivity at Federal Mill," *Pulp & Paper,* January 1985, pp. 32-34.

Taylor, S.S., "Clay Company & Mills Benefit From SPQC," *American Papermaker,* January 1989, pp. 2829.

Taylor, S. S., "Building in Quality Project by Project," *Southern Pulp & Paper,* September 1986, pp. 2933.

Index

What is TAPPI?

Founded in 1915, TAPPI is the world's largest professional society of executives, operating managers, engineers, scientists, and technologists serving the paper and related industries. Total membership is approximately 32,000 with some 80% residing in the United States. The remainder live in 76 other countries.

TAPPI is renowned for its industry publications. Members produce technical books, reports, conference proceedings, course notes, home study courses, and videotapes through TAPPI PRESS. *Tappi Journal,* distributed monthly to all members, is the leading publication for technical information on the manufacture and use of pulp, paper, packaging, and converted products. Through TAPPI, Association members develop, update, and publish test methods and technical information sheets on which much of the industry depends to analyze its products and processes.

TAPPI sponsors a variety of technical conferences, seminars, and short courses to foster worldwide technical information exchange and enhance the professional development of members.

For membership information, to order any of TAPPI's professional development products, or to register for a meeting, please call TAPPI's toll-free Service Line:

1-800-332-8686 (U.S.)
1-800-446-9431 (Canada).

TAPPI's Vision

We are a global community of motivated individuals who lead the technical advancement of the paper and related industries.

Together...

- We provide outstanding educational and professional growth opportunities.

- We serve as a worldwide forum to exchange technical information, promote research, and recognize individual achievement.

- We create success by the quality, timeliness and innovativeness of our products and services.

Integrity and fellowship characterize our association.

Units of Measurement and Conversion Factors

The reproduction of this Technical Information Sheet (TIS 0800-01) is supplied solely for the convenience of the readers of this publication.

This Technical Information Sheet deals with the application of the International System of Units (abbreviated "SI") within the field of pulp, paper, and paperboard, often called the "SI metric units." TAPPI regulations require the use of the SI units as the preferred units in TAPPI Testing Procedures and other TAPPI publications.

Details of the SI are given in the various parts of ISO Standard 31 and ISO Standard 1000. Application of this information is not always easy, however. For instance, some quantities can be expressed in different units, all within the SI. Such variations can lead to confusion in reporting test results and quoting property values.

In order to overcome such problems within the pulp, paper, and paperboard field, ISO/TC6 has recommended appropriate units in ISO Standard 5651. These are reproduced in this Technical Information Sheet, together with other information and recommendations.

Scope

This Technical Information Sheet states the units recommended for use in expressing the properties of pulp, paper, and paperboard and other quantities found in the pulp, paper, and paperboard documents. It also provides the conversion factors for converting customary units to metric units or other recommended units, and marks with an asterisk (*) when the conversion factor is exact.

Units recommended in ISO 5651 are marked (+) or (#), with the latter indicating that the recommended unit for pulp, paper, and paperboard is not a preferred SI unit.

Mechanics of use

In converting from the customary units to the recommended form, multiply the test value expresses in customary units by the conversion factor to obtain the test value in the recommended form. Suppose that the property of interest is the thickness of a sheet of paper, and that this has been determined to be 5.3 mils. Examination of the table shows that in Section 1.2, Thickness, the conversion factor from mils to micrometers is exactly 25.4. Multiplying 5.3 by 25.4, the test value in the recommended form is 134.62 micrometers.

Generally, the converted value should be rounded to the same number of significant figures or one more than is in the value in customary units, depending, respectively, on whether the first digit of the new value is more or less than that of the original value. Thus, in this case, the reported value would be 135 micrometers. *(Note: The rounding rule given here is easy to remember and is "safe" in the sense that it does not result in the loss of significant information.)*

If the property of interest had been the thickness of corrugated board of, say, 180 mils, the value would be converted to millimeters, not micrometers. The conversion factor is 0.0254 and the value in the recommended form is 4.57 mm or 4.6 mm, depending on whether the zero in the last digit of 180 is significant or simply holding the decimal place.

Generally, if the table allows a choice in the recommended form, choose that form that would have no more than three significant figures in front of the decimal point for most of the values. However, use, for example, 0.13 mm, not 130 μm, if the zero in 130 is not significant.

Notation

The quantities given in square brackets in the second and fourth columns are symbols, not abbreviations. Hence, do not use abbreviation marks. When two of these symbols are multiplied, as in Part 20. of Table 10, Dynamic viscosity in pascal seconds, use a raised dot between the symbols, this Pa•s.

The same symbol is used for the plural; so adding an "s" for the plural is incorrect. Thus, write fifty-two grams as 52 g. Also, the symbol for per is a slash (/), not the letter p, and scientific notation is used for squares, cubes, and higher powers. Thus, write fifty-two grams per square meter as 52 g/m^2, not gpsm nor gsm.

The symbols and the names of units are not mixed in the same expression. Therefore, g per square meter, grams/square meter, grams/m^2, g/sq m, and g per m^2 are all incorrect.

In the test, write out the name of the unit, except usually when preceded by a numerical value. Thus use "determine the mass in grams to the nearest 0.1 g." However, expressions like "one gram" or "two to three grams" may be used when the value as well as the name is written out.

Table 1. General properties

PROPERTY and TAPPI Test Method (if any) [ISO Standard]		To convert values expressed in CUSTOMARY UNITS	Multiply by	To obtain values expressed in RECOMMENDED FORM
		Name [Symbol]	*exactly	Name [Symbol]
1.1	Mass per unit area:	pounds per ream, 17 x 22 - 500	3.7597	+grams per square meter [g/m^2]
	Basis weight (mass on ream basis)	pounds per ream, 24 x 36 - 500	1.6275	grams per square meter [g/m^2]
	Grammage (mass in grams per square meter)	pounds per ream, 25 x 38 - 500	1.4801	grams per square meter [g/m^2]
	(See T 410 or TIS 0808-01 for other conversions)	pounds per ream, 25 x 40 - 500	1.4061	grams per square meter [g/m^2]
	T 410, T 220	pounds per 1000 square feet [lb/1000 ft^2]	4.8824	grams per square meter [g/m^2]
	[ISO 536, 3039,5270,5638]	pounds per 3000 square feet [lb/3000 ft^2]	1.6275	grams per square meter [g/m^2]
1.2	Thickness, or	mils (or points or thousandths of an inch)	25.4	+micrometers [μm]
	Caliper	mils [mil or 0.001 in]	0.0254	+millimeters [mm]
	T 411	inches [in]	25.4	millimeters [mm]
	[ISO 534, 3034]			
1.3	Bulking thickness T 411 (Note 4 of T 411 os-76), T 220 [ISO 438]	mils [mil or 0.001 in]	25.4	+micrometers [μm]
1.4	Apparent density (also see 10.9)	pounds per cubic foot [lb/ft^3]	16.01846	kilograms per cubic meter [kg/m^3]
	[ISO 438,534]	grams per cubic centimeter [g/cm^3]	* 1000	kilograms per cubic meter [kg/m^3]
		grams per cubic centimeter [g/cm^3]	*1	# grams per cubic centimeter [g/cm^3]
1.5	Dimensional change after water immersion [ISO 5635, 5637]	percent [%]	*1	+percent [%]
1.6	Hygroinstability	percent [%]	*1	+percent [%]

* Exactly
+ Unit recommended in ISO Standard 5651-1978
# Unit recommended in ISO 5651, although not preferred SI unit
Conversions to 6 signifigant maximum

Table 2. Strength properties

PROPERTY and TAPPI Test Method (if any) [ISO Standard]	To convert values expressed in CUSTOMARY UNITS	Multiply by	To obtain values expressed in RECOMMENDED FORM
	Name [Symbol]	*exactly	Name [Symbol]
2.1 Tensile strength T 494, T 404, T 456, T 220, T 813 [ISO 1924, 3781, 5270]	pounds-force per inch [lbf/in]	0.175127	+kilonewtons per meter [kN/m]
	pounds-force per 15 millimeter width [lbf/15 mm]	0.29655	kilonewtons per meter [kN/m]
	ounce-force per inch [ozf/in]	10.945	newtons per meter [N/m]
	kilograms-force per 15 millimeter width [kgf/15 mm]	0.65378	kilonewtons per meter [kN/m]
	kilograms-force per 25 millimeter width [kgf/25 mm]	0.39227	kilonewtons per meter [kN/m]
	kilograms-force per centimeter [kgf/cm]	0.980665	kilonewtons per meter [kN/m]
	grams-force per millimeter [gf/mm]	9.80665	newtons per meter [N/m]
	newtons per 15 millimeter width [N/15 mm]	66.6667	newtons per meter [N/m]
2.2 Tensile strength of wax T 644	pounds-force per square inch [lbf/in^2]	6.89476	kilonewtons per sq meter [kN/m^2]
2.3 Tensile index [ISO 5270]	newton meters per gram [N · m/g]	*1	#newton meters per gram [N · m/g]
	kilometers breaking length [km]	*9.80665	newton meters per gram [N · m/g]
2.4 Breaking length (to be replaced by tensile index) T 220, T231, T494 [ISO 1924]	meters [m]	*0.001	+kilometers [km]
	kilometers [km]	*1	kilometers [km]
2.5 Stretch at rupture (elongation) T 220, T 404, T 494	percent [%]	*1	+percent [%]
2.6 Tensile energy absorption (TEA) T 494	foot pounds-force per square foot [ft · lbf/ft^2]	14.5939	+joules per square meter [J/m^2]
	inch pounds-force per square inch [in · lbf/in^2]	175.127	joules per square meter [J/m^2]
	kilogram-force meters per square meter [kgf · m/m^2]	9.80665	joules per square meter [J/m^2]
	joules per square meter [J/m^2]	*1	joules per square meter [J/m^2]
2.7 T.E.A. index	millijoules per gram [mJ/g]	*1	#millijoules per gram [mJ/g]
2.8 Tearing strength T 220, T 414, T 496 [ISO 1974, 5270]	grams-force [gf]	9.80665	+millinewtons [mN]
2.9 Tear index (formerly tear factor) T 220 [ISO 5270]	Tear factor computed as: 100 grams-force per (gram per square meter) [100 gf/(g/m^2)]	0.0980665	#millinewton sq meters per gram [mN·m^2/g]
2.10 Edge tearing resistance (Finch) T 470	pounds-force [lbf]	4.44822	newtons [N]
	kilograms-force [kgf]	9.80665	newtons [N]
2.11 Bursting strength T 403, T 807, T 810 [ISO 2758, 2759, 3689]	pounds-force per square inch [psi]	6.89476	+kilopascals [kPa]
	points	6.89476	kilopascals [kPa]
	kilograms-force per square centimeter [kgf/cm^2]	98.0665	kilopascals [kPa]
	kilonewtons per square meter [kN/m^2]	*1	kilopascals [kPa]
2.12 Burst index (formerly burst factor) T 220 [ISO 2758, 2759, 5270]	Burst factor computed as: Grams-force per square centimeter per (gram per square meter) [(gf/cm^2)/(g/m^2)]	0.0980665	#kilopascal sq meters per gram [kPa · m^2/g]
2.13 Puncture resistance T 803 [ISO 3036]	centimeter kilograms-force [cm · kgf]	0.0980665	+joules [J]
	scale units (= 0.305 cm · kgf)	0.0299	joules [J]
	foot pounds-force [ft · lbf]	1.35582	joules [J]
	inch ounces-force [in · ozf]	7.06155	millijoules [mJ]
	inch pounds-force [in · lbf]	0.112985	joules [J]
2.14 Adhesion strength of glue bonds of corrugated fibreboard T 813	pounds-force per inch [lbf/in]	0.175127	+kilonewtons per meter [kN/m]
	kilograms-force per millimeter [kgf/mm]	9.80665	kilonewtons per meter [kN/m]
2.15 Z-direction strength properties (units as appropriate) T 506	pounds-force per square inch [lbf/in^2]	6.89476	+kilopascals [kPa]
	kilograms-force per square centimeter [kgf/cm^2]	98.0665	kilopascals [kPa]
	pounds-force per inch [lbf/in]	0.175127	+kilonewtons per meter [kN/m]
	foot pounds-force per square inch [ft·lbf/in^2]	2101.5	+joules per square meter [J/m^2]
	foot pounds-force per square inch [ft·lbf/in^2]	2.1015	kilojoules per square meter [kJ/m^2]

Table 3. Folding, bending, and compression properties

PROPERTY and TAPPI Test Method (if any) [ISO Standard]	To convert values expressed in CUSTOMARY UNITS	Multiply by	To obtain values expressed in RECOMMENDED FORM
	Name [Symbol]	*exactly	Name [Symbol]
3.1 Static bending force [ISO 2493]	pounds-force [lbf] pounds-force [lbf] milligrams-force [mgf] (or Gurley units)	4.44822 4448.22 9.80665	+newtons [N] +millinewtons [mN] micronewtons [µN]
3.2 Bending stiffness T 489, T 535, T 820	gram-force centimeters [gmf · cm] (or Taber units) gram-force centimeters [gmf · cm] (or Taber units) pound-force inches [lbf · in]	98.0665 0.0980665 0.112985	+micronewton meters [µN · m] +millinewton meters [mN · m] +newton meters [N · m]
3.3 Bending strength (modulus of rupture) T 655, T 1003	pounds-force per square inch [lbf/in²]	6.89478	kilopascals [kPa]
3.4 Fold number T 220, T 423, T 511 [ISO 5270, 5626]	numerical value (number of double folds)	*1	+numerical value
3.5 Folding endurance T 423, T 511 [ISO 5270, 5626]	log to the base 10 of the number of double folds	*1	+$\log_{10}$ (number of double folds)
3.6 Concora medium test (flat crush) T 809	pounds-force [lbf]	4.44822	#newtons (CMT) [N(CMT)]
3.7 Ring crush T 818	pounds-force (for a 6-inch length) [lbf/6 in] ** kilograms-force (for a 6-inch length) [kgf/6 in] ** newtons (for a 6-inch length) [N/6 in] kilograms-force per centimeter [kgf/cm]	0.02919 0.06435 0.006562 0.980665	+kilonewtons per meter [kN/m] kilonewtons per meter [kN/m] kilonewtons per meter [kN/m] kilonewtons per meter [kN/m]
3.8 Edgewise compressive strength T 811 [ISO 3037]	pounds-force per inch [lbf/in] ** pounds-force per inch [lbf/in] ** kilograms-force per inch [kgf/in] **	0.175127 175.127 0.38609	+kilonewtons per meter [kN/m] newtons per meter [N/m] kilonewtons per meter [kN/m]
3.9 Flat crush resistance of corrugated board T 808 [ISO 3035]	pounds-force per square inch [lbf/in²] ** kilograms-force per square centimeter [kgf/cm²]	6.89476 98.0665	+kilopascals [kPa] kilopascals [kPa]

**Usually expressed simply as "pounds" or "kilograms"

Table 4. Surface properties

PROPERTY and TAPPI Test Method (if any) [ISO Standard]	To convert values expressed in CUSTOMARY UNITS	Multiply by	To obtain values expressed in RECOMMENDED FORM
	Name [Symbol]	*exactly	Name [Symbol]
4.1 Roughness, general	microns or micrometers [µm]	*1	+micrometers [µm]
4.2 Smoothness, Bekk T 479	seconds [s]	*1	#seconds (Bekk) [s(Bekk)]
4.3 Roughness, Bendtsen [ISO 2494]	milliliters per minute [mL/min]	*1	#milliliters per minute (Bendtsen) [mL/min (Bendtsen)]
4.4 Roughness, Sheffield T 538 [ISO 2494]	Sheffield units	*1	#Sheffield units
4.5 Surface strength: Picking velocity, IGT T 499 [ISO 3782, 3783]	millimeters per second [mm/s] feet per minute [ft/min] feet per minute [ft/min]	*1 5.080 0.00508	+millimeters per second [mm/s] millimeters per second [mm/s] meters per second [m/s]
4.6 Surface strength: Viscosity-velocity-product (VVP), IGT T 514	kilopoise centimeters per second [kP · cm/s] poise meters per second [P · m/s] pascal-seconds meters-per-second [(Pa · s)(m/s)]	*1 *0.1 *1	newtons per meter [N/m] newtons per meter [N/m] newtons per meter [N/m]
4.7 Specific external surface of pulp T 226	square centimeters per gram [cm²/g]	*0.1	square meters per kilogram [m²/kg]

Table 5. Permeability and absorption properties

PROPERTY and TAPPI Test Method (if any) [ISO Standard]	To convert values expressed in CUSTOMARY UNITS	Multiply by	To obtain values expressed in RECOMMENDED FORM
	Name [Symbol]	*exactly	Name [Symbol]
5.1 Air permeance, general T 251 [ISO 5636]	cubic feet per minute square foot 0.5-inch water [ft^3/(min • ft^2 • 0.5 inH$_2$O)]	0.04083	millimeters per pascal second [mm/(Pa • s)]
	cubic feet per minute square foot 0.5-inch water [ft^3/(min • ft^2 • 0.5 inH$_2$O)]	40.83	micrometers per pascal second [μm/(Pa • s)]
5.2 Air permeance, Bendtsen [ISO 5636/3]	milliliters per minute [mL/min]	* 1	milliliters per minute (Bendtsen) [mL/min(Bendtsen)]
	milliliters per minute [mL/min]	0.01134	micrometers per pascal second [μm/(Pa • s)]
5.3 Air permeance, Sheffield [ISO 5636/4]	Sheffield units	* 1 (1)	milliliters per minute (Sheffield) [mL/min(Sheffield)] millimeters per pascal second [mm/(Pa • s)]
5.4 Air resistance			+pascal seconds per meter [Pa • s/m]
5.5 Air resistance, GUrley T 460, T 536 [ISO 3687, 5270,5636/5]	seconds [s] sometimes expressed as seconds per 100 milliliters 1 inch sq test area	* 1	#seconds (Gurley) [s(Gurley)]
5.6 Water vapor transmission rate T 448, T 464, T 523 [ISO 2528]	grams per square meter day [g/(m^2 • d)]	* 1	+grams per square meter day [g/(m^2 • d)]
5.7 Ink absorbency, K and N	"K and N" units	* 1	#"K and N" units
5.8 Water absorbency and absorption			
5.9 Area basis T 441 [ISO 535, 5637]	grams per square meter [g/m^2]	* 1	+grams per square meter [g/m^2]
5.10 Mass basis T 491	percent [%] (of initial weight)	* 1	+percent [%] (of initial weight)
5.11 Capillary rise	inches [in]	25.4	+millimeters [mm]

Note (1): RGF. should be made to Journal article or standard to obtain best level of known conversion

Table 6. Optical properties **

PROPERTY and TAPPI Test Method (if any) [ISO Standard]	To convert values expressed in CUSTOMARY UNITS	Multiply by	To obtain values expressed in RECOMMENDED FORM
	Name [Symbol]	*exactly	Name [Symbol]
6.1 Reflectance factor T 442, T 452, T 525, T 534, T 646 [ISO 2469, 2470, 3688]	Percent (relative to MgO = 100%) [%(relative, MgO = 100%)] (conversion factor depends on instrument geometry and wave length)		+percent (absolute) [%(absolute)]
6.2 Opacity T 425, T 519 [ISO 2471]	percent [%]	* 1	+percent [%]
6.3 Gloss T 480, T 653	percent [%] or numerical value	* 1	+percent [%] or numerical value
6.4 Reflection (optical) density	numerical value	* 1	+numerical value
6.5 Transmission (optical) density	numerical value	* 1	+numerical value
6.6 Light absorption power	numerical value	* 1	+numerical value
6.7 Light scattering power	numerical value	* 1	+numerical value
6.8 Light absorption coefficient	square centimeters per gram [cm^2/g]	* 0.1	+square meters per kilogram [m^2/kg]
6.9 Light scattering coefficient T 220	square centimeters per gram [cm^2/g]	* 0.1	+square meters per kilogram [m^2/kg]

**See also Section 10.22

Table 7. Electrical properties

PROPERTY and TAPPI Test Method (if any) [ISO Standard]	To convert values expressed in CUSTOMARY UNITS	Multiply by	To obtain values expressed in RECOMMENDED FORM
	Name [Symbol]	*exactly	Name [Symbol]
7.1 Surface resistivity	ohms [Ω]	* 1	+ohms [Ω]
7.2 Volume resistivity	ohm meters [Ω · m]	* 1	+ohm meters [Ω·m]
7.3 Electrical strength	kilovolts per millimeter [kV/mm] kilovolts per millimeter [kV/mm]	* 1 * 1	#kilovolts per millimeter [kV/mm] megavolts per meter [MV/m]
7.4 Electrical conductivity of extracts T 252 [ISO 6587]	millisiemens per meter [mS/m] millisiemens per meter [mS/m] microsiemens per centimeter [µS/cm] micromhos per centimeter [µΩ^{-1}/cm]	* 1 * 1 * 0.1 * 0.1	+microsiemens per meter [µS/m] millisiemens per meter [mS/m] millisiemens per meter [mS/m] millisiemens per meter [mS/m]

Table 8. Composition properties**

PROPERTY and TAPPI Test Method (if any) [ISO Standard]	To convert values expressed in CUSTOMARY UNITS	Multiply by	To obtain values expressed in RECOMMENDED FORM
	Name [Symbol]	*exactly	Name [Symbol]
8.1 Moisture content (or dry matter content) T 264, T 208, T 220, T 258, T 412, T 671 [ISO 287, 638]	percent [%] (of total weight)	* 1	+percent [%] (of total weight)
8.2 Stock concentration T 240 [ISO 4119]	percent [%] (of total weight)	* 1	percent [%] (of total weight)
8.3 Ash T 211, T 413 [ISO 1762, 2144]	percent [%] (of dry weight)	* 1	+percent [%] (of dry weight)
8.4 Dirt: Area basis T 213, T 437	parts per million [ppm]	* 1	square millimeters per sq meter [mm^2/m^2]
8.5 Dirt: Mass Basis T 246	square millimeters per kilogram [mm^2/kg]	* 1	square millimeters per kilogram [mm^2/kg]
8.6 Other major constituents or coatings: Mass per unit area T 405, T 497, T 531, T 532, T 688, T 690, T 691	grams per square meter [g/m^2] pounds per 1000 square feet [lb/1000 ft^2] pounds per ream (see 1.1 for conversion factors) grains per square yard	* 1 4.88243 0.07750	+grams per square meter [g/m^2] grams per square meter [g/m^2] grams per square meter [g/m^2] grams per square meter [g/m^2]
8.7 Other major constituents or coatings: Relative mass T 203, T 204, T 207, T 212, T 222, T 235, T 249, T 250, T 261, T 405, T 406, T 408, T 418, T 419, T 428, T 438, T 493, T 504, T 612, T 627, T 688, T 691 [ISO 624, 692, 699]	percent [%] (of dry weight)	* 1	+percent [%] (of dry weight)
8.8 Minor constituents T 241, T 242, T 243, T 244, T 245, T 247, T 434 [ISO 776, 777, 778, 779, 1830]	parts per million [ppm] (of dry weight)	* 1	+milligrams per kilogram [mg/kg]

**Also see Table 9

Table 9. Pulp properties**

PROPERTY and TAPPI Test Method (if any) [ISO Standard]	To convert values expressed in CUSTOMARY UNITS	Multiply by	To obtain values expressed in RECOMMENDED FORM
	Name [Symbol]	*exactly	Name [Symbol]

9.1   Saleable mass
      T 210
      [ISO 801]
	pounds [lb]	0.45359	+kilograms [kg]

9.2   Drainability of pulp (freeness)
      T 227
      [ISO 5267/II]
	Canadian standard freeness	* 1	milliliters (CSF) [mL (CSF)]

9.3 Degree of delignification			

9.4   Kappa number
      T 236
      [ISO 302]
	numerical value	* 1	+numerical value

9.5   Chlorine consumption (hypo number)
      T 253
      [ISO 3260]
	number	* 1	+percent [%] (of dry weight)

**See also Table 8

Table 10. General units found in pulp, paper, and board testing documents

PROPERTY and TAPPI Test Method (if any) [ISO Standard]	To convert values expressed in CUSTOMARY UNITS	Multiply by	To obtain values expressed in RECOMMENDED FORM
	Name [Symbol]	*exactly	Name [Symbol]
10.1 Length	angstroms [Å] microns mils [mil or 0.001 in] inches [in] feet [ft] miles [mi]	* 0.1 * 1 0.0254 25.4 0.3048 1.609	nanometers [nm] micrometers [μm] millimeters [mm] millimeters [mm] meters [m] kilometers [km]
10.2 Fiber length T 232, T 233	millimeters [mm]	* 1	millimeters [mm]
10.3 Area	square inches [in²] square feet [ft²] square yards [yd²] acres hectares [ha] square miles [mi²]	6.4516 0.0929030 0.8361274 4046.86 * 0.01 2.58999	square centimeters [cm²] square meters [m²] square meters [m²] square meters [m²] square kilometers [km²] square kilometers [km²]
10.4 Volume, fluid (preferred)	ounces [oz] gallons [gal] milliliters [mL] liters [L]	29.5735 3.785412 * 1 * 1	milliliters [mL] liters [l or L] (L preferred, USA) milliliters [mL] liters [L]
10.5 Volume, solid (or fluid)	cubic inches [in³] cubic feet [ft³] cubic yards [yd³] microliters [μL] milliliters [mL] liters [L] liters [L]	16.38706 0.0283169 0.764555 * 1 * 1 * 1 * 0.001	cubic centimeters [cm³] cubic meters [m³] cubic meters [m³] cubic millimeters [mm³] cubic centimeters [cm³] cubic decimeters [dm³] cubic meters [m³]
10.6 Mass	ounces [oz] pounds [lb] tons (= 2000 lb)	28.3495 0.453592 0.907185	grams [g] kilograms [kg] metric tons (tonne) [t] (= 1000 kg)
10.7 Mass per unit length T 234	milligrams per 100 meters [mg/100 m] or decigrex milligrams per 100 meters [mg/100 m] or decigrex	* 10 * 0.01	micrograms per meter [μg/m] milligrams per meter [mg/m]
10.8 Mass per unit area (also see 1.1, 8.6)	tons per 100 square feet [ton/100 ft²]	0.092903	tonne per square meter [t/m²]

10.9   Mass per unit volume, or Density
       (also see 1.4)
       T 258, T 694
	ounces per gallon [oz/gal]
	pounds per gallon [lb/gal]
	grams per liter [g/L]
	pounds per cubic inch [lb/in³]
	pounds per cubic foot [lb/ft³]
		7.48915
		0.119826
		* 1
		27.6799
		16.0184
			grams per liter [g/L]
			kilograms per liter [kg/L]
			kilograms per cubic meter [kg/m³]
			megagrams per cubic meter [Mg/m³]
			kilograms per cubic meter [kg/m³]

Table 10. General units found in pulp, paper, and board testing documents (continued)

PROPERTY and TAPPI Test Method (if any) [ISO Standard]	To convert values expressed in CUSTOMARY UNITS	Multiply by	To obtain values expressed in RECOMMENDED FORM
	Name [Symbol]	*exactly	Name [Symbol]
10.10 Time	microseconds [µs]	* 1	microseconds [µs]
	milliseconds [ms]	* 1	milliseconds [ms]
	seconds [s]	* 1	seconds [s]
	minutes [min]	* 1	minutes [min] = 60 seconds [s]
	hours [h]	* 1	hours [h] = 3.6 kiloseconds [ks]
	days [d]	* 1	days [d] = 86.4 kiloseconds [ks]
10.11 Speed	feet per second [ft/s]	0.30480	meters per second [m/s]
	feet per minute [ft/min or fpm]	5.080	millimeters per second [mm/s]
10.12 Volume flow rate	gallons per minute [gal/min]	3.78541	liters per minute [L/min]
	cubic feet per second [ft³/s]	28.31685	liters per second [L/s]
	cubic feet per second [ft³/s]	0.0283169	cubic meters per second [m³/s]
	cubic feet per minute [ft³/min or cfm]	1.69901	cubic meters per hour [m³/h]
	cubic yards per second [yd³/s]	0.76455	cubic meters per second [m³/s]
	gallons per day [gal/d]	0.00378541	cubic meters per day [m³/d]
10.13 Force	pounds-force [lbf] (factor exactly 4.448 221 615 260 5)	4.44822	newtons [N]
	ounces-force [ozf]	0.278014	newtons [N]
	kilograms-force [kgf]	9.80665	newtons [N]
	dynes [dyn]	* 0.01	millinewtons [mN]
10.14 Force per unit length T 517	grams-force per millimeter [gf/mm]	9.80665	newtons per meter [N/m]
	pounds-force per inch [lbf/in]	0.1751268	kilonewtons per meter [kN/m]
10.15 Surface tension	dynes per centimeter [dyn/cm]	* 1	millinewtons per meter [mN/m]
10.16 Pressure; stress; force per unit area	pounds-force per square inch [lbf/in² or psi]	6.89476	kilopascals [kPa]
	pounds-force per square foot [lbf/ft²]	47.8803	pascals [Pa]
	feet of water (39.2°F) [ftH$_2$O]	2.98898	kilopascals [kPa]
	inches of water (60°F) [inH$_2$O]	0.24884	kilopascals [kPa]
	inches of mercury (32°F) [inHg]	3.38638	kilopascals [kPa]
	inches of mercury (60°F) [inHg]	3.37685	kilopascals [kPa]
	millimeters of mercury (0°C) [mmHg]	0.133322	kilopascals [kPa]
	atmospheres [atm]	0.101325	megapascals [MPa]
	grams-force per square centimeter [gf/cm²]	98.0665	pascals [Pa]
	newtons per square meter [N/m²]	* 1	pascals [Pa]
	bars [bar]	* 100	kilopascals [kPa]
10.17 Torque, or bending moment	pound-force inches [lbf • in]	0.11298	newton meters [N • m]
	pound-force feet [lbf • ft]	1.35582	newton meters [N • m]
	kilogram-force centimeters [kgf • cm]	0.0980665	newton meters [N • m]
	dyne centimeters [dyn • cm]	* 0.1	micronewton meters [µN • m]
10.18 Energy	foot pounds-force [ft • lbf]	1.35582	joules [J]
	meter kilograms-force [m • kgf]	9.80665	joules [J]
	centimeter grams-force [cm • gf]	0.0980665	millijoules [mJ]
	British thermal units, Int. [Btu]	1.05506	kilojoules [kJ]
	horsepower hours [hp • h]	2.68452	megajoules [MJ]
	kilowatt hours [kW • h or kWh]	3.600	megajoules [MJ]
	kilocalories, Int. Table [kcal]	4.1868	kilojoules [kJ]
	meter newtons [m • N]	* 1	joules [J]
10.19 Power	foot pounds-force per second [ft • lbf/s]	1.35582	watts [W]
	horsepower [hp] (= 550 foot pounds-force per second)	745.700	watts [W]
	horsepower [hp]	0.74570	kilowatts [kW]
	metric horsepower	735.499	watts [W]
10.20 Viscosity, dynamic T 230, T 254, T 648, T 666, T 677, T 687	poise [P]	* 0.1	pascal seconds [Pa • s]
	centipoise [cP]	* 1	millipascal seconds [mPa • s]
10.21 Viscosity, kinematic	centistokes [cSt]	* 1	square millimeters per second [mm²/s]
10.22 Illumination	footcandles [fc]	10.7639	lumen per square meter [lm/m²]
	footcandles [fc]	10.7639	lux [lx]
10.23 Temperature T 630, T 634, T 652, T 662, T 675, T 814	degrees Fahrenheit [°F] $T_c = \frac{5}{9}(T_F - 32)$		degrees Celsius [°C]
10.24 Temperature interval	Fahrenheit degrees change [∆F]	0.555556	Celsius degrees change [∆C]
10.25 Frequency	cycles per second [s⁻¹]	* 1	hertz [Hz]
10.26 pH T 252, T 435, T 509, T 529, T 667 [ISO 6588]	pH units	* 1	pH units

■